A DICTIONARY OF CIVIL, WATER RESOURCES & ENVIRONMENTAL ENGINEERING

A DICTIONARY OF CIVIL, WATER RESOURCES & ENVIRONMENTAL ENGINEERING

Harry C. Friebel, Ph.D., P.E.

GOLDEN RATIO PUBLISHING

Published by Golden Ratio Publishing, LLC, Berlin, New Jersey.

Library of Congress Control Number: 2012952387

Library of Congress subject headings:
Civil engineering – Dictionaries
Water – Dictionaries
Water resources – Dictionaries
Hydrology – Dictionaries
Hydraulic engineering – Dictionaries
Environmental engineering – Dictionaries

To the love of my life,
April
And our children,
Nathan and Juliet - you
are my everything.

TESTIMONIALS

HERE'S WHAT PEOPLE ARE SAYING ABOUT DR. FRIEBEL AS A PE CIVIL REVIEW INSTRUCTOR AND THE NEED TO BRING A DICTIONARY TO THE EXAM:

"With regards to a dictionary, Dr. Friebel, you were correct. It gave me the answer to at least 4 problems by definition only, and another 2 questions by showing which answers were incorrect, so I definitely recommend continuing to push any test takers to pick up a separate dictionary of some sort."

"Dr. Friebel provided some great test taking tips, which I will use on the exam. Although I am not taking the depth in Water, he was able to provide great tips to complete the breadth section along with tips to pass any depth section. Great Instructor!!!!!"

"Some 3-4 questions were answered straight out of the dictionary - thanks Dr. Friebel."

"I had the NCEES water resources topics for the morning exam and material and Dr. Friebel covered all subjects. He explained subjects like we were in High School, I like it...That's how you do it. He is very fresh in his ideas."

"Thank you Dr. Friebel, I brought in a dictionary and I used it twice!"

"Dr. Friebel is the best in my opinion. He was clear, concise and demonstrated profound knowledge of the subject area. He explained the problems with emphasis on how to think through and solve them with a level that I never would have been able to see on my own. I'm going into this exam very, very confident in my ability to solve Water Resource problems."

"Bring a good dictionary. I was able to look up a few environmental terms that helped me on a few problems - easy points!"

"Professor Friebel's presentation was very straight-forward and helpful, a great review as someone who has a background in the field and will be taking it for the PM depth section. Thank you!"

"Dr. Friebel said, "Take an engineering dictionary." Guess what? There were about three problems that were answered by simply reading a definition from the dictionary."

"Not only did Dr. Friebel know the material, he was able to teach it to the students so we understood it. He applied everything we learned to the PE test. He gave us test taking tips. He was by far one of the best teachers I have had."

"Dr. Friebel recommended I bring a good dictionary and I am glad I did, I needed it."

"This is my problem area and Dr. Friebel really helped me not be "fearful of the subject."

"Subject material was very hard for me personally, but Dr. Friebel was able to break it down in a way that made it much easier. Great instructor! He's really there to help the students."

"Great advice to bring a dictionary to the exam; I answered three questions correctly because of your advice. Many times an exam question will simply ask for the definition of a word. All you have to do is look it up in the dictionary!"

"I am taking the Water Resources/Env. depth examination. Dr. Friebel was incredibly helpful. His teaching methods were fantastic."

"Dr. Friebel knew the material very well, communicated it to the class well and made the class enjoyable. I have not done any water resource work since college and I now feel like I have a very good general understanding of the subject and will do well with any questions on the exam."

"Dr. Friebel was very straight-forward and understandable. I feel very confident and ready to ace the Water Resource/Env. section."

"Dr. Friebel's lecture style was excellent. He assumed that the topics he covered were new to all of us, giving us a chance to better understand the key topics."

"I thought Dr. Friebel was an excellent instructor. He taught in such a manner which made water resource/environmental easy to understand. He was GREAT!"

ACKNOWLEDGMENTS

I credit my knowledge and experience regarding Water Resources and Environmental Engineering to my colleagues and former academic mentors. I am very grateful for all the patience, advice and guidance I have received over the years. I have been taught so much, by so many people - you have all inspired me!

Since I began helping others pass the Water Resources and Environmental Engineering portion of the PE Civil exam, I have benefited enormously from discussions with former students, fellow PE review course instructors and readers of my website (www.goldenratiopublishing.com). Special acknowledgement is made to my colleagues at School of PE (www.schoolofpe.com.) In particular, I would like to thank Vinod Dega, owner of School of PE, for offering me an opportunity to become an instructor and Jeffrey S. MacKay of NTM Engineering, Inc. (my former School of PE instructor) for providing such excellent guidance when I was studying to pass the exam. If not for either of these gentlemen, I certainly would not be creating new resources to help pass the PE Civil exam and instead be heading down another path in life.

Many more colleagues and friends, from a multitude of disciplines, have contributed to this book. I owe my sincerest gratitude to: my cousin Robert Anthony for designing the logo of Golden Ratio Publishing; Edward St John of Woodworth & St. John, LLC for providing endless accounting advice yet only asking for friendship in return; Rita Stevens of Planning Matters LLC for helping me think outside the box and broaden my goals and horizons; and Glendon Stevens (my former supervisor) for encouraging me to pursue my dreams all while insisting simplicity is best.

To my family and especially my Mother, thank you for your lifetime encouragement. To my children Nathan and Juliet, thanks for your patience - I sacrificed way too much family time to finish this dictionary. The last word is for my loving wife April - thank you for your unconditional love and support.

H. C. F.

ABOUT THE AUTHOR

Harry C. Friebel, Ph.D., P.E., is the founder and president of the publishing company Golden Ratio Publishing, LLC. Dr. Friebel is a hydraulic engineer with the U.S. Army Corps of Engineers (views expressed herein are the authors own and do not represent the views of the U.S. Army Corps of Engineers or Federal government). He is also a Civil - Water Resources and Environmental Engineering P.E. review course instructor and maintains a related website at www.goldenratiopublishing.com.

Dr. Friebel formerly taught Hydraulic Design to both undergraduate and graduate students as an adjunct professor at Rowan University, in Glassboro, New Jersey. He earned both a B.S. and M.S. engineering degree from Florida Institute of Technology and a Ph.D. from Stevens Institute of Technology. He is a registered Professional Engineer (P.E.) and has served as principle investigator on numerous water research projects sponsored by federal, state and local government agencies.

For more information about Dr. Friebel or Golden Ratio Publishing, please visit: www.goldenratiopublishing.com.

If you have any comments regarding this book or Golden Ratio Publishing, please contact: info@goldenratiopublishing.com.

AFTER YOU PASS, WRITE ME

Please, after your pass, write me your success story! I'd love to post it on my website (www.goldenratiopublishing.com) or even feature it in my next book.

I welcome your feedback and suggestions. If this book helps, please let me know. If I can improve it in any way or there is a term you would like to see included in a future edition, let me know that too. I only write and teach because I love it - I experienced some exhilarating teachers in my career and now it is time for me to give back.

I must admit, I do enjoy the positive feedback I receive from students. Through their comments I was inspired to produce this dictionary. Your input will help me improve future editions, so please send me your thoughts:

info@goldenratiopublishing.com

or visit:

www.goldenratiopublishing.com

There you will find updates to this book and announcements of future publications. I built it for you, so please take advantage of it.

PREFACE

Think you don't need a dictionary, think again!

This dictionary was written with the Professional Engineering (PE) Civil licensure exam in mind. Typically, a PE examinee spends well over a thousand dollars on review books, courses and sample exams. Many previous examinees believe they did not pass the exam the first time due to a single problem or two. What if one of those questions could have been answered correctly with a dictionary? What is it worth to increase your chances of getting additional problems correct?

Think of this dictionary as insurance! Hopefully you won't need it, but what if you do? Be wary of anyone stating a dictionary is not necessary for the exam. If you don't think the good people that put the exam together are not aware of the exact terms defined (and NOT defined) in your review manual glossary, you're kidding yourself. I have received testimonial after testimonial from former students thanking me for recommending that they bring a dictionary to the exam. Terms can (and do) appear on the exam that you may not be familiar with. What if they are not defined in your reference books, what then? What's familiar to you may not be familiar to the person sitting next to you and vice versa. This dictionary may very well be the difference in passing the PE Civil exam.

I sincerely hope you find this dictionary an invaluable resource. I spent many years gathering, writing, and re-writing the terms found in this dictionary to provide a treasure of material for passing the PE Civil exam without breaking anyone's piggy bank. I wrote this dictionary with only one objective, to help you pass the PE Civil exam!

H. C. FRIEBEL

Berlin, New Jersey

January, 2013

A

A AMS Arctic Air Mass.

A INDEX Daily index of geomagnetic activity calculated as the average of the eight 3-hourly a indices.

A&E (A/E) Architectural and Engineering.

AAAS American Association for the Advancement of Science.

AAC All American Canal.

AAHU Average Annual Habitat Units.

AAM Annual Arithmetic Mean.

AAR Alkali-Aggregate Reaction.

AAWU Alaskan Aviation Weather Unit.

AB Assembly Bill.

ABANDONED WELL Well that is no longer being used or cannot be used because of its poor condition; well whose use has been permanently discontinued or that is in a state of such disrepair that it cannot be used for its intended purpose.

ABATEMENT Reducing the degree or intensity of, or eliminating, pollution.

ABATEMENT DEBRIS Waste from remediation activities.

ABC Association of Boards of Certification.

ABERRANT Atypical; departing from the normal type or structure.

ABIOTIC Absence of living organisms.

ABLATION Depletion of snow and ice from any mass of frozen water through processes of melting, sublimation and evaporation, wind erosion, and calving; wasting and removal from a rock mass of material by physical processes such as wind erosion or by chemical processes.

ABNDT Abundant.

ABRASION Reduction of rock fragments or surfaces by the wearing, friction, grinding, or rubbing action of other rock particles or of the transport medium.

ABSOLUTE PRESSURE Atmospheric pressure plus gauge pressure.

ABSOLUTELY STABLE AIR Atmospheric condition that exists when the environmental lapse rate is less than the moist adiabatic lapse rate.

ABSOLUTELY UNSTABLE AIR Atmospheric condition that exists when the environmental lapse rate is greater than the dry adiabatic lapse rate.

ABSORBED DOSE Amount of a substance that enters the body of an exposed organism through physical or biological processes.

ABSORBED WATER Water held mechanically in a soil or rock mass and having physical properties not substantially different from ordinary water at the same temperature and pressure.

ABSORPTION Entrance of water into the soil or rocks by all natural processes including infiltration of precipitation or snowmelt, gravity flow of streams into the valley alluvium into sinkholes or other large openings, and the movement of atmospheric moisture; uptake of water, other fluids, or dissolved chemicals by a cell or an organism; taking in of fluids or other substances through cells or tissues; passing of a substance into the circulatory system of the body; entry of toxicants through the skin; dissolving or mixing of a substance in gaseous, liquid or solid form with groundwater; process in which incident radiant energy is retained by a substance by conversion to some other form of energy.

ABSORPTION BARRIER Any of the exchange sites of the body (skin, lung tissue, and gastrointestinal-tract wall) that permit uptake of various substances at different rates.

ABSORPTION FACTOR Fraction of a chemical making contact with an organism that is absorbed by the organism.

ABT About.

ABUTMENT Part of a valley or canyon wall against which a dam is constructed where left and right abutments of dams are defined with the observer viewing the dam looking in the downstream direction; part of a dam that contacts the riverbank; structure that supports the ends of a dam or bridge; part of a structure that is the terminal point or receives thrust or pressure; action or place of abutting.

ABUTMENT SEEPAGE Reservoir water that moves through seams or pores in the natural abutment material and exits as seepage.

ABV Above.

AC Acre.

ACCAS (AltoCumulus CAStellanus)
Mid-level clouds (bases generally 8 to 15 thousand feet) often taller than wide, of which at least a fraction of their upper parts show cumulus-type development indicating instability aloft that may precede the rapid development of thunderstorms.

ACCELERATION Time rate of change of the velocity vector, either of magnitude or direction or both.

ACCELEROGRAM Record from an accelerometer showing acceleration as a function of time.

ACCEPTABLE DAILY INTAKE (ADI)
Estimate of the largest amount of a chemical to which a person can be exposed on a daily basis that is not anticipated to result in adverse effects (usually expressed in mg/kg/day).

ACCESS CHARGE Charge levied on a power supplied, or its customer, for access to a utility's transmission or distribution system; charge for the right to send electricity over another's wires.

ACCESS CONTROL POINT Location staffed to restrict entry of unauthorized personnel into a risk area during emergency and/or disaster events involving use of vehicles, barricades, or other measures to deny access.

ACCESS SHAFT Concrete portion of an outlet works between the shaft house and the gate chamber; provides vertical access to the gates.

ACCESSORY CLOUD Cloud that is dependent on a larger cloud system for development and continuance (e.g., roll clouds, shelf clouds, and wall clouds).

ACCIDENT ASSESSMENT Evaluation of the nature, severity, and impact of an accident.

ACCIDENT SITE Location of an unexpected occurrence, failure or loss, either at a plant or along a transportation route, resulting in a release of hazardous materials.

ACCLIMATION Adjustment of an organism to a new habitat or environment.

ACCLIMATIZATION Physiological and behavioral adjustments of an organism to changes in its environment.

ACCRETION Process of growth whereby material is added to the outside of non-living matter; growth of a precipitation particle by the collision of a frozen particle with a supercooled liquid water droplet that freezes upon impact; gradual increase in flow of a stream attributable to seepage, groundwater discharge, or tributary inflow.

ACCUMS Accumulation.

ACCURACY Degree of conformity of a measure to a standard or true value; how close a predicted or measured value is to the true value.

ACEC Area of Critical Environmental Concern.

ACEQUIA Gravity-driven waterways that can be from simple ditches with dirt banks to lined with concrete; similar in concept to a flume,.

ACER Assistant Commissioner - Engineering and Research.

ACES Automated Coastal Engineering System.

AC-FT Acre-foot, acre-feet.

ACGIH American Conference of Governmental Industrial Hygienist.

ACHP Advisory Council on Historic Preservation.

ACID Corrosive solution with a pH value between 0 and 7; more free hydrogen ions (H+) than hydroxyl ions (OH-); pH modifier used in the U.S. Fish and Wildlife Service wetland classification system.

ACID AEROSOL Acidic liquid or solid particles small enough to become airborne.

ACID DEPOSITION Complex chemical and atmospheric phenomenon that occurs when emissions of sulfur and nitrogen compounds and other substances are transformed by chemical processes into the atmosphere and then deposited on Earth in either wet forms through rain "acid rain," snow, or fog or dry forms through acidic gases or particulates.

ACID MINE DRAINAGE Drainage of water from areas that have been mined for coal or other mineral ores, the water has a low pH because of its contact with sulfur-bearing material and is harmful to aquatic organisms.

ACID NEUTRALIZING CAPACITY (ANC) Measure of ability of a base (e.g., water or soil) to resist changes in pH, the equivalent sum of all bases or base-producing materials, solutes plus particulates, in an aqueous system that can be titrated with acid to an equivalence point.

ACID PRECIPITATION Precipitation, such as rain, snow or sleet, containing relatively high concentrations of acid-forming chemicals that have been released into the atmosphere and combined with water vapor; harmful to the environment.

ACID RAIN Precipitation that has been rendered acidic by airborne pollutants that is harmful to the environment; result of sulfur dioxide (SO2) and nitrogen oxides (NOx) reacting in the atmosphere with water and returning to Earth as rain, fog, or snow.

ACIDIC Condition of water or soil that contains a sufficient amount of acid substances to lower the pH below 7.0.

ACIDIC DEPOSITION Transfer by either by wet-deposition processes (rain, snow, dew, fog, frost, hail) or by dry deposition (gases, aerosols, or fine to coarse particles) of acidic or acidifying substances from the atmosphere to the surface of the Earth or to objects on its surface.

ACIDIFIED Addition of an acid (usually nitric or sulfuric) to a sample to lower the pH below 2.0.

ACLD Above Cloud Level.

ACPY Accompany.

ACR Alkali-Carbonate Reaction.

ACRE (AC) Unit for measuring land, equal to 43,560 ft^2, 4840 yds^2, or 160 rds^2.

ACRE-FOOT (AC-FT, ACRE-FT) Unit of volume, commonly used to measure quantities of water used or stored, equivalent to the volume of water required to cover one acre to a depth of one foot; equals 43,560 ft^3, 326,851 gallons, or 1,233 m^3.

ACRM Assistant Commissioner-Resources Management.

ACRS Across.

ACTION LEVEL Level of lead or copper that, if exceeded, triggers treatment or other requirements that a water system must follow; regulatory levels recommended by EPA for enforcement by FDA and USDA when pesticide residues occur in food or feed commodities for reasons other than the direct application of the pesticide. As opposed to “tolerances” that are established for residues occurring as a direct result of proper usage, action levels are set for inadvertent residues resulting from previous legal use or accidental contamination; in the Superfund program, the existence of a contaminant concentration in the environment high enough to warrant action or trigger a response under SARA and the National Oil and Hazardous Substances Contingency Plan.

ACTION STAGE Stage that, when reached by a rising stream, represents the level where the National Weather Service or a partner/user needs to take some type of mitigation action in preparation for possible significant hydrologic activity.

ACTIVATED CARBON Used to remove dissolved organic matter from waste drinking water; highly adsorbent form of carbon used to remove odors and toxic substances from liquid or gaseous emissions; adsorptive particles or granules of carbon usually obtained by heating carbon (such as wood) where particles or granules have a high capacity to selectively remove certain trace and soluble materials from water.

ACTIVATED SLUDGE Product that results when primary effluent is mixed with bacteria-laden sludge and then agitated and aerated to promote biological treatment, speeding the breakdown of organic matter in raw sewage undergoing secondary waste treatment.

ACTIVATOR Chemical added to a pesticide to increase its activity.

ACTIVE Solar activity levels with at least one geophysical event or several larger radio events (10cm) per day (Class M Flares).

ACTIVE (USABLE) STORAGE CAPACITY
Total amount of reservoir capacity normally available for release from a reservoir below the maximum storage level (volume of water between the outlet works and the spillway crest); total or reservoir capacity minus inactive storage capacity.

ACTIVE BED Layer of material between the bed surface and a hypothetical depth at which no transport will occur for the given gradation of bed material and flow conditions.

ACTIVE CAPACITY Reservoir capacity normally usable for storage and regulation of reservoir inflows to meet established reservoir operating requirements (irrigation, power, municipal and industrial use, fish and wildlife, recreation, water quality, etc.); extends from the highest of either the top of exclusive flood control capacity, the top of joint use capacity, or the top of active conservation capacity, to the top of inactive capacity; total capacity less the sum of the inactive and dead capacities.

ACTIVE CONSERVATION CAPACITY
See ACTIVE STORAGE CAPACITY.

ACTIVE CONSERVATION STORAGE
Portion of water stored in a reservoir that can be released for all useful purposes such as municipal water supply, power, irrigation, recreation, fish, wildlife, etc.; volume of water stored between the inactive pool elevation and flood control stage.

ACTIVE DARK FILAMENT (ADF)
Active prominence seen on the disk.

ACTIVE EARTH PRESSURE Minimum value of earth pressure; exists when a soil mass is permitted to yield sufficiently to cause its internal shearing resistance along a potential failure surface to be completely mobilized.

ACTIVE FAULT Fault, because of its present tectonic setting, can undergo movement from time to time in the immediate geologic future; fault that has moved during the recent geologic past and may move again.

ACTIVE INGREDIENT Component that kills, or otherwise controls, target pests in pesticide product.

ACTIVE LAYER Depth of material from bed surface to equilibrium depth continually mixed by flow.

ACTIVE LONGITUDE Approximate center of a range of heliographic longitudes in that "active regions" are more numerous and more flare-active than the average.

ACTIVE PROMINENCE Prominence displaying material motion and changes in appearance over a few minutes of time.

ACTIVE PROMINENCE REGION (APR) Portion of the solar limb displaying active prominences.

ACTIVE REGION (AR) Localized, transient volume of the solar atmosphere in which plages, sunspots, faculae, flares, etc. may be observed.

ACTIVE STORAGE CAPACITY Total amount of reservoir capacity normally available for release from a reservoir below the maximum storage level; total or reservoir capacity minus inactive storage capacity; volume of water between the outlet works and the spillway crest.

ACTIVE SURGE REGION (ASR) Active Region that exhibits a group or series of spike-like surges that rise above the limb.

ACTIVE TRANSPORT Energy-expending mechanism by which a cell moves a chemical across the cell membrane from a point of lower concentration to a point of higher concentration, against the diffusion gradient.

ACTIVITY Ratio of the plasticity index to the percent by dry mass of soil particles finer than 0.002 mm (2 microns) in size.

ACTIVITY PLANS Written procedures in a school's asbestos-management plan that detail the steps a Local Education Agency (LEA) will follow in performing the initial and additional cleaning, operation and maintenance-program tasks, periodic surveillance, and re-inspection required by the Asbestos Hazard Emergency Response Act (AHERA).

ACTS Action Correspondence Tracking System.

ACUTE Occurring over a short period of time; used to describe brief exposures and effects that appear promptly after exposure; stimulus severe enough to rapidly induce an effect; effect observed in 96- hours or less is typically considered acute in aquatic toxicity tests.

ACUTE EFFECT Adverse effect on any living organism that results in severe symptoms that develop rapidly and often subside after the exposure stops.

ACUTE EXPOSURE Exposure to a toxic substance that occurs in a short or single time period; single exposure to a toxic substance that results in severe biological harm or death; usually characterized as lasting no longer than a day.

ACUTE HEALTH EFFECT Immediate (i.e., within hours or days) effect that may result from exposure to certain drinking water contaminants (e.g. pathogens).

ACUTE TOXICITY Ability of a substance to cause poisonous effects resulting in severe biological harm or death soon after a single exposure or dose; any severe poisonous effect resulting from a single short-term exposure to a toxic substance; LD50 of a substance (the lethal dose at which 50 percent of test animals succumb to the toxicity of the chemicals) is typically used as a measure of its acute toxicity.

ACUTE-CHRONIC RATIO (ACR) Ratio of the acute toxicity of an effluent or a toxicant to its chronic toxicity; used as a factor for estimating chronic toxicity on the basis of acute toxicity data, used for estimating acute toxicity on the basis of chronic toxicity data.

ACUTELY TOXIC CONDITIONS Those acutely toxic to aquatic organisms following their short-term exposure within an affected area.

ACWA Association of California Water Agencies.

ADAPTATION Adjustment to environmental conditions; changes in an organism's physiological structure or function or habits that allow it to survive in new surroundings.

ADAPTATION PARAMETER Data related to a specific WSR-88D unit that may consist of meteorological or hydrological parameters or of geographic boundaries, political boundaries, system configuration, telephone numbers (auto dial), or other like data.

ADAS Automated Data Acquisition System.

ADDITIVE DATA Group of coded remarks that includes pressure tendency, amount of precipitation, and maximum/minimum temperature during specified periods of time.

ADDITIVE EFFECT Combined effect of two or more chemicals equal to the sum of their individual effects.

ADDITIVITY Characteristic property of a mixture of toxicants that exhibits a total toxic effect equal to the arithmetic sum of the effects of the individual toxicants.

ADD-ON CONTROL DEVICE Air pollution control device such as carbon absorber or incinerator that reduces the pollution in an exhaust gas.

ADDS Aviation Digital Data Service.

ADENOSINE TRIPHOSPHATE (ATP) Organic, phosphaterich compound important in the transfer of energy in organisms; excellent indicator of the presence of living material in water providing a sensitive and rapid estimate of biomass reported in micrograms per liter.

ADEQUATELY WET Asbestos containing material that is sufficiently mixed or penetrated with liquid to prevent the release of particulates.

ADHESION Shearing resistance between soil and another material under zero externally applied pressure.

ADI Acceptable Daily Intake.

ADIABAT Line on a thermodynamic chart relating the pressure and temperature of a substance (such as air) that is undergoing a transformation in which no heat is exchanged with its environment.

ADIABATIC Changes in temperature caused by the expansion (cooling) or compression (warming) of a body of air as it rises or descends in the atmosphere, with no exchange of heat with the surrounding air.

ADIABATIC LAPSE RATE Rate of decrease of temperature experienced by a parcel of air when it is lifted in the atmosphere under the restriction that it cannot exchange heat with its environment. For parcels that remain unsaturated during lifting, the (dry adiabatic) lapse rate is ≈ 9.8 °C per kilometer.

ADIABATIC PROCESS Process which occurs with no exchange of heat between a system and its environment.

ADIRONDACK TYPE SNOW SAMPLING SET
Snow sampler consisting of a 5-foot fiberglass tube, 3 inches in diameter, 60-inch snow depth capacity, with a serrated-edge steel cutter at one end and a twisting handle at the other.

ADIT Nearly horizontal underground excavation in an abutment having an opening in only one end; opening in the face of a dam for access to galleries or operating chambers.

ADJ Adjacent.

ADJUSTED DISCHARGE Discharge data that have been mathematically adjusted (i.e., to remove the effects of a daily tide cycle or reservoir storage).

ADMINISTERED DOSE Amount of a substance given to a test subject (human or animal) to determine dose-response relationships.

ADMINISTRATIVE ORDER Legal document signed by EPA directing an individual, business, or other entity to take corrective action or refrain from an activity, describes the violations and actions to be taken, and can be enforced in court.

ADMINISTRATIVE ORDER ON CONSENT
Legal agreement signed by EPA and an individual, business, or other entity through which the violator agrees to pay for correction of violations, take the required corrective or cleanup actions, or refrain from an activity; describes the actions to be taken, may be subject to a comment period, applies to civil actions, and can be enforced in court.

ADMINISTRATIVE PROCEDURES ACT
Law that spells out procedures and requirements related to the promulgation of regulations.

ADMINISTRATIVE RECORD All documents which EPA considered or relied on in selecting the response action at a Superfund site, culminating in the record of decision for remedial action or, an action memorandum for removal actions.

ADMS Accessibility Data Management System.

ADP Automated Data Processing.

ADPC Acoustic Doppler Current Profiler.

ADR Alternative Dispute Resolution.

ADSORBATE Material being removed by the adsorption process.

ADSORBED WATER Water in a soil or rock mass, held by physicochemical forces, having physical properties substantially different from absorbed water or chemically combined water, at the same temperature and pressure.

ADSORBENT Material (e.g. activated carbon) that is responsible for removing the undesirable substance in the adsorption process.

ADSORPTION Removal of a pollutant from air or water by collecting the pollutant on the surface of a solid material (e.g., activated carbon removes organic matter from wastewater); Bonding of chemicals to soil particles or other surfaces; adherence of a gas, liquid, or dissolved material on the surface of solids; increase in concentration of gas or solute at the interface of a two-phase system.

ADSORPTION ISOTHERM graphical representation of the relationship between the solute concentration and the mass of the solute species adsorbed on the aquifer sediment or rock.

ADSS Advanced Decision Support System.

ADT average daily traffic.

ADULTERANTS Chemical impurities or substances that by law do not belong in a food, or pesticide.

ADULTERATED Any pesticide whose strength or purity falls below the quality stated on its label; A food, feed, or product that contains illegal pesticide residues.

ADVANCED DECISION SUPPORT SYSTEM (ADSS) Computer software designed to provide easy access to and allow efficient use of methods of analysis and information management.

ADVANCED TREATMENT Level of wastewater treatment more stringent than secondary treatment; requires an 85-percent reduction in conventional pollutant concentration or a significant reduction in non-conventional pollutants.

ADVANCED VERY HIGH RESOLUTION RADIOMETER (AVHRR) Main sensor on U.S. polar orbiting satellites.

ADVANCED WASTEWATER TREATMENT Any treatment of sewage that goes beyond the secondary or biological water treatment stage and includes the removal of nutrients such as phosphorus and nitrogen and a high percentage of suspended solids.

ADVANCED WEATHER INTERACTIVE PROCESSING SYSTEM (AWIPS) Allows the operator to overlay meteorological data from a variety of sources; this system replaced the Automation of Field Operations and Services (AFOS).

ADVECTION (ADVCTN) Transport of an atmospheric property by the wind; process by which solutes are transported by moving groundwater.

ADVECTION FOG Fog that forms when warm air flows over a cold surface and cools from below until saturation is reached.

ADVERSE CONSEQUENCES Negative impacts that may result from the failure of a dam where the primary concerns are loss of human life, economic loss (including property damage), lifeline disruption, and environmental impact.

ADVERSE EFFECT Action that has an apparent direct or indirect negative effect on the conservation and recovery of an ecosystem component listed as threatened or endangered.

ADVERSE EFFECTS DATA
FIFRA requires a pesticide registrant to submit data to EPA on any studies or other information regarding unreasonable adverse effects of a pesticide at any time after its registration.

ADVIS Program which combines the Antecedent Precipitation Index (API) method of estimating runoff with unit hydrograph theory to estimate streamflow for a headwater basin.

ADVISORY Highlights special weather conditions that are less serious than a warning; a non-regulatory document that communicates risk information to those who may have to make risk management decisions.

ADVISORY COUNCIL ON HISTORIC PRESERVATION (ACHP)
Executive agency responsible for ensuring requirements of National Historic Preservation Act and 36 CFR Part 800 are met.

AEOLIAN (EOLIAN) Materials carried, deposited, produced, or eroded by the wind.

AEOLIAN DEPOSITS Wind-deposited material such as dune sands and loess deposits.

AERATE Impregnate with gas (e.g. air); to supply air to water, soil, or other media.

AERATED LAGOON Holding and/or treatment pond that speeds up the natural process of biological decomposition of organic waste by stimulating the growth and activity of bacteria that degrade organic waste.

AERATION Process of adding air to water by either passing air through water or passing water through air and may be passive (as when waste is exposed to air), or active (as when a mixing or bubbling device introduces the air); process which promotes biological degradation of organic matter in water.

AERATION TANK Chamber used to inject air into water; portion of the lithosphere in which the functional interstices of permeable rock or earth are not filled with water under hydrostatic pressure where the interstices either are not filled with water or are filled with water that is no held by capillarity.

AERIAL COVER Ground area circumscribed by the perimeter of the branches and leaves of a given plant or group of plants (measure of relative density).

AEROALLERGENS Any of a variety of allergens such as pollens, grasses, or dust carried by winds.

AEROBIC Condition in which free (atmospheric) or dissolved oxygen is present in water; opposite of anaerobic; life or processes that require, or are not destroyed by, the presence of oxygen.

AEROBIC TREATMENT Process by which microbes decompose complex organic compounds in the presence of oxygen and use the liberated energy for reproduction and growth.

AEROBIC TREATMENT UNIT (ATU)
Mechanical wastewater treatment unit that provides secondary wastewater treatment for a single home, a cluster of homes, or a commercial establish-

ment by mixing air (oxygen) and aerobic and facultative microbes with the wastewater.

AEROSOL Finely divided material suspended in air or other gaseous environment; pressurized gas used to propel substances out of a container; larger than a molecule and can be filtered from the air.

AF Acre-feet.

AFFECTED ENVIRONMENT Existing biological, physical, social, and economic conditions of an area subject to change, both directly and indirectly, as the result of a proposed human action.

AFFECTED LANDFILL Under the Clean Air Act, landfills that meet criteria for capacity, age, and emissions rates set by the EPA and are required to collect and combust their gas emissions.

AFFECTED PUBLIC People who live and/or work near a hazardous waste site; human population adversely impacted following exposure to a toxic pollutant in food, water, air, or soil.

AFOS Automation of Field Operations and Services.

AFTERBAY (TAILRACE) Body of water immediately downstream from a power plant or pumping plant; reservoir or pool that regulates fluctuating discharges from a hydroelectric power plant or a pumping plant; short stretch of stream or conduit, or to a pond or reservoir.

AFTERBAY DAM Dam located downstream from a large hydroelectric power plant used to regulate discharges downstream.

AFTERBURNER Burner located so that the combustion gases are made to pass through its flame in order to remove smoke and odors that may be attached to or separated from the incinerator proper.

AFTN Afternoon.

AFTR After.

AFWA Air Force Weather Agency.

AFY (AF/YR) Acre-feet per year.

AG Agriculture; association of governments.

AGC Automatic Generation Control; Automatic Gain Control.

AGDISP Particular atmospheric dispersion model used for treating the transport and diffusion of aerially sprayed pest control agents in agricultural applications.

AGE TANK (DAY TANK) Tank used to store a chemical solution of known concentration for feed to a chemical feeder.

AGENT Any physical, chemical, or biological entity that can be harmful to an organism.

AGENT ORANGE Toxic herbicide and defoliant.

AGFS Aviation Gridded Forecast System.

AGGLOMERATE Ice cover of floe formed by the freezing together of various forms of ice.

AGGRADATION Geologic process wherein streambeds, floodplains, sandbars, and the bottom of water bodies are raised in elevation by the deposition of sediment; opposite of degradation.

AGGREGATE Crushed rock or gravel screened to sizes for use in road surfaces, concrete, or bituminous mixes; mass or cluster of soil particles, often having a characteristic shape.

AGL Above Ground Level.

AGM Annual Geometric Mean.

AGN Again.

AGRICULTURAL DRAINAGE Process of directing excess water away from root zones by natural or artificial means; water drained away from irrigated farmland.

AGRICULTURAL DRAINAGE WELLS Class V wells that receive agricultural runoff including improved sinkholes, abandoned drinking water wells, wells that recharge aquifers with agricultural tail waters, and wells that drain flood irrigation.

AGRICULTURAL POLLUTION Farming wastes, including runoff and leaching of pesticides and fertilizers; erosion and dust from plowing; improper disposal of animal manure and carcasses; crop residues, and debris.

AGRICULTURAL WASTE Poultry and livestock manure, and residual materials in liquid or solid form generated from the production and marketing of poultry, livestock or furbearing animals; also includes grain, vegetable, and fruit harvest residue.

AGROCHEMICAL Synthetic chemicals (pesticides and fertilizers) used in agricultural production.

AGROECOSYSTEM Land used for crops, pasture, and livestock; the adjacent uncultivated land that supports other vegetation and wildlife; and the associated atmosphere, the underlying soils, groundwater, and drainage networks.

AHD Ahead.

AHERA DESIGNATED PERSON (ADP) Person designated by a Local Education Agency to ensure that the AHERA requirements for asbestos management and abatement are properly implemented.

AHOS Automatic Hydrologic Observing System.

AHOS-S Automatic Hydrologic Observing System - Satellite.

AHOS-T Automatic Hydrologic Observing System - Telephone.

AHS Archaeological and Historical Services.

AIPC American Indian Program Council.

AIR Mixture of gases comprising the Earth's atmosphere.

AIR BINDING Situation where air enters the filter media and harms both the filtration and backwash processes.

AIR CHANGES PER HOUR (ACH) Movement of a volume of air in a given period of time; if a building has one air change per hour, it means that the air in the building will be replaced in a one-hour period.

AIR CLEANING Indoor-air quality-control strategy to remove various airborne particulates and/or gases from the air where the most common methods are particulate filtration, electrostatic precipitation, and gas sorption.

AIR CONTAMINANT Particulate matter, gas, or combination thereof, other than water vapor.

AIR CURTAIN Air bubbling through a perforated pipe causes an upward water flow that slows the spread of oil; used to stop fish from entering polluted water.

AIR EXCHANGE RATE Rate at which outside air replaces indoor air in a given space.

AIR GAP Open vertical gap or empty space that protects and separates drinking water supply from contamination by backflow or back siphonage by another water system in a treatment plant.

AIR HANDLING UNIT Type of heating and/or cooling distribution that channels warm or cool air to different parts of a building; equipment includes a blower or fan, heating and/or cooling coils, and related equipment such as controls, condensate drain pans, and air filters.

AIR MASS Body of air covering a relatively wide area and exhibiting horizontally uniform properties.

AIR MASS THUNDERSTORM Thunderstorm not associated with a front or other type of synoptic-scale forcing mechanism.

AIR PADDING Pumping dry air into a container to assist with the withdrawal of liquid or to force a liquefied gas out of the container.

AIR PERMEABILITY Permeability of soil with respect to air.

AIR PLENUM Any space used to convey air in a building, furnace, or structure.

AIR POLLUTION Presence of contaminants or pollutant substances in the air that interfere with human health or welfare, or produce other harmful environmental effects.

AIR POLLUTION CONTROL DEVICE Mechanism or equipment that cleans emissions generated by a source by removing pollutants that would otherwise be released to the atmosphere.

AIR POLLUTION EPISODE Period of abnormally high concentration of air pollutants, often due to low winds and temperature inversion, that can cause illness and death.

AIR POLLUTION POTENTIAL Meteorological potential for air pollution problems, considered without regard to the presence or absence of actual pollution sources.

AIR QUALITY Measure of the health-related and visual characteristics of the air, often derived from quantitative measurements of the concentrations of specific injurious or contaminating substances.

AIR QUALITY CRITERIA Levels of pollution and lengths of exposure above which adverse health and welfare effects may occur.

AIR QUALITY MODEL Mathematical or conceptual model used to estimate present or future air quality.

AIR QUALITY STANDARDS Level of pollutants prescribed by regulations that are not be exceeded during a given time in a defined area.

AIR RELEASE VALVE Valve, usually manually operated, that is used to release air from a pipe or fitting.

AIR SIDE ECONOMIZER Air economizer cycle reduces the load on the chilled water system by increasing the flow of outside air above the minimum required for ventilation when the outside air temperature is favorable in comparison to return air temperature.

AIR SLAKING Process of breaking up or sloughing when an indurated soil is exposed to air.

AIR SPARGING Injecting air or oxygen into an aquifer to strip or flush volatile contaminants as air bubbles up through the groundwater and is captured by a vapor extraction system.

AIR STAGNATION Meteorological situation in which there is a major buildup of air pollution in the atmosphere.

AIR STAGNATION ADVISORY National Weather Service product issued when major buildups of air pollution, smoke, dust, or industrial gases are expected near the ground for a period of time.

AIR STRIPPING Treatment system that removes volatile organic compounds (VOCs) from contaminated groundwater or surface water by forcing an airstream through the water and causing the compounds to evaporate.

AIR TOXICS Any air pollutant for which a national ambient air quality standard (NAAQS) does not exist (i.e., excluding ozone, carbon monoxide, PM-10, sulfur dioxide, nitrogen oxide) that may reasonably be anticipated to cause cancer; respiratory, cardiovascular, or developmental effects, reproductive dysfunctions, neurological disorders, heritable gene mutations, or other serious or irreversible chronic or acute health effects in humans.

AIR TRANSPORTABLE MOBILE UNIT Modularized transportable unit containing communications and observational equipment necessary to support a meteorologist preparing on-site forecasts at a wildfire or other incident.

AIR WAVES Air borne vibrations caused by explosions.

AIR/OIL TABLE Surface between the vadose zone and ambient oil; the pressure of oil in the porous medium is equal to atmospheric pressure.

AIRBORNE PARTICULATES Total suspended particulate matter found in the atmosphere as solid particles or liquid droplets.

AIRBORNE RELEASE Release of any pollutant into the air.

AIRBORNE SNOW SURVEY PROGRAM Center (NOHRSC) that makes airborne snow water equivalent and soil moisture measurements over large areas of the country that are subject to severe and chronic snow melt flooding.

AIRBORNE SNOW WATER EQUIVALENT MEASUREMENT THEORY Theory based on the fact that natural terrestrial gamma radiation is emitted from the potassium, uranium, and thorium radioisotopes in the upper eight inches of the soil.

AIRFA American Indian Religious Freedom Act.

AIRMET AIRman's METeorological advisory.

AIR-SPACE RATIO Ratio of volume of water that can be drained from a saturated soil or rock under the action of force of gravity to total volume of voids.

AIR-VOID RATIO Ratio of the volume of airspace to the total volume of voids in a soil mass.

AIS Automated Information System.

AISC American Institute of Steel Construction.

AIV Aviation Impact Variables.

ALACHLOR Herbicide, marketed under the trade name Lasso, used mainly to control weeds in corn and soybean fields.

ALAR Trade name for daminozide, a pesticide that makes fruits color deeper, firmer, and less likely to drop off before harvesting.

ALASKA CURRENT North Pacific Ocean current flowing counterclockwise in the Gulf of Alaska; northward flowing (warm) division of the Aleutian Current.

ALBEDO Reflectivity; the fraction of radiation striking a surface that is reflected by that surface.

ALBERTA CLIPPER Fast moving low pressure system that moves southeast out of Canadian Province of Alberta (southwest Canada) through the Plains, Midwest, and Great Lakes region usually during the winter; low pressure area is usually accompanied by light snow, strong winds, and colder temperatures; sometimes referred to as "Saskatchewan Screamer".

ALC Agency Location Code.

ALDICARB Insecticide sold under the trade name Temik. It is made from ethyl isocyanate.

ALERT (AUTOMATED LOCAL EVALUATION IN REAL TIME) Flood warning system consisting of remote sensors, data transmission by radio, and a computer software package developed by the National Weather Service; generic term used for a decision making software package.

ALERT FLOOD WARNING SYSTEM Cooperative, community-operated flood warning system.

ALERT STAGE Elevation, or stage, of a stream at which need-to-know officials are notified of the threat of possible flooding; synonymous with monitor stage or caution stage.

ALEUTIAN CURRENT Eastward flowing North Pacific Ocean current which lies north of the North Pacific Current.

ALEUTIAN LOW Semi-permanent, subpolar area of low pressure located in the Gulf of Alaska near the Aleutian Islands.

ALEVIN Young fish which has not yet absorbed its yolk sac.

ALGAE Simple plants containing chlorophyll that produce oxygen during sunlight hours and use oxygen during the night hours; microscopic plants that contain chlorophyll and live floating or suspended in water; affect water quality adversely by lowering the dissolved oxygen in the water; food for fish and small aquatic animals.

ALGAL BLOOM Rapid and flourishing growth of algae (typically green or bluegreen algae) in and on a body of water as a result of high phosphate concentration from farm fertilizers and detergents; affect water quality adversely and indicate potentially hazardous changes in local water chemistry.

ALGAL GROWTH POTENTIAL (AGP) Maximum algal dry weight biomass that can be produced in a natural water sample under standardized laboratory conditions.

ALGICIDE Substance or chemical used specifically to kill or control algae.

ALGORITHM Computer program (or set of programs) that is designed to systematically solve a certain kind of problem; step by step procedure for solving a problem.

ALIASING Process by which frequencies too high to be analyzed with the given sampling interval appear at a frequency less than the Nyquist frequency.

ALIQUOT Portion of a sample where one or more aliquots make up a sample.

ALKALI Soluble salt obtained from the ashes of plants.

ALKALI AGGREGATE REACTION (AAR) Deterioration of concrete by which the alkali in the cement paste in the concrete reacts chemically with the silica or carbonate present in some aggregates.

ALKALI CARBONATE REACTION (ACR) Reaction of alkalis which occurs between certain argillaceous dolomitic limestones and the alkaline pore solution in the concrete and causes expansion and extensive cracking.

ALKALINE Having a pH of 7.0 or above; condition of water or soil which contains a sufficient amount of alkali substances to raise the pH above 7.0; quality of being bitter due to alkaline content; pH modifier in the U.S. Fish and Wildlife Service wetland classification system.

ALKALINITY Capacity of bases, water or solutes to neutralize acids.

ALKALI-SILICA REACTION (ASR) Reaction of alkalis with aggregate with various forms of poorly crystalline reactive silica: opal, chert, flint and chalcedony and also tridymite, crystoblite and volcanic glasses.

ALKALI-SILICATE/SILICA REACTION (ASSR) Reaction of alkalis with strained quartz is thought to be one reactive component of aggregates causing this reaction.

ALLELOPATHY Influence of plants upon each other caused by products of metabolism.

ALLERGEN Substance that causes an allergic reaction in individuals sensitive to it.

ALLOCHTHONOUS Exotic species of a given area; deposits of material that originated elsewhere.

ALLOPATRIC Having separate and mutually exclusive areas of geographical distribution.

ALLOWABLE BEARING CAPACITY Maximum pressure that can be permitted on foundation soil, giving consideration to all pertinent factors, with adequate safety against rupture of the soil mass or movement of the foundation of such magnitude that the structure is impaired.

ALLOWABLE PILE BEARING LOAD Maximum load that can be permitted on a pile with adequate safety against movement of such magnitude that the structure is endangered.

ALLUVIAL Relating to and/or sand deposited by flowing water, such as in a riverbed, flood plain, or delta; alluvium deposited by a stream or flowing water.

ALLUVIAL AQUIFER Water-bearing deposit of unconsolidated material (sand and gravel) left behind by a river or other flowing water.

ALLUVIAL DEPOSIT Clay, silt, sand, gravel, or other sediment deposited by the action of running or receding water.

ALLUVIAL FAN Large fan-shaped accumulation of sediment deposited by streams where they emerge at the front of a mountain range.

ALLUVIAL REACH Reach of river with a sediment bed composed of the same type of sediment material as that moving in the stream.

ALLUVIAL STREAM Stream whose channel boundary is composed of appreciable quantities of the sediments transported by the flow, and which generally changes its bed forms as the rate of flow changes.

ALLUVIUM Material transported and deposited by flowing water; sediments deposited by erosional processes usually in the beds of rivers and streams, on a flood plain, on a delta, or at the base of a mountain.

ALONG-SLOPE WIND SYSTEM Closed, thermally driven diurnal mountain wind circulation whose lower branch blows up or down the sloping sidewalls of a valley or mountain and the upper branch blows in the opposite direction closing the circulation.

ALPINE SNOW GLADE Marshy clearing between slopes above the timberline in mountains.

ALTERNATE METHOD Any method of sampling and analyzing for an air or water pollutant that is not a reference or equivalent method but that has been demonstrated in specific cases-to EPA's satisfaction-to produce results adequate for compliance monitoring.

ALTERNATING CURRENT (AC) Electric current that reverses its direction (positive/negative values) at regular intervals.

ALTERNATIVE COMPLIANCE Policy that allows facilities to choose among methods for achieving emission-reduction or risk-reduction instead of command-and control regulations that specify standards and how to meet them; use of a theoretical emissions bubble over a facility to cap the amount of pollution emitted while allowing the company to choose where and how (within the facility) it complies.

ALTERNATIVE FUELS Substitutes for traditional liquid, oil-derived motor vehicle fuels like gasoline and diesel including mixtures of alcohol-based fuels with gasoline, methanol, ethanol, compressed natural gas, and others.

ALTERNATIVE REMEDIAL CONTRACT STRATEGY CONTRACTORS Government contractors who provide project management and technical services to support remedial response activities at National Priorities List sites.

ALTERNATIVES Courses of action that may meet the objectives of a proposal at varying levels of accomplishment, including the most likely future conditions without the project or action.

ALTIMETER Instrument that indicates the altitude of an object above a fixed level.

ALTIMETER SETTING Correction of the station pressure to sea level used by aviation.

ALTOCUMULUS Cloud of a class characterized by globular masses or rolls in layers or patches, the individual elements being larger and darker than those of cirrocumulus and smaller than those of stratocumulus. These clouds are of medium altitude, about 8000-20,000 ft (2400-6100 m).

ALTOSTRATUS Cloud of a class characterized by a generally uniform gray sheet or layer, lighter in color than nimbostratus and darker than cirrostratus. These clouds are of medium altitude, about 8000 to 20,000 ft (2400-6100 m).

ALUM Aluminum Sulfate.

AMA Adaptive Management Area.

AMALGAMATION Dissolving or blending of a metal (commonly gold and silver) in mercury to separate it from its parent material.

AMBIENT Surrounding natural conditions or environment at a given place and time; environmental or surrounding conditions.

AMBIENT AIR Any unconfined portion of the atmosphere: open air, surrounding air.

AMBIENT MEASUREMENT Measurement of the concentration of a substance or pollutant within the immediate environs of an organism; taken to relate it to the amount of possible exposure.

AMBIENT MEDIUM Material surrounding or contacting an organism (e.g., outdoor air, indoor air, water, or soil) through which chemicals or pollutants can reach the organism.

AMBIENT MONITORING All forms of monitoring conducted beyond the immediate influence of a discharge pipe or injection well and may include sampling of sediments and living resources.

AMBIENT TEMPERATURE Temperature of the surrounding air or other medium.

AMBIENT TOXICITY Measured by a toxicity test on a sample collected from a water body.

AMBURSEN DAM Buttress dam in which the upstream part is a relatively thin flat slab usually made of reinforced concrete.

AMCR Alternative Management Control Review.

AMMONIA Compound of nitrogen and hydrogen (NH_3) that is a common by-product of animal waste; readily converts to nitrate in soils and streams.

AMMONIUM One form of nitrogen that is usable by plants.

AMP See AMPERE; Adaptive Management Plan.

AMPERAGE Strength of an electric current measured in amperes; amount of electric current flow.

AMPERE (AMP) Unit of electric current or rate of flow of electrons. One volt across 1 ohm of resistance causes a current flow of 1 ampere.

AMPEROMETRIC Based on the electric current that flows between two electrodes in a solution.

AMPHIBIAN Vertebrate animals that have life stages both in water and on land (e.g., salamanders, frogs, and toads); animals capable of living either in water or land.

AMPLIFICATION Modification of the input bedrock ground motion by the overlying un-consolidation materials; causes the amplitude of the surface ground motion to be increased in some range of frequencies and decreased in others; function of the shear wave velocity and damping of the unconsolidated materials, its thickness and geometry, and the strain level of the input rock motion.

AMPLIFIER Device used to increase the strength of an analog signal.

AMPLITUDE Maximum magnitude of a quantity; maximum height of a wave.

AMPROMETRIC TITRATION Way of measuring concentrations of certain substances in water using an electric current that flows during a chemical reaction.

AMR Acquisition Management Review.

AMS American Meteorological Society.

ANABRANCH Diverging branch of a river which reenters the mainstream.

ANADROMOUS Fish that migrate from salt water to freshwater to breed; going up rivers to spawn.

ANADROMOUS FISH Migratory species that are born in freshwater, live mostly in estuaries and ocean water, and return to freshwater to spawn.

ANAEROBIC Condition in which free (atmospheric) or dissolved oxygen is not present in water; life or process that occurs in, or is not destroyed by, the absence of oxygen; pertaining to, taking place in, or caused by the absence of oxygen; opposite of aerobic.

ANAEROBIC DECOMPOSITION Reduction of the net energy level and change in chemical composition of organic matter caused by microorganisms in an oxygen-free environment.

ANALOG Class of measuring devices in which the output varies continuously as a function of the input (non-digital); historical instance of a given meteorological scenario or feature that is used for comparison with another scenario or feature.

ANALOG SIGNAL Signal that varies in a continuous manner.

ANALYTICAL MODEL Mathematical model generally assuming homogeneous aquifer properties, uniform flow direction and hydraulic gradient, uniform aquifer thickness, with simple upper and lower boundaries, and lateral boundaries are placed at an infinite distance.

ANBURS AlphaNumeric BackUp Replacement System.

ANCHOR BLOCK See THRUST BLOCK.

ANCHOR ICE Submerged Frazil ice attached or anchored to the river bottom, irrespective of its formation; ice in the bed of a stream or upon a submerged body or structure.

ANCHOR ICE DAM Accumulation of anchor ice which acts as a dam and raises the water level.

ANCILLARY DATA Other categories of data critical to interpreting water-quality data and formulating courses of action.

ANCILLARY SERVICES Other energy-related services that are required to control system frequency, to meet changing scheduling requirements, to react to changing loads and unexpected contingencies, and to ensure system stability (i.e. preventing blackouts).

ANDESITE Fine-grained, medium gray volcanic rock of intermediate composition between rhyolite and basalt.

ANEMOMETER Instrument used for measuring the speed of the wind.

ANEROID BAROMETER Instrument for measuring atmospheric pressure in which a needle, attached to the top of an evacuated box, is deflected as changes in atmospheric pressure cause the top of the box to bend in or out.

ANGELS Radar echoes caused by birds, insects, and localized refractive index discontinuities.

ANGLE OF EXTERNAL FRICTION (ANGLE OF WALL FRICTION) Angle between the abscissa and the tangent of the curve representing the relationship of shearing resistance to normal stress acting between soil and surface of another material.

ANGLE OF INTERNAL FRICTION (ANGLE OF SHEAR RESISTANCE) Angle between the axis of normal stress and the tangent to the Mohr envelope at a point representing a given failure-stress condition for solid material.

ANGLE OF OBLIQUITY Angle between the direction of the resultant stress or force acting on a given plane and the normal to that plane.

ANGLE OF REFLECTION Angle at which a reflected ray of energy leaves a reflecting surface that is measured between the outgoing ray and a perpendicular to the surface at the point of incidence (i.e. where the ray strikes).

ANGLE OF REPOSE Angle between the horizontal and the maximum slope that a particular soil or geologic material assumes through natural processes.

ANGLE OF WALL FRICTION Angle between the abscissa and the tangent of the curve representing the relationship of shearing resistance to normal stress acting between soil and surface of another material.

ANGLER-DAY Time spent fishing by one person for any part of a day.

ANGSTROM Unit of length equal to 10^{-8} cm.

ANIMAL DANDER Tiny scales of animal skin commonly considered an indoor air pollutant.

ANIMAL FEEDING OPERATION (AFO) Lot or facility (other than an aquatic animal production facility) where the following conditions are met: animals (other than aquatic animals) have been, are, or will be stabled or confined and fed or maintained for a total of 45 days or more in any 12-month period, and crops, vegetation, forage growth, or post-harvest residues are not sustained in the normal growing season over any portion of the lot or facility.

ANIMAL SPECIALTIES Water use associated with the production of fish in captivity except fish hatcheries, fur-bearing animals in captivity, horses, rabbits, and pets.

ANIMAL STUDIES Investigations using animals as surrogates for humans with the expectation that the results are pertinent to humans.

ANION Negatively charged ion in an electrolyte solution, attracted to the anode under the influence of a difference in electrical potential.

ANISOTROPIC MASS Mass having different properties in different directions at any given point.

ANISOTROPY Flow conditions vary with direction; conditions under which one or more hydraulic properties (e.g. hydraulic conductivity) of an aquifer vary from a reference point.

ANNEX (FUNCTIONAL) Emergency operations plan element that describes the jurisdiction's plan for functioning in that component area of activity during emergencies.

ANNUAL 7-DAY MINIMUM Lowest mean value for any 7-consecutive-day period in a year whereas the year may be either the calendar year or water year (October 1 through September 30).

ANNUAL ENERGY COST Variable costs relating to energy production in a year, usually expressed in mills per kilowatt-hour.

ANNUAL FAILURE PROBABILITY Probability of the load multiplied by the probability of failure.

ANNUAL FLOOD Maximum discharge peak during a given water year (October 1 - September 30).

ANNUAL FLOOD SERIES List of annual floods.

ANNUAL INSPECTION (AI) Yearly inspections of a dam and appurtenant facilities conducted by the local operating office that address both O&M and dam safety issues.

ANNUAL LOAD FACTOR Factor equal to energy generated in a year divided by the product of the peak demand for that year and the number of total hours in a year.

ANNUAL OPERATING COST Includes all annual operation and maintenance expense, wheeling, purchased power, etc.

ANNUAL RUNOFF Total quantity of water that is discharged ("runs off") from a drainage basin in a year that may be reported as a volume, a discharge per unit of drainage area, or a depth of water on the drainage basin.

ANNUAL WORK PLAN Annual budget document that describes proposed work to be performed and details the amount of funds required.

ANNUALIZED LOSS OF LIFE Sum of the probability of dam failure multiplied by the annual probability of the loading and the estimated number of lives that would be lost for each dam failure scenario under a particular loading category.

ANNULAR Ring-shaped.

ANNULAR SPACE Ring-shaped space located between two circular objects (e.g. two pipes).

ANNULUS Space between the casing of a well and the well bore or space between the tubing and casing of a well.

ANODE Positive pole or electrode of an electrolytic system; attracts negatively charged particles or ions (anions).

ANOMALIES Externally visible skin or subcutaneous disorders, including deformities, eroded fins, lesions, and tumors.

ANOMALOUS PROPAGATION (AP) Radar term for false (non-precipitation) echoes resulting from non-standard propagation of the radar beam under certain atmospheric conditions.

ANOMALY Deviation of a measurable unit over a period in a given region from the long-term average.

ANOXIC Without oxygen.

ANSI American National Standards Institute.

ANTAGONISM Situation where two chemicals interfere with each other's actions, or one chemical interferes with the action of the other; characteristic property of a mixture of toxicants that exhibits a less-than-additive total toxic effect.

ANTARCTIC "OZONE HOLE" Seasonal depletion of ozone in the upper atmosphere above a large area of Antarctica.

ANTECEDENT CONDITIONS Watershed conditions prevailing prior to an event; normally used to characterize basin wetness (e.g. soil moisture).

ANTECEDENT FLOOD Flood or series of floods assumed to occur prior to the occurrence of an inflow design flood (IDF).

ANTECEDENT PRECIPITATION INDEX (API) Index of moisture stored within a drainage basin before a storm.

ANTENNA GAIN Measure of effectiveness of a directional antenna as compared to an isotropic radiator where the maximum value is called antenna gain by convention.

ANTHELION Luminous white spot that appears on the parhelic circle at the same altitude as the Sun and 180 degrees from it in azimuth.

ANTHROPOGENIC Created, occurring because of, or influenced by human activity.

ANTHROPOGENIC SOURCE Pollutant source caused or produced by humans.

ANTI-BACKSLIDING Provision in U.S. Federal Regulations that requires a reissued permit to be as stringent as the previous permit with some exceptions.

ANTICLINE Fold in the Earth's crust that curves upward (convex upward) whose core contains stratigraphically older rock; upward fold in rock layers that creates an arched or domelike uplift of sedimentary layers.

ANTICYCLOGENESIS Formation or intensification of an anticyclone or high pressure center.

ANTICYCLONE Large-scale circulation of winds around a central region of high atmospheric pressure, clockwise in the Northern Hemisphere, counterclockwise in the Southern Hemisphere.

ANTICYCLONE (ACYC) A large-scale circulation of winds around a central region of high atmospheric pressure, clockwise in the Northern Hemisphere, counterclockwise in the Southern Hemisphere.

ANTICYCLONIC ROTATION Rotation in the opposite sense as the Earth's rotation; opposite of cyclonic rotation.

ANTIDEGRADATION Policies which ensure protection of water quality for a particular water body where the water quality exceeds levels necessary to protect fish and wildlife propagation and recreation on and in the water; special protection of waters designated as outstanding natural resource waters; typically adopted by each state to minimize adverse effects on water.

ANTI-DEGRADATION CLAUSE Part of federal air quality and water quality requirements prohibiting deterioration where pollution levels are above the legal limit.

ANTILLES CURRENT Current that originates in the vicinity of the Leeward Islands as part of the Atlantic North Equatorial Current.

ANTI-MICROBIAL Agent that kills microbes.

ANTI-WIND Upper or return branch of an along-valley wind system, as confined within a valley, and blowing in a direction opposite to the winds in the lower altitudes of the valley.

ANVIL Flat, spreading top of a cumulonimbus cloud, often shaped like an anvil.

ANVIL CRAWLER Lightning discharge occurring within the anvil of a thunderstorm, characterized by one or more channels that appear to crawl along the underside of the anvil.

ANVIL DOME Large overshooting top or penetrating top.

ANVIL ROLLOVER Slang for circular or semicircular lip of clouds along the underside of the upwind part of a back-sheared anvil, indicating rapid expansion of the anvil.

ANVIL ZITS Slang for frequent (often continuous or nearly continuous), localized lightning discharges occurring from within a thunderstorm anvil.

AO Area Office.

AOP Annual Operating Plan.

AP Analysis Plan.

AP INDEX Averaged planetary A Index based on data from a set of specific stations.

APCD Air Pollution Control District.

APD Average wave period (seconds) of all waves during the 20-minute period.

APE Area of Potential Effect.

APHELION Point on the annual orbit of a body (about the Sun) that is farthest from the Sun; at present; opposite of perihelion.

API METHOD Statistical method to estimate the amount of surface runoff which will occur from a basin from a given rainstorm based on the antecedent precipitation index, physical characteristics of the basin, time of year, storm duration, rainfall amount, and rainfall intensity.

APOGEE Farthest distance between the Moon and Earth or the Earth and Sun.

APPARENT TEMPERATURE Measure of human discomfort due to combined heat and humidity (e.g. heat index).

APPARENT WIND Speed and true direction from which the wind appears to blow with reference to a moving point (i.e. relative wind).

APPENDIX Emergency operations plan element attached to a functional annex to provide information on special approaches or requirements generated by unique characteristics of specified hazards of particular concern to the jurisdiction.

APPLICABLE OR RELEVANT AND APPROPRIATE REQUIREMENTS (ARARS) State or federal statute that pertains to protection of human life and the environment in addressing specific conditions or use of a particular cleanup technology at a Superfund site.

APPLICATION EFFICIENCY Ratio of the average depth of irrigation water infiltrated and stored in the root zone to the average depth of irrigation water applied that is typically expressed as a percent.

APPLIED DOSE Amount of a substance in contact with the primary absorption boundaries of an organism (e.g., skin, lung tissue, gastrointestinal track) and available for absorption.

APPLIED WATER (DELIVERED WATER) Water delivered to a user; may be used for either inside uses or outside watering, does not include precipitation or distribution losses and may apply to metered or unmetered deliveries.

APPRAISAL ESTIMATE Estimate used in an appraisal study as an aid in selecting the most economical plan by comparing alternative features or for determining whether more detailed investigations of a potential project are economically justified; used to obtain approximate costs in a short period of time with inadequate data; not to be used for project authorization.

APPRAISAL LEVEL OF DETAIL Level of detail necessary to facilitate making decisions on whether or not to proceed with a detailed study and evaluation of any alternative.

APPRAISAL STUDY (APPRAISAL REPORT) Study incorporating an appraisal level of detail.

APPROACH CHANNEL Channel upstream from that portion of the spillway having a concrete lining or concrete structure; channel upstream from intake structure of an outlet works; channel that is generally unlined, excavated in rock or soil, with or without riprap, soil cement or other types of erosion protection.

APPROPRIATION Amount of water legally set apart or assigned to a particular purpose or use.

APPROPRIATION DOCTRINE System for allocating water to private individuals used in most Western states whereas the doctrine of Prior Appropriation was in common use throughout the arid west as early settlers and miners began to develop the land that was based on the concept of "First in Time, First in Right."

APPROPRIATIVE Water rights to or ownership of a water supply which is acquired for the beneficial use of water by following a specific legal procedure.

APPURTENANT STRUCTURE Ancillary features of a dam such as outlets, spillways, power plants, tunnels, bridges, drain systems, tunnels, towers, etc.

APRCH Approach.

APRCHG Approaching.

APRNT Apparent.

APRON (FORE APRON) Section of concrete or riprap constructed upstream or downstream from a control structure to prevent undercutting of the structure; floor or lining of concrete, timber, or other suitable material at the toe of a dam, discharge side of a spillway, a chute, or other discharge structure, to protect the waterway from erosion from falling water or turbulent flow; short ramp with a slight pitch.

APST Aviation Products and Services Team.

AQIA Air Quality Impact Analysis.

AQMD Air Quality Management District.

AQMP Air Quality Management Plan.

AQUACULTURE Science of farming organisms (plants and animals) that live in water, such as fish, shellfish, and algae.

AQUACULTURE WELLS Class V wells that dispose of water used for the cultivation of marine and freshwater animals and plants.

AQUATIC Living, growing, or occurring in or on the water.

AQUATIC ALGAE Microscopic plants that grow in sunlit water containing phosphates, nitrates, and other nutrients.

AQUATIC COMMUNITY Association of interacting populations of aquatic organisms in a given water body or habitat.

AQUATIC ECOSYSTEM Stream channel, lake or estuary bed, water, and (or) biotic communities and the habitat features that occur therein.

AQUATIC GUIDELINES Specific levels of water quality which, if reached, may adversely affect aquatic life; non-enforceable guidelines issued by a governmental agency or other institution.

AQUATIC HABITAT Environments characterized by the presence of standing or flowing water.

AQUATIC-LIFE CRITERIA Water-quality guidelines for protection of aquatic life; criteria established by the U.S. Environmental Protection Agency for protection of aquatic organisms.

AQUEDUCT Pipe, man-made canal, conduit, or channel designed to transport water from a remote source, usually by gravity.

AQUEOUS Something made up of, similar to, or containing water; watery.

AQUEOUS SOLUBILITY Maximum concentration of a chemical that will dissolve in pure water at a reference temperature.

AQUICLUDE Layer of clay which limits the movement of groundwater; formation which contains water but cannot transmit it rapidly enough to furnish a significant supply to a well or spring.

AQUIFER Natural underground bed or layer (i.e., soil, sand, gravel, rock or porous storage) that contains water usually capable of yielding a large amount (i.e. supply) of water; water-bearing stratum of permeable rock, sand, or gravel; underground water-bearing geologic formation or structure that provides a groundwater reservoir; geologic formation, group of formations, or part of a formation that stores and transmits water and yields significant quantities of groundwater to wells and springs; permeable underground layers that hold or transmit groundwater below the water table that will yield water to a well in sufficient quantities for beneficial use.

AQUIFER (CONFINED) Aquifer bounded above and below by confining beds (e.g. impermeable material) in which the hydraulic head is above the top of the aquifer; aquifer under pressure so that when the aquifer is penetrated by a well, the water will rise to a level associated with the pressure.

AQUIFER (UNCONFINED) Aquifer whose hydraulic head surface (i.e. water table) is at equilibrium with atmospheric pressure.

AQUIFER RECHARGE/RECOVERY WELLS
Class V wells that are used to inject fluids to recharge an aquifer and may have secondary purposes such as saline intrusion prevention, subsidence control, or aquifer storage and recovery (ASR).

AQUIFER REMEDIATION WELLS
Class V wells that are used to clean up, treat, or prevent contamination of underground sources of drinking water (USDWs).

AQUIFER TEST Test to determine hydraulic properties of an aquifer.

AQUIFUGE Geological formation which has no interconnected openings and cannot hold or transmit water.

AQUITARD Geological formation that may contain groundwater but is not capable of transmitting significant quantities of it under normal hydraulic gradients; may function as confining bed.

ARABLE Having soil or topographic features suitable for cultivation (e.g. farming).

ARABLE LAND Land which when farmed that provides a reasonable return to the farm's expenses.

ARAM Aviation, Range, and Aerospace Meteorology.

ARAR Applicable or Relevant and Appropriate Requirement.

ARCH DAM Concrete or masonry dam which is curved upstream in plan so as to transmit the major part of the water load to the abutments and to keep the dam in compression; solid concrete dam curved upstream in plan where arch provides resistance to movement and both the weight and shape of the structure provide great resistance to the pressure of water; used in sites where the ratio of width between abutments to height is not great and where the foundation at the abutments is solid rock capable of resisting great forces.

ARCH FILAMENT SYSTEM (AFS) Bright, compact plage crossed by a system of small, arched filaments, which is often a sign of rapid or continued growth in an Active Region.

ARCHAIC Cultural stage following the earliest known human occupation in the New World (about 5,500 B.C. to A.D. 100) characterized by a generalized hunting and gathering lifestyle and seasonal movement to take advantage of a variety of resources.

ARCH-BUTTRESS DAM (CURVED BUTTRESS DAM) Buttress dam which is curved in plan.

ARCHEOLOGY Study of human cultures through the recovery and analysis of their material relics.

ARCH-GRAVITY DAM An arch dam which is only slightly thinner than a gravity dam.

ARCHIMEDEAN SCREW Ancient water-raising device attributed to Archimedes, made up of a spiral tube coiled about a shaft or of a large screw in a cylinder, revolved by hand; pump consisting of an inclined, revolving, corkscrew-shaped shaft tightly enclosed in a pipe.

ARCHING Transfer of stress from a yielding part of a soil or rock mass to adjoining less-yielding or restrained parts of the mass.

ARCHITECTURAL COATINGS Coverings such as paint and roof tar that are used on exteriors of buildings.

ARCTIC Region within the Arctic Circle, or, loosely, northern regions in general, characterized by very low temperatures.

ARCTIC FRONT Boundary or front separating deep, cold arctic air from shallower, relatively less cold polar air.

ARCTIC OSCILLATION (AO) Pattern in which atmospheric pressure at polar and middle latitudes fluctuates between negative and positive phases.

ARCTIC SEA SMOKE Steam fog, but often specifically applied to steam fog rising from small open water within sea ice.

ARCUS Low, horizontal cloud formation associated with the leading edge of thunderstorm outflow (i.e., the gust front).

AREA FORECAST DISCUSSION National Weather Service product intended to provide a well-reasoned discussion of the meteorological thinking which went into the preparation of the Zone Forecast Product that at the end of the discussion includes a list of all advisories, non-convective watches, and non-convective warnings.

AREA OF INFLUENCE Area covered by the drawdown curves of a given pumping well or combination of wells at a particular time.

AREA OF INFLUENCE OF A WELL
Area surrounding a well within which the piezometric surface has been lowered when pumping has produced a maximum steady rate of flow (i.e. zone of influence).

AREA OF REVIEW (AOR) Area around a deep injection well that must be checked for artificial penetrations, such as other wells, to determine if flow between aquifers will be induced by the injection operation before a permit is issued.

AREA SOURCE Any source of air pollution that is released over a relatively small area but which cannot be classified as a point source (e.g., vehicles, small engines, small businesses, household activities, biogenic sources, etc.); array of pollutant sources, so widely dispersed and uniform in strength that they can be treated in a dispersion model as an aggregate pollutant release from a defined area at a uniform rate.

AREA WIDE HYDROLOGIC PREDICTION SYSTEM (AWHPS)
Computer system which automatically ingests areal flash flood guidance values and WSR-88D products and displays this data and other hydrologic information on a map background.

AREA-CAPACITY CURVE Graph showing the relation between the surface area of the water in a reservoir, the corresponding volume, and elevation.

AREA-CAPACITY TABLE Table giving reservoir storage capacity, and sometimes surface areas, in terms of elevation increments.

ARID Term describing a climate or region in which precipitation is so deficient in quantity or occurs so infrequently that intensive agricultural production is not possible without irrigation.

ARMORING Process of progressive coarsening of the bed layer by removal of fine particles until it becomes resistant to scour; formation of a resistant layer of relatively large particles resulting from removal of finer particles by erosion.

ARMS Automated Records Management System.

AROCLOR Registered trademark for a group of polychlorinated biphenyls that were manufactured by the Monsanto Company prior to 1976.

AROMATICS Type of hydrocarbon (e.g., benzene or toluene) with a specific type of ring structure.

ARP Address Resolution Protocol.

ARROYO Small, deep, flat-floored channel or gully of an ephemeral or intermittent stream, usually with nearly vertical banks cut, into unconsolidated material; gully or channel cut by an intermittent stream; water-carved channel or gulley in an arid area, usually rather small in cross section with steep banks, dry much of the time due to infrequent rainfall and the depth of the cut which does not penetrate below the level of permanent groundwater.

ARSENICALS Pesticides containing arsenic.

ARTESIAN Ground water held under pressure in porous rock or soil confined by impermeable geological formations.

ARTESIAN WATER Ground water that is under pressure when tapped by a well and is able to rise above the level at which it is first encountered.

ARTESIAN WELL Well drilled into a confined aquifer with enough hydraulic pressure for the water to flow to the surface without pumping; free flowing well.

ARTIFACT Any human-made or used object, intact or in pieces, 50 years or older.

ARTIFICIAL CONTROL Weir or other man-made structure which serves as the control for a stream-gaging station.

ARTIFICIAL DRAINS Man-made or constructed drains.

ARTIFICIAL RECHARGE Process where water is put back into groundwater storage from surface-water supplies such as irrigation, or induced infiltration from streams or wells, or by pumping water directly into an aquifer; addition of surface water to a groundwater reservoir by human activity.

ARTIFICIAL SUBSTRATE Device that purposely is placed in a stream or lake for colonization of organisms.

ARTS Automated Resources Training System.

ARWRI American River Water Resources Investigation.

AS Assistant Secretary.

ASAP AHOS SHEF Automatic Processing System, As Soon As Possible.

ASAPTRAN Software component of ASAP.

ASBESTOS Mineral fiber that can pollute air or water and cause cancer or asbestosis when inhaled.

ASBESTOS ABATEMENT Procedures to control fiber release from asbestos-containing materials in a building or to remove them entirely, including removal, encapsulation, repair, enclosure, encasement, and operations and maintenance programs.

ASBESTOS ASSESSMENT Evaluation of the physical condition and potential for damage of all friable asbestos containing materials and thermal insulation systems.

ASBESTOS PROGRAM MANAGER Building owner or designated representative who supervises all aspects of the facility asbestos management and control program.

ASBESTOS-CONTAINING WASTE MATERIALS (ACWM) Mill tailings or any waste that contains commercial asbestos and is generated by a source covered by the Clean Air Act Asbestos NESHAPS.

ASBESTOSIS Disease associated with inhalation of asbestos fibers that makes breathing progressively more difficult and can be fatal.

ASC Administrative Service Center.

A-SCALE SOUND LEVEL Measurement of sound approximating the sensitivity of the human ear, used to note the intensity or annoyance level of sounds.

ASCE American Society of Civil Engineers.

ASCII American Standard Code for Information Interchange.

ASCMP Administrative Support Career Management Program.

ASCS Agricultural Stabilization and Conservation Service.

ASH Mineral content of a product remaining after complete combustion.

ASH MASS Mass or amount of residue present after the residue from a dry-mass determination has been ashed in a muffle furnace at a temperature of 500 °C for 1 hour.

ASHFALL ADVISORY Advisory issued for conditions associated with airborne ash plume resulting in ongoing deposition at the surface that may originate directly from a volcanic eruption, or indirectly by wind suspending the ash.

ASL Above Sea Level.

ASOS IDS Four character identifier assigned to each Automated Surface Observing System.

ASPECT Direction toward which a slope faces with respect to the compass.

ASPHYXIANTS Chemicals that starve the cells of an individual from the life-giving oxygen needed to sustain metabolism.

ASR Alkali-Silica Reaction.

ASR-9 Airport Surveillance Radar (FAA)

ASSAY Test for a specific chemical, microbe, or effect.

ASSE American Society of Safety Engineers.

ASSESSED WATERS Water bodies for which the state is able to make use-support decisions based on actual information.

ASSESSMENT ENDPOINT Explicit expression of the environmental value to be protected.

ASSIMILATION Ability of a body of water to purify itself of pollutants.

ASSIMILATIVE CAPACITY Capacity of a natural body of water to receive wastewaters or toxic materials without deleterious effects and without damage to aquatic life or humans who consume the water.

ASSOCIATED FACILITY Term used to describe those facilities examined by the respective regional or area office including most carriage, distribution, and drainage systems, small diversion works, small pumping plants and power plants, open and closed conduits, tunnels, siphons, small regulating reservoirs, waterways, and class B bridges.

ASSOCIATED PRINCIPAL USER User with dedicated communications to a WSR-88D unit.

ASSOCIATION OF BOARDS OF CERTIFICATION (ABC) International organization representing boards which certify the operators of waterworks and wastewater facilities.

ASSR Alkali-Silicate/Silica Reaction.

ASTM American Society for Testing and Materials.

ASTRONOMICAL DAWN Time at which the Sun is 18 degrees below the horizon in the morning; point in time at which the Sun starts lightening the sky; prior to this time, the sky is completely dark.

ASTRONOMICAL DUSK Time at which the Sun is 18 degrees below the horizon in the evening; the Sun no longer illuminates the sky.

ASTRONOMICAL UNIT (AU) Mean Earth-Sun distance; equal to 1.496×10^{13} cm; equal to 214.94 solar radii.

ASTS Above Ground Storage Tanks.

ASYMMETRIC Not similar in size, shape, form or arrangement of parts on opposite sides of a line, point or plane.

ATMOSPHERE Air surrounding and bound to the Earth.

ATMOSPHERIC BOUNDARY LAYER Layer of air adjacent to a bounding surface where the effects of friction are significant; roughly the lowest one or two kilometers of Earth's atmosphere.

ATMOSPHERIC CIRCULATION MODEL Mathematical model for quantitatively describing, simulating, and analyzing the structure of the circulation in the atmosphere and the underlying causes.

ATMOSPHERIC DEPOSITION Transfer of substances from the air to the surface of the Earth either in wet form (rain, fog, snow, dew, frost, hail) or in dry form (gases, aerosols, particles).

ATMOSPHERIC PRESSURE Pressure exerted by the Earth's atmosphere at any given point; determined by taking the product of the gravitational acceleration at the point and the mass of the unit area column of air above the point; equal to 14.7 pounds per square inch at sea level or 29.92 inches of mercury as measured by a standard barometer.

ATMOSPHERIC RADIATION Infrared radiation (energy in the wavelength interval of 3- 80 micrometer) emitted by or being propagated through the atmosphere; consists of both upwelling and downwelling components.

ATMP On a buoy report, the air temperature (Celsius).

AT-REST EARTH PRESSURE Value of the earth pressure when the soil mass is in its natural state without having been permitted to yield or without having been compressed.

ATTAINMENT AREA Area considered to have air quality as good as or better than the national ambient air quality standards as defined in the Clean Air Act; there may be an attainment area for one pollutant and a nonattainment area for others.

ATTENUATION Reduction in the peak of a hydrograph resulting in a more broad, flat hydrograph; decrease in the amplitude of a flood wave due to channel geometry and energy loss as it progresses downstream; process by which a compound is reduced in concentration over time, through absorption, adsorption, degradation, dilution, and/or transformation; decrease with distance of sight caused by attenuation of light by particulate pollution; reduction of the radar beam power due to the reflection or absorption of energy when it strikes a target; decrease in amplitude of the seismic waves with distance due to geometric spreading, energy absorption, and scattering.

ATTERBERG LIMITS (CONSISTENCY LIMITS) Boundaries (determined by laboratory tests) of moisture content in a soil between the liquid state and plastic state (known as liquid limit), between the plastic state and the semisolid state (known as the plastic limit), and between the semisolid state and the solid state (known as the shrinkage limit).

ATTM AT This Time.

ATTRACTANT Chemical or agent that lures insects or other pests by stimulating their sense of smell.

ATTRIBUTE SURVEY Survey to determine the important components of the recreational experience.

ATTRITION Wearing or grinding down of a substance by friction; dust from such processes contributes to air pollution.

ATV All-Terrain Vehicle.

ATYPICAL Not typical.

AUGER Rotating drill having a screw thread that carries cuttings away from the face.

AUGMENTED REPORT Meteorological report prepared by an automated surface weather observing system for transmission with certified observers signed on to the system to add information to the report.

AUM Animal Unit Months.

AUO Administratively Uncontrollable Overtime.

AURORA Faint visual phenomenon associated with geomagnetic activity that occurs mainly in the high-latitude night sky; typical 100 to 250 km above the ground.

AURORA AUSTRALIS Also known as the southern lights; the luminous, radiant emission from the upper atmosphere over middle and high latitudes, and centered around the Earth's magnetic poles; same as Aurora Borealis, but in the Southern Hemisphere.

AURORA BOREALIS Luminous, radiant emission from the upper atmosphere over middle and high latitudes, and centered around the Earth's magnetic poles.

AURORAL OVAL Oval band around each geomagnetic pole which is the locus of structured aurorae.

AUTHORIZATION Act by the Congress of the United States which authorizes use of public funds to carry out a prescribed action.

AUTHORIZED PROGRAM OR AUTHORIZED STATE State, Territorial, Tribal, or interstate NPDES program which has been approved or authorized by EPA.

AUTHORIZED RECLAMATION PROJECT Congressionally approved Bureau of Reclamation project that has been authorized for specific purposes.

AUTHOR'S SIGNATURE Signature of the person or persons with primary responsibility for writing the document.

AUTOCLAVE Device that uses steam to sterilize equipment and deactivate bacteria, viruses, fungi, and spores.

AUTOMATED EVENT-REPORTING GAUGE For river stage gauges, IFLOWS pressure transducer type gauges can be programmed to report if water surface rises or falls by a predetermined amount.

AUTOMATED LOCAL EVALUATION IN REAL TIME (ALERT) Local flood warning system where river and rainfall data are collected via radio signals in real-time at an ALERT base station.

AUTOMATED MUTUAL ASSISTANCE VESSEL RESCUE SYSTEM (AMVER) System operated by the U.S. Coast Guard which computes the nearest available rescue vessels for vessels in distress using vessel track and position reports supplied by participating vessels.

AUTOMATED REPORT Meteorological report prepared by an automated surface weather observing system for transmission, and with no certified weather observers signed on to the system.

AUTOMATED SURFACE OBSERVING SYSTEM (ASOS) Joint effort of the National Weather Service (NWS), the Federal Aviation Administration (FAA), and the Department of Defense (DOD) to serve as the nation's primary surface weather observing network.

AUTOMATED TONE DIAL TELEPHONE DATA COLLECTION SYSTEM (ATDTDCS) Data collection system where cooperative observers collect precipitation, stage, and temperature data then transmit the data to the National Weather Service ATDTDCS computer through the telephone lines.

AUTOMATIC GAIN CONTROL Method of automatically controlling the gain of a receiver, particularly one that holds the output level constant regardless of the input level.

AUTOMATIC GENERATION CONTROL (AGC) Computerized power system regulation to maintain scheduled generation within a prescribed area in response to changes in transmission system operational characteristics.

AUTOMATIC RADIOTHEODOLITE (ART) Ground-based radio direction finder that automatically tracks a balloon-borne radiosonde.

AUTOMATION OF FIELD OPERATIONS AND SERVICES (AFOS) Computer system linking National Weather Service offices for the transmission of weather data.

AUTUMN Season of the year that is the transition period from summer to winter, occurring as the Sun approaches the winter solstice. Meteorological autumn (different from standard/astronomical autumn) begins September 1 and ends November 30.

AUTUMNAL EQUINOX Equinox at which the Sun approaches the Southern Hemisphere, marking the start of astronomical autumn in the Northern Hemisphere; time of this occurrence is approximately September 22.

AUXILIARY EQUIPMENT Accessory equipment necessary for the operation of a generating station.

AUXILIARY SPILLWAY Secondary spillway designed to operate only during exceptionally large flood flows; allows inflows from large storms to be released from the reservoir before the water level raises high enough to overtop the dam; any secondary spillway which is designed to be operated very infrequently and possibly in anticipation of some degree of structural damage or erosion to the spillway during operation.

AVADS Automated Vacancy Announcement Distribution System.

AVAILABILITY SESSION Informal meeting at a public location where interested citizens can talk with public officials on a one-to-one basis.

AVAILABLE CAPACITY Amount of water held in the soil that is available to the plants.

AVAILABLE CHLORINE Measure of the amount of chlorine available in chlorinated lime, hypochlorite compounds, and other materials used as a source of chlorine when compared with that of liquid or gaseous chlorines.

AVALANCHE Mass of snow, rock, and/or ice falling down a mountain or incline.

AVALANCHE ADVISORY Preliminary notification that conditions may be favorable for the development of avalanches in mountain regions.

AVERAGE Arithmetic mean; sum of the values divided by the number of values.

AVERAGE ANNUAL RUNOFF Average value of annual runoff amounts calculated for a selected period of record that represents average hydrologic conditions.

AVERAGE DEGREE OF CONSOLIDATION Ratio of the total volume change in a soil mass at a given time to the total volume change anticipated in the soil mass due to primary consolidation.

AVERAGE DISCHARGE Arithmetic average of all complete water years of record of surface water discharge whether consecutive or not where the term "average" is generally reserved for average of record and "mean" is used for averages of shorter periods, namely, daily mean discharge.

AVERAGE END CONCEPT Averaging of the two end cross sections of a reach in order to smooth the numerical results.

AVERAGE ENERGY Total power generation produced by a power plant during all of the years of its actual or simulated operation divided by the number of years of actual or simulated operation.

AVERAGE MONTHLY DISCHARGE LIMITATIONS Highest allowable average of daily discharges over a calendar month, calculated as the sum of all daily discharges measured during that month divided by the number of days on which monitoring was performed (except in the case of fecal coliform).

AVERAGE WAVE PERIOD (AVP) Average period (seconds) of the highest one-third of the wave observed during a 20 minute sampling period.

AVERAGE WEEKLY DISCHARGE LIMITATION Highest allowable average of daily discharges over a calendar week, calculated as the sum of all daily discharges measured during a calendar week divided by the number of daily discharges measured during that week.

AVERAGE YEAR SUPPLY Average annual supply of a water development system over a long period.

AVERAGE YEAR WATER DEMAND Demand for water under average hydrologic conditions for a defined level of development.

AVERAGING PERIOD Period of time over which the receiving water concentration is averaged for comparison with criteria concentrations; limits the duration of concentrations above the criteria.

AVIAN Having to do with birds.

AVIATION MODEL (AVN) 120-hour numerical model of the atmosphere.

AVOIDED COST Cost a utility would incur to generate the next increment of electric capacity using its own resources.

AVOIRDUPOIS WEIGHT English and American system of weights based on a pound of 16 ounces.

AWARDS Agricultural WAter Resources Decision Support.

AWOL Absence WithOut Leave.

AWTR Advanced Water Treatment Research.

AWWA American Water Works Association.

AXIS Straight line around which a shaft or body revolves.

AXIS OF DAM Vertical plane or curved surface, appearing as a line in plan or cross section, to which horizontal dimensions can be referred, coincident with the upstream face at the top of the dam.

AZIMUTH Direction in terms of a 360° compass where North is 0°, east is 90°, south is 180°, and west is 270°.

AZIMUTH ANGLE Direction or bearing toward which a sloping surface faces.

AZORES CURRENT One of the currents of the North Atlantic subtropical gyre.

AZORES HIGH Alternate term for Bermuda High; semi-permanent, subtropical area of high pressure in the North Atlantic Ocean off the East Coast of North America that migrates east and west with varying central pressure.

B

B Abbreviation used on long-term climate outlooks issued by CPC to indicate areas that are likely to be below normal for a given parameter (temperature, precipitation, etc.).

B/C Benefit-Cost ratio.

BA Biological Assessment.

BACK DOOR COLD FRONT Cold front moving south or southwest along the Atlantic seaboard and Great Lakes.

BACK PRESSURE Pressure that can cause water to backflow into the water supply when a user's water system is at a higher pressure than the public water system.

BACK-BUILDING THUNDERSTORM Thunderstorm in which new development takes place on the upwind side, such that the storm seems to remain stationary or propagate in a backward direction.

BACKFILL Material used to fill an excavated trench, or the process of such refilling.

BACKFILL CONCRETE Concrete used in refilling excavation in lieu of earth material.

BACKFIRE Fire started to stop an advancing fire by creating a burned area in its path.

BACKFLOW Backing up of water through a conduit or channel in the direction opposite to normal flow; reverse flow condition, created by a difference in water pressures, which causes water to flow back into the distribution system.

BACKFLOW/BACK SIPHONAGE Reverse flow condition created by a difference in water pressures that causes water to flow back into the distribution pipes of a drinking water supply from any source other than the intended one.

BACKFURROW First cut of a plow, from which the slice is laid on undisturbed soil.

BACKGROUND CONCENTRATION Concentration of a substance in a particular environment that is indicative of minimal influence by human (anthropogenic) sources.

BACKGROUND LEVEL Concentration of a substance in an environmental media (air, water, or soil) that occurs naturally or is not the result of human activities; in exposure assessment, the concentration of a substance in a defined control area, during a fixed period of time before, during, or after a data-gathering operation.

BACKING (BCKG) Counterclockwise shift in wind direction.

BACKING WINDS Winds which shift in a counterclockwise direction with time at a given location or change direction in a counterclockwise sense with height; opposite of veering winds.

BACKSCATTER Portion of power scattered back in the incident direction.

BACK-SHEARED ANVIL Slang for thunderstorm anvil which spreads upwind, against the flow aloft; often implies a very strong updraft and a high severe weather potential.

BACKSIGHT Rod reading taken on a point of known elevation, a benchmark or a turning point; added to the known elevation to arrive at the height of the instrument.

BACKSIPHONAGE Form of backflow caused by a negative or below atmospheric pressure within a water system.

BACKWASHING Reversing the flow of water back through the filter media to remove entrapped solids.

BACKWATER Water backed up or retarded in its course as compared with its normal or natural condition of flow; body of water in which the flow is slowed or turned back by an obstruction such as a bridge or dam, an opposing current, or the movement of the tide; small, generally shallow body of water with little or no current of its own; stagnant water in a small stream or inlet.

BACKWATER CURVE Longitudinal profile of the water surface in an open channel where the depth of flow has been increased by an obstruction, an increase in channel roughness, a decrease in channel width, or a flattening of the bed slope.

BACKWATER EFFECT Effect which a dam or other obstruction or construction has in raising the surface of the water upstream from it.

BACKWATER FLOODING Upstream flooding caused by downstream conditions such as channel restriction and/or high flow in a downstream confluence stream.

BACKWATER PROFILE Longitudinal profile of the water surface in a stream where the water surface is raised above its normal level by a natural or artificial obstruction.

BACKYARD COMPOSTING Diversion of organic food waste and yard trimmings from the municipal waste stream by composting hem in one's yard through controlled decomposition of organic matter by bacteria and fungi into a humus-like product.

BACT Best Available Control Technology

BACTERIA Single-celled microscopic organisms; microscopic unicellular organisms, typically spherical, rod-like, or spiral and threadlike in shape, often clumped into colonies; microscopic living organisms that can aid in pollution control by metabolizing organic matter in sewage, oil spills or other pollutants; microscopic living organisms that can cause human, animal and plant health problems.

BACTERICIDE Pesticide used to control or destroy bacteria, typically in the home, schools, or hospitals.

BADA Bay Area Dischargers Association.

BAFFLE Flat board or plate, deflector, guide, or similar device constructed or placed in flowing water or slurry systems to cause more uniform flow velocities to absorb energy and to divert, guide, or agitate liquids.

BAFFLE BLOCK (DENTATE) Block, usually of concrete, constructed in a channel or stilling basin to dissipate the energy of water flowing at high velocity; one of a series of upright obstructions designed to dissipate energy as in the case of a stilling basin or drop structure.

BAFFLE CHAMBER Chamber designed to promote the settling of fly ash and coarse particulate matter by changing the direction and/or reducing the velocity of the gases produced by the combustion of the refuse or sludge.

BAGHOUSE FILTER Large fabric bag, usually made of glass fibers, used to eliminate intermediate and large (greater than 20 PM in diameter) particles.

BAILER 10- to 20-foot-long pipe equipped with a valve at the lower end, used to remove slurry from the bottom or side of a well as it is being drilled, or to collect groundwater samples from wells or open boreholes; tube of varying length.

BALANCED HEAD CONDITION Condition in which the water pressure on the upstream and downstream sides of an object are equal (such as an emergency or regulating gate).

BALING Compacting solid waste into blocks to reduce volume and simplify handling.

BALLISTIC SEPARATOR Machine that sorts organic from inorganic matter for composting.

BALL-MILLING Repeated churning action of cobbles, gravel, and sand caused by the force of water in a stilling basin or other structure by which severe concrete abrasion can occur.

BAM Binary Angular Measure.

BAND APPLICATION Spreading of chemicals over, or next to, each row of plants in a field.

BANDPASS FILTER Filter whose frequencies are between given upper and lower cutoff values, while substantially attenuating all frequencies outside these values.

BANDWIDTH Frequency range between the lowest and highest frequencies that are passed through a component, circuit, or system with acceptable attenuation.

BANK Margins of a channel; sloping ground that borders a stream and confines the water in the natural channel when the water level, or flow, is normal; called right or left as viewed facing in the direction of the flow.

BANK FULL Established river stage at a given location along a river which is intended to represent the maximum safe water level that will not overflow the river banks or cause any significant damage within the river reach.

BANK SEDIMENT RESERVOIR Portion of the alluvium on the sides of a channel.

BANK STORAGE Water that has infiltrated from a reservoir into the surrounding land where it remains in storage until water level in the reservoir is lowered; water absorbed and stored in the void in the soil cover in the bed and banks of a stream, lake, or reservoir, and returned in whole or in part as the level of water body surface falls; change in the amount of water stored in an aquifer adjacent to a surface-water body resulting from a change in stage of the surface-water body.

BANKFULL Water level, or stage, at which a stream, river or lake is at the top of its banks and any further rise would result in water moving into the flood plain.

BANKFULL STAGE Maximum stage of a stream before it overflows its banks; established river stage at a certain point along a river which is intended to represent the maximum safe water level which will not overflow the river banks or cause any significant damage within the reach of the river; stage at which a stream first overflows its natural banks formed by floods with 1- to 3-year recurrence intervals.

BANKING System for recording qualified air emission reductions for later use in bubble, offset, or netting transactions.

BANNER CLOUD Cloud plume often observed to extend downwind behind isolated mountain peaks, even on otherwise cloud-free days.

BAR An obstacle formed at the shallow entrance to the mouth of a river or bay.

BAR SCREEN Device used to remove large solids in wastewater treatment.

BARBER POLE Slang for thunderstorm updraft with a visual appearance including cloud striations that are curved in a manner similar to the stripes of a barber pole.

BAROCLINIC LEAF SHIELD Cloud pattern on satellite images - frequently noted in advance of formation of a low pressure center.

BAROCLINIC ZONE Region in which a temperature gradient exists on a constant pressure surface.

BAROCLINITY Measure of the state of stratification in a fluid in which surfaces of constant pressure (isobaric) intersect surfaces of constant density (isosteric).

BAROGRAM Analog record of pressure produced by a barograph.

BAROGRAPH Barometer that records its observations continuously.

BAROMETER Instrument that measures atmospheric pressure.

BAROMETRIC PRESSURE Pressure of the atmosphere as indicated by a barometer.

BAROTROPIC SYSTEM Weather system in which temperature and pressure surfaces are coincident (i.e., temperature is uniform or no temperature gradient) on a constant pressure surface.

BAROTROPY State of a fluid in which surfaces of constant density (or temperature) are coincident with surfaces of constant pressure; it is the state of zero baroclinity.

BARRAGE Artificial obstruction placed in water to increase water level or divert it.

BARREL Volumetric unit of measure for crude oil and petroleum products equivalent to 42 U.S. gallons.

BARREL SAMPLER Open-ended steel tube used to collect soil samples.

BARRIER BAR Elongate offshore ridge, submerged at least at high tide, built up by the action of waves or currents.

BARRIER BEACH Narrow, elongate sandy ridge rising slightly above the high-tide level and extending generally parallel with the mainland shore, but separated from it by a lagoon.

BARRIER COATING Layer of a material that obstructs or prevents passage of something through a surface that is to be protected (e.g., grout, caulk, or various sealing compounds).

BARRIER JET Jet-like wind current that forms when a stably-stratified low-level airflow approaches a mountain barrier and turns to the left to blow parallel to the longitudinal axis of the barrier.

BARTEL'S ROTATION NUMBER Serial number assigned to 27-day rotation periods of solar and geophysical parameters. Rotation 1 in this sequence was assigned arbitrarily by Bartel to begin in January 1833.

BASAL APPLICATION Application of a chemical on plant stems or tree trunks just above the soil line.

BASALT Fine-grained, dark-colored volcanic rock rich in iron-bearing minerals; consistent year-round energy use of a facility; minimum amount of electricity supplied continually to a facility.

BASCULE GATE See FLAP GATE.

BASE Substance that has a pH value between 7 and 14; less free hydrogen ions (H^+) than hydroxyl ions (OH^-).

BASE COURSE Layer of specified or selected material of planned thickness constructed on the subgrade or subbase for the purpose of serving one or more functions such as distributing load, providing drainage, minimizing frost action, etc.

BASE DATA Digital fields of reflectivity, mean radial velocity, and spectrum width data in spherical coordinates provided at the finest resolution available from the radar.

BASE DISCHARGE (FOR PEAK DISCHARGE) Discharge value, determined for selected stations, above which peak discharge data are published and selected so that an average of about three peak flows per year will be published.

BASE EXCHANGE Physicochemical process whereby one species of ions adsorbed on soil particles is replaced by another species.

BASE FLOOD Flood having a one percent chance of being equaled or exceeded in any given year; national standard for floodplain management is the base, or one percent chance flood that has at least one chance in 100 of occurring in any given year (100-year flood).

BASE FLOOD PLAIN Flood plain inundated by the 100-year flood.

BASE PRODUCTS Products that present some representation of the base data.

BASE REFLECTIVITY One of the three fundamental quantities (along with base [radial] velocity and spectrum width) that a Doppler radar measures.

BASE RUNOFF Sustained or fair weather runoff composed largely of groundwater effluent.

BASE SAFETY CONDITION (BSC)
Level of loading above which a dam failure does not contribute an incremental loss of life.

BASE STATION Computer that accepts radio signals from ALERT gaging sites, decodes the data, places the data in a database, and makes the data available to other users.

BASE THICKNESS Maximum thickness or width of the dam measured horizontally between upstream and downstream faces and normal to the axis of the dam, but excluding projections for outlets or other appurtenant structures; thickness is used for gravity or arch dams, and width is used for other dams.

BASE WIDTH Time duration of a unit hydrograph.

BASEFLOW Sustained low flow of a stream in the absence of direct runoff; ground-water inflow to the stream channel; portion of stream discharge that is derived from natural storage; not attributable to short-term surface run off, precipitation or snow melt events; includes natural and human-induced streamflows.

BASELINE (CONDITION OR ALTERNATIVE)
Conditions that would prevail if no actions were taken.

BASELINE PROFILE Used for a survey of the environmental conditions and organisms existing in a region prior to unnatural disturbances.

BASELOAD Minimum load in a power system over a given period of time; minimum constant amount of load connected to the power system over a given time period, usually on a monthly, seasonal, or yearly basis.

BASELOAD PLANT Power plant normally operated to carry baseload; consequently, it operates essentially at a constant load; plant, usually housing high-efficiency steam-electric units, which is normally operated to take all or part of the minimum load of a system, and which consequently produces electricity at an essentially constant rate and runs continuously; operated to maximize system mechanical and thermal efficiency and minimize system operating costs.

BASELOADING Running water through a power plant at a roughly steady rate, thereby producing power at a steady rate.

BASIC Opposite of acidic; water that has a pH of greater than 7.

BASIC FIXED SITES Sites on streams at which streamflow is measured and samples are collected for temperature, salinity, suspended sediment, major ions and metals, nutrients, and organic carbon to assess the broad-scale spatial and temporal character and transport of inorganic constituents of streamwater in relation to hydrologic conditions and environmental settings.

BASIC HYDROLOGIC DATA Includes inventories of features of land and water that vary spatially (topographic and geologic maps are examples), and records of processes that vary with both place and time.

BASIC HYDROLOGIC INFORMATION Includes surveys of the water resources of particular areas and a study of their physical and related economic processes, interrelations and mechanisms.

BASIC-STAGE FLOOD SERIES See PARTIAL-DURATION FLOOD SERIES.

BASIN An area having a common outlet for its surface runoff.

BASIN AND RANGE PHYSIOGRAPHY Region characterized by a series of generally north-trending mountain ranges separated by alluvial valleys.

BASIN BOUNDARY Topographic dividing line around the perimeter of a basin, beyond which overland flow (i.e. runoff) drains away into another basin.

BASIN LAG Time it takes from the centroid of rainfall for the hydrograph to peak.

BASIN RECHARGE Rainfall that adds to the residual moisture of the basin in order to help recharge the water deficit (i.e.; water absorbed into the soil that does not take the form of direct runoff).

BASIN RUNOFF MODEL Any one of the computer programs that mathematically models basin characteristics to forecast reservoir inflow from rainfall and/or streamflow data.

BASIN STATES Arizona, California, Colorado, New Mexico, Utah, Wyoming.

BAT Best Available Treatment technologies.

BATHYMETRY Science of measuring depths of the oceans, lakes, seas, etc.

BATTER Inclination from the vertical; pile driven at an angle to widen the area of support and to resist thrust.

BAY-DELTA Sacramento/San Joaquin River Delta and San Francisco Bay.

BCM Become.

BCMNG Becoming.

BD Blowing Dust.

BEA U.S. Bureau of Economic Analysis.

BEACH EROSION Movement of beach materials by some combination of high waves, currents and tides, or wind.

BEACHING Action of water waves by which beach materials settle into the water because of removal of finer materials.

BEAM FILLING Measure of variation of hydrometeor density throughout the radar sampling volume.

BEAM WIDTH Angular width of antenna pattern; usually width where the power density is one-half that of the axis beam (Half-Power or 3 dB point).

BEAN SHEET Common term for a pesticide data package record.

BEAR'S CAGE Slang for region of storm-scale rotation, in a thunderstorm, which is wrapped in heavy precipitation.

BEAR-TRAP GATE Any of a family of crest gates consisting of two leaves, an upstream leaf hinged and sealed at the upstream edge and a downstream leaf hinged and sealed on its downstream edge.

BEAVER('S) TAIL Slang for particular type of inflow band with a relatively broad, flat appearance suggestive of a beaver's tail.

BED ELEVATION Height of streambed above a specified level.

BED FORMS Irregularities found on the bottom (bed) of a stream that are related to flow characteristics; related to the transport of sediment and interact with the flow because they change the roughness of the stream bed; given names such as "dunes", "ripples", and "antidunes."

BED LAYER Flow layer, several grain diameters thick (usually taken as two grain diameters thick), immediately above the bed.

BED LOAD Bed material that moves in continuous contact with the bed layer; sediment that moves by rolling or sliding along the bed; particles have a density or grain size such as to preclude movement far above or for a long distance out of contact with the stream bed under natural conditions of flow.

BED LOAD DISCHARGE Quantity of bed load passing a cross section in a unit of time (i.e.; the rate).

BED MATERIAL Sediment mixture of which a streambed, lake, pond, reservoir, or estuary bottom is composed.

BED MATERIAL DISCHARGE Total rate (tons/day) at which bed material is transported by a given flow at a given location on a stream.

BED MATERIAL LOAD Total rate (tons/day) at which bed material is transported by a given location on a stream.

BED SEDIMENT Material that temporarily is stationary in the bottom of a stream or other watercourse.

BED SEDIMENT AND TISSUE STUDIES Assessment of concentrations and distributions of trace elements and hydrophobic organic contaminants in streambed sediment and tissues of aquatic organisms to identify potential sources and to assess spatial distribution.

BED SEDIMENT CONTROL VOLUME Source-sink component of sediment sources in a river system (the other component is the suspended sediment in the inflowing discharge); user-defined dimensions are the movable bed width and depth, and the average reach length.

BEDDING Ground, or layer of such, for support purposes on which pipe is laid; soil placed beneath and beside a pipe to support the load on the pipe.

BEDDING PLANE Separation or weakness between two layers of rock, caused by changes during the building up of the rock-forming material.

BEDLOAD Sediment that moves on or near the bottom of a streambed and is in almost continuous contact with the bed.

BED-LOAD DISCHARGE Quantity of bed load passing a cross section of a stream in a unit of time.

BED-MATERIAL DISCHARGE Part of the total sediment discharge which is composed of grain sizes found in the bed; assumed equal to the transport capability of the flow.

BEDROCK Solid rock at the surface or underlying other surface materials; rock of relatively great thickness and extent in its native location; any solid rock, not exhibiting soil-like properties, that underlies soil or other unconsolidated, bed material.

BEDROCK MOTION PARAMETERS Numerical values representing vibratory ground motion, such as particle acceleration, velocity, and displacement, frequency content, predominant period, spectral intensity, and a duration that define a design earthquake.

BEGINNING OF FREEZUP Date on which ice forming a stable winter ice cover is first observed on the water surface.

BEGINNING OF THE BREAKUP Date of definite breaking, movement, or melting of ice cover or significant rise of water level.

BEHAVIOR Reaction of an animal to its environment.

BELL Expanded, or enlarged, end of a pipe section, into which the next pipe fits.

BEN EPA's computer model for analyzing a violator's economic gain from not complying with the law.

BENCH Working level or step in a cut.

BENCHMARK (BM) Permanent or temporary monument of known elevation above sea level, used for vertical control at a construction site; point of known or assumed elevation used as a reference in determining other elevations; permanent reference point (elevation) whose known elevation is tied to a national network used in a survey.

BENCH-SCALE TESTS Laboratory testing of potential cleanup technologies.

BEND Change of direction in piping.

BENEFICIAL USE Water loss through use for the betterment of society (e.g., irrigation or municipal use).

BENEFICIAL USES Management objectives.

BENEFICIARY Any individual, entity, or governmental agency (local, state, or federal) that benefits from a Reclamation project.

BENEFIT-COST ANALYSIS Economic method for assessing the benefits and costs of achieving alternative health-based standards at given levels of health protection.

BENEFIT-COST RATIO (B/C) Ratio of the present value of project benefits to the present value of the project costs, used in economic analysis.

BENTHIC Plants or animals that live on the bottom of lakes, streams, or oceans; Bottom of rivers, lakes, or oceans; organisms that live on the bottom of water bodies; bottom- or depth-inhabiting.

BENTHIC FAUNA (OR BENTHOS) Organisms attached to or resting on the bottom or living in the bottom sediments of a water body.

BENTHIC INVERTEBRATES Insects, mollusks, crustaceans, worms, and other organisms without a backbone that live in, on, or near the bottom of lakes, streams, or oceans.

BENTHIC ORGANISM Form of aquatic life (organisms) that inhabits the bottom of an aquatic environment including bacteria, fungi, insect larvae and nymphs, snails, clams, and crayfish; useful as indicators of water quality.

BENTHOS Organisms living in or on the bottom of a lake, pond, ocean, stream, etc.

BENTONITE Colloidal clay, expansible when moist, commonly used to provide a tight seal around a well casing.

BENTONITIC CLAY (BENTONITE) Clay with a high content of the mineral montmorillonite, usually characterized by high swelling on wetting and shrinkage on drying.

BERGERON PROCESS Process by which ice crystals in a cloud grow at the expense of supercooled liquid water droplets.

BERGY BIT Piece of ice which has broken away from an iceberg, extending 1-5 meters above the sea surface and 100-300 square meters in area; remains of a melting iceberg.

BERM Horizontal strip or shelf built into an embankment or cut to break the continuity of the slope, usually for the purpose of reducing erosion or to increase the thickness of the embankment at a point of change in a slope or defined water surface elevation; horizontal step in the sloping profile of an embankment dam; shelf that breaks the continuity of a slope, or artificial ridge of earth; ledge or shoulder, as along the edge of a road or canal; artificial ridge of earth.

BERMUDA HIGH Semi-permanent, subtropical area of high pressure in the North Atlantic Ocean off the East Coast of North America that migrates east and west with varying central pressure.

BERYLLIUM Metal hazardous to human health when inhaled as an airborne pollutant that is discharged by machine shops, ceramic and propellant plants, and foundries.

BEST Budget Estimate System Tools.

BEST AVAILABLE CONTROL MEASURES (BACM) Term used to refer to the most effective measures (according to EPA guidance) for controlling small or dispersed particulates and other emissions from sources such as roadway dust, soot and ash from woodstoves and open burning of rush, timber, grasslands, or trash.

BEST AVAILABLE CONTROL TECHNOLOGY (BACT) Currently available technology producing the greatest reduction of air pollutant emissions, taking into account energy, environmental, economic, and other costs.

BEST AVAILABLE TECHNOLOGY Water treatment that EPA certifies to be the most effective for removing a contaminant.

BEST AVAILABLE TECHNOLOGY ECONOMICALLY ACHIEVABLE (BAT) Technology-based standard established by the Clean Water Act (CWA) as the most appropriate means available on a national basis for controlling the direct discharge of toxic and nonconventional pollutants to navigable waters.

BEST CONVENTIONAL POLLUTANT CONTROL TECHNOLOGY (BCT) Technology-based standard for the discharge from existing industrial point sources of conventional pollutants including BOD, TSS, fecal coliform, pH, oil and grease.

BEST DEMONSTRATED AVAILABLE TECHNOLOGY (BDAT) Most effective commercially available means of treating specific types of hazardous waste.

BEST GATE Gate opening where the peak efficiency of a turbine occurs at a particular head.

BEST MANAGEMENT PRACTICE (BMP) Agricultural practice that has been determined to be an effective, practical means of preventing or reducing non-point-source pollution.

BEST MANAGEMENT PRACTICES (BMPS) Schedules of activities, prohibitions of practices, maintenance procedures, and other management practices to prevent or reduce the discharge of pollutants to waters of the United States.

BEST PRACTICABLE CONTROL TECHNOLOGY CURRENTLY AVAILABLE (BPT) First level of technology-based standards established by the CWA to control pollutants discharged to waters of the U.S.

BEST PROFESSIONAL JUDGMENT (BPJ) Method used by permit writers to develop technology-based NPDES permit conditions on a case-by-case basis using all reasonably available and relevant data.

BEST TRACK Subjectively-smoothed representation of a tropical cyclone's location and intensity over its lifetime.

BFR Before.

BGN Begin.

BHND Behind.

BIA U.S. Bureau of Indian Affairs.

BIAS Systematic difference between an estimate of and the true value of a parameter.

BIENNIAL Plant which produces seeds during its second year and then dies.

BIFURCATION Point where a stream channel splits into two distinct channels.

BIG BASIN Large retention, detention basins or ponds that detain and slow stormwater, allowing sediment, chemicals, and trash to be filtered out before the water is released into receiving waters.

BILLOW CLOUD Cloud consisting of broad parallel bands oriented perpendicular to the wind.

BIMETAL Beverage containers with steel bodies and aluminum tops; handled differently from pure aluminum in recycling.

BIN Radar sample volume.

BIND To exert a strong chemical attraction.

BINDER Portion of soil passing a No. 40 (0.425 mm) United States Standard sieve.

BINOMIAL Scientific name of plants or animals which has two parts: a genus and a species name.

BINOVC Breaks IN OVerCast.

BIOACCUMULANTS Substances that increase in concentration in living organisms as they take in contaminated air, water, or food because the substances are very slowly metabolized or excreted.

BIOACCUMULATE Net uptake of a material by an organism from food, water, and (or) respiration that results in elevated internal concentrations.

BIOACCUMULATION Biological sequestering of a substance at a higher concentration than that at which it occurs in the surrounding environment or medium, intake and retention of non-food substances by a living organism from its environment, resulting in a build-up of the substances in the organism; process whereby a substance enters organisms through the gills, epithelial tissues, dietary, or other sources

BIOACCUMULATION FACTOR (BAF) Ratio of a substance's concentration in tissue versus its concentration in ambient water, in situations where the organism and the food chain are exposed.

BIOASSAY Test which determines the effect of a chemical on a living organism.

BIOASSIMILATION Accumulation of a substance within a habitat.

BIOAVAILABILITY Capacity of a chemical constituent to be taken up by living organisms either through physical contact or by ingestion; degree of ability to be absorbed and ready to interact in organism metabolism; measure of the physicochemical access that a toxicant has to the biological processes of an organism.

BIOCHEMICAL Chemical processes that occur inside or are mediated by living organisms.

BIOCHEMICAL OXYGEN DEMAND (BOD) Measure of the amount of oxygen consumed in the biological processes that break down organic matter in water; measurement of the amount of oxygen utilized by the decomposition of organic material, over a specified time period (usually 5 days) in a wastewater sample; measurement of the readily decomposable organic content of a wastewater; greater the BOD, the greater the degree of pollution.

BIOCHEMICAL PROCESS Process characterized by, produced by, or involving chemical reactions in living organisms.

BIOCONCENTRATION Accumulation of a chemical in tissues of a fish or other organism to levels greater than in the surrounding medium; process by which a compound is absorbed from water through gills or epithelial tissues and is concentrated in the body.

BIOCONCENTRATION FACTOR (BCF) Ratio of a substance's concentration in tissue versus its concentration in water, in situations where the food chain is not exposed or contaminated.

BIODEGRADABLE Capable of decomposing quickly through the action of microorganisms under natural conditions.

BIODEGRADATION Transformation of a substance into new compounds through biochemical reactions or the actions of microorganisms such as bacteria.

BIODEGRADATION, AEROBIC Decomposition of organic matter by microorganisms in the presence of free oxygen where the decomposition end-products are carbon dioxide and water.

BIODEGRADATION, ANAEROBIC decomposition of organic matter by microorganisms in the absence or near absence of free oxygen where the decomposition end-products are enriched in carbon.

BIODIVERSITY Variety and variability among living organisms and the ecological complexes in which they occur.

BIOLOGICAL ASSESSMENT Evaluation of the biological condition of a water body by using biological surveys and other direct measurements of a resident biota in surface water.

BIOLOGICAL CONTAMINANTS Living organisms or derivates (e.g., viruses, bacteria, fungi, and mammal and bird antigens) that can cause harmful health effects when inhaled, swallowed, or otherwise taken into the body.

BIOLOGICAL CONTROL Use of animals and organisms that eat or otherwise kill or out-compete pests.

BIOLOGICAL CRITERIA (OR BIOCRITERIA) Numerical values or narrative expressions that describe the reference biological integrity of aquatic communities that inhabit water of a given designated aquatic life use.

BIOLOGICAL DIVERSITY Number and kinds of organisms per unit area or volume; composition of species in a given area at a given time.

BIOLOGICAL GROWTH Activity and growth of any and all living organisms.

BIOLOGICAL INTEGRITY Ability to support and maintain balanced, integrated, functionality in the natural habitat of a given region; functionally defined as the condition of the aquatic community that inhabits unimpaired water bodies of a specified habitat as measured by community structure and function; concept applied primarily in drinking water management.

BIOLOGICAL MAGNIFICATION . The substances become concentrated in tissues or internal organs as they move up the chain.

BIOLOGICAL MAGNIFICATION (BIOMAGNIFICATION) Step by step concentration of substances in successive levels of food chains; process whereby certain substances such as pesticides or heavy metals move up the food chain, work their way into rivers or lakes, and are eaten by aquatic organisms such as fish, which in turn are eaten by large birds, animals or humans; enhancement of a substance (usually a contaminant) in a food web such that the organisms eventually contain higher concentrations of the substance than their food sources.

BIOLOGICAL MEASUREMENT Measurement taken in a biological medium.

BIOLOGICAL MEDIUM One of the major components of an organism (e.g., blood, fatty tissue, lymph nodes or breath) in which chemicals can be stored or transformed.

BIOLOGICAL MONITORING (BIOMONITORING) Use of living organisms in water quality surveillance to indicate compliance with water quality standards or effluent limits and to document water quality trends; use of a biological entity as a detector and its response as a measure to determine environmental conditions; toxicity tests and biological surveys are common biomonitoring methods.

BIOLOGICAL OPINION (BO) Document stating the U.S. Fish and Wildlife Service (FWS) and the National Marine Fisheries Service (NMFS) opinion as to whether a federal action is likely to jeopardize the continued existence of a threatened or endangered species or result in the destruction or adverse modification of critical habitat.

BIOLOGICAL OXIDATION Decomposition of complex organic materials by microorganisms; occurs in self-purification of water bodies and in activated sludge wastewater treatment.

BIOLOGICAL OXYGEN DEMAND (BOD) See BIOCHEMICAL OXYGEN DEMAND.

BIOLOGICAL PESTICIDES Certain microorganism, including bacteria, fungi, viruses, and protozoa that are effective in controlling pests; usually do not have toxic effects on animals and people and do not leave toxic or persistent chemical residues in the environment.

BIOLOGICAL PROCESSES Processes characteristic of, or resulting from, the activities of living organisms.

BIOLOGICAL STRESSORS Organisms accidentally or intentionally dropped into habitats in which they do not evolve naturally (e.g., gypsy moths, Dutch elm disease, certain types of algae, and bacteria).

BIOLOGICAL SURVEY (OR BIOSURVEY) Collecting, processing, and analyzing a representative portion of the resident aquatic community to determine the community structure and function.

BIOLOGICAL TREATMENT Treatment technology that uses bacteria to consume organic waste.

BIOLOGICALLY EFFECTIVE DOSE Amount of a deposited or absorbed compound reaching the cells or target sites where adverse effect occur, or where the chemical interacts with a membrane.

BIOLOGICALS Vaccines, cultures and other preparations made from living organisms and their products, intended for use in diagnosing, immunizing, or treating humans or animals, or in related research.

BIOLOGY Scientific study of life.

BIOMAGNIFICATION Tendency of certain chemicals to become concentrated as they move into and up the food chain.

BIOMASS Total mass or amount of living organisms in a particular area or environment; amount of living matter, in the form of organisms, present in a particular habitat, usually expressed as weight-per-unit area.

BIOME Entire community of living organisms in a single major ecological area.

BIOMONITORING Measurement of biological parameters in repetition to assess the current status and changes in time of the parameters measured; use of living organisms to test the suitability of effluents for discharge into receiving waters and to test the quality of such waters downstream from the discharge; analysis of blood, urine, tissues, etc. to measure chemical exposure in humans.

BIOREMEDIATION Use of living organisms to clean up oil spills or remove other pollutants from soil, water, or wastewater; use of organisms such as non-harmful insects to remove agricultural pests or counteract diseases of trees, plants, and garden soil.

BIOSENSOR Analytical device comprising a biological recognition element (e.g., enzyme, receptor, DNA, antibody, or microorganism) in intimate contact with an electrochemical, optical, thermal, or acoustic signal transducer that together permit analyses of chemical properties or quantities.

BIOSOLIDS Sewage sludge that is used or disposed through land application, surface disposal, incineration, or disposal in a municipal solid waste landfill; defined as solid, semisolid, or liquid untreated residue generated during the treatment of domestic sewage in a treatment facility.

BIOSPHERE Portion of Earth and its atmosphere that can support life.

BIOSTABILIZER Machine that converts solid waste into compost by grinding and aeration.

BIOTA All living organisms of an area.

BIOTECHNOLOGY Techniques that use living organisms or parts of organisms to produce a variety of products (from medicines to industrial enzymes) to improve plants or animals or to develop microorganisms to remove toxics from bodies of water, or act as pesticides.

BIOTIC COMMUNITY Naturally occurring assemblage of plants and animals that live in the same environment and are mutually sustaining and interdependent.

BIOTIC POTENTIAL Inherent capacity of an organism to reproduce and survive, which is pitted against limiting influences of the environment.

BIOTIC PYRAMID Set of all food chains or hierarchic arrangements of organisms as eaters and eaten in a prescribed area when tabulated by numbers or by biomasses, usually takes the form of an inverted pyramid.

BIOTOPE Smallest geographical unit of a habitat, characterized by a high degree of uniformity in the environment and its plant and animal life.

BIOTRANSFORMATION Conversion of a substance into other compounds by organisms; includes biodegredation.

BIOTYPE Genetically homogeneous population composed only of closely similar individuals; a genotypic race or group of organisms.

BIP Budget Input Program.

BITUMINOUS Containing asphalt or tar.

BKN Broken.

BLACK ICE Transparent ice formed in rivers and lakes, or on roads and bridges.

BLACKBODY Hypothetical "body" that absorbs all of the electromagnetic radiation striking it - it does not reflect or transmit any of the incident radiation; not only absorbs all wavelengths, but emits at all wavelengths with the maximum possible intensity for any given temperature.

BLACKBODY RADIATION Electromagnetic radiation emitted by an ideal blackbody adhering to the radiation laws; theoretical maximum amount of electromagnetic radiation of all wavelengths that can be emitted by a body at a given temperature.

BLACKOUT Disconnection of the source of electricity from all the electrical loads in a certain geographical area brought about by an emergency forced outage or other fault in the generation, transmission, or distribution system serving the area.

BLACKWATER Water that contains animal, human, or food waste and the carriage water generated through toilet usage.

BLADE Usually a part of an excavator which digs and pushes dirt but does not carry it.

BLANKET DRAIN Layer of pervious material placed to facilitate drainage of the foundation and/or embankment.

BLAST Loosen or move rock or soil by means of explosives or an explosion.

BLASTING MATS Blanket usually composed of woven cable or interlocked rings placed over a blast to reduce flyrock.

BLD Build.

BLDUP Buildup.

BLIZZARD Means that the following conditions are expected to prevail for a period of 3 hours or longer: Sustained wind or frequent gusts to 35 miles an hour or greater; and considerable falling and/or blowing snow (i.e., reducing visibility frequently to less than 1/4 mile).

BLIZZARD WARNING Issued for winter storms with sustained or frequent winds of 35 mph or higher with considerable falling and/or blowing snow that frequently reduces visibility to 1/4 of a mile or less where these conditions are expected to prevail for a minimum of 3 hours.

BLM U.S. Bureau of Land Management.

BLO Below.

BLOCKED FLOW Flow approaching a mountain barrier that is too weak or too stable to be carried over the barrier.

BLOCKLOADING Providing a consistent amount of electrical power in a stated period of time.

BLOOD PRODUCTS Any product derived from human blood, including but not limited to blood plasma, platelets, red or white corpuscles, and derived licensed products such as interferon.

BLOOM Proliferation of algae and/or higher aquatic plants in a body of water; often related to pollution, especially when pollutants accelerate growth.

BLOWING Descriptor used to amplify observed weather phenomena whenever the phenomena are raised to a height of 6 feet or more above the ground.

BLOWING DUST OR SAND Strong winds over dry ground, that has little or no vegetation, can lift particles of dust or sand into the air; airborne particles that can reduce visibility, cause respiratory problems, and have an abrasive effect on machinery.

BLOWING SNOW Wind-driven snow that reduces surface visibility.

BLOWING SNOW ADVISORY Issued when wind driven snow reduces surface visibility, possibly, hampering traveling.

BLOWOUT Small saucer- or trough-shaped hollow or depression formed by wind erosion on a pre-existing dune or other sand deposit.

BLUE TOPS Grade stakes whose tops indicate finish grade level.

BLUE WATCH OR BLUE BOX Slang for severe thunderstorm watch.

BLUE-BABY SYNDROME Condition most common in young infants and certain elderly people that can be caused by ingestion of high amounts of nitrate, which results in the blood losing its ability to effectively carry oxygen.

BLUE-GREEN ALGAE (CYANOPHYTA) Group of phytoplankton and periphyton organisms with a blue pigment in addition to a green pigment called chlorophyll; cause nuisance water-quality conditions in lakes and slow-flowing rivers; however, they are found commonly in streams throughout the year.

BLUSTERY Same as Breezy; 15 to 25 mph winds.

BLV Before.

BMP Best Management Practices.

BNDRY Boundary.

BO Biological Opinion.

BOD Biochemical Oxygen Demand.

BOD5 Amount of dissolved oxygen consumed in five days by biological processes breaking down organic matter.

BODY BURDEN Amount of a chemical stored in the body at a given time, especially a potential toxin in the body as the result of exposure.

BODY WAVE Waves propagated in the interior of the Earth (i.e., the compression and shear waves of an earthquake).

BOG Type of wetland that accumulates appreciable peat deposits; nutrient-poor, acidic wetland dominated by a waterlogged, spongy mat of sphagnum moss that ultimately forms a thick layer of acidic peat; generally has no inflow or outflow; fed primarily by rain water.

BOILER Vessel designed to transfer heat produced by combustion or electric resistance to water that provide hot water or steam.

BOILER FEED WATER Input water utilized by a boiler.

BOILING POINT Temperature at which a liquid will start to become a gas, and boil.

BOLSON Extensive, flat, saucer-shaped, alluvium-floored basin or depression, almost or completely surrounded by mountains and from which drainage has no surface outlet; term used in the desert regions of the Southwestern United States.

BOM U.S. Bureau of Mines.

BOMB Popular expression of a rapid intensification of a cyclone (low pressure) with surface pressure expected to fall by at least 24 millibars in 24 hour.

BON Basis Of Negotiation.

BONNET Upper portion or the cover of a gate valve body.

BONNEVILLE POWER ADMINISTRATION (BPA) One of five federal power marketing administrations that sell low cost electric power produced by federal hydroelectric dams to agricultural and municipal users that serves Idaho, Oregon, and Washington as well as parts of Nevada and Wyoming.

BOOM Floating device used to contain oil on a body of water; piece of equipment used to apply pesticides from a tractor or truck.

BOR U.S. Bureau of Reclamation.

BORA Regional downslope wind whose source is so cold that it is experienced as a cold wind, despite compression warming as it descends the lee slope of a mountain range.

BORDER ICE Ice sheet in the form of a long border attached to the bank or shore; shore ice.

BORDERLINE SOILS Soils that have the characteristics of two of the classification groups in the Unified Soil Classification System.

BOREAL Climatic zone having a definite winter with snow and a short summer that is generally hot, and which is characterized by a large annual range of temperature.

BOREHOLE Hole made with drilling equipment.

BOREHOLE EXTENSOMETERS (SINGLE-POINT OR MULTI-POINT) Instrument designed to measure axial displacement of a fixed point or points along its length; can be a rod-type, or a wire-type and are usually grouted into an uncased borehole; can be installed horizontally, vertically, or at any angle.

BORING Rotary drilling.

BORROW Material excavated from one area to be used as fill material in another area.

BORROW AREA Area from which natural materials, such as rock, gravel or soil, used for construction purposes is excavated.

BORROW PIT Specific site within a borrow area from which material is excavated for use.

BOSQUE Dense growth of trees and underbrush.

BOTANICAL PESTICIDE Pesticide whose active ingredient is a plant-produced chemical such as nicotine or strychnine (plant-derived pesticide).

BOTTLE BILL Proposed or enacted legislation which requires a returnable deposit on beer or soda containers and provides for retail store or other redemption.

BOTTOM ASH Non-airborne combustion residue from burning pulverized coal in a boiler; the material which falls to the bottom of the boiler and is removed mechanically; a concentration of non-combustible materials, which may include toxics.

BOTTOM LAND See FLOODPLAIN.

BOTTOM LAND HARDWOODS Forested freshwater wetlands adjacent to rivers in the southeastern United States, especially valuable for wildlife breeding, nesting and habitat.

BOTTOM-LAND FOREST Low-lying forested wetland found along streams and rivers, usually on alluvial flood plains.

BOULDER Rock fragment, usually rounded by weathering or abrasion, with an average dimension of 12 inches or more that will not pass a 12-inch screen; rock which is too heavy to be lifted readily by hand.

BOULDER CLAY Glacial drift that has not been subjected to the sorting action of water and therefore contains particles from boulders to clay sizes.

BOUNDARY CONDITION Mathematical statement specifying the dependent variable (e.g. hydraulic head of concentration) at the boundaries of the modeled domain which contain the equations of the mathematical model; known or hypothetical conditions at the boundary of a problem that govern its solution.

BOUNDARY LAYER Layer of air adjacent to a bounding surface where the effects of friction are significant; roughly the lowest one or two kilometers of Earth's atmosphere.

BOUNDED WEAK ECHO REGION (BWER) Radar signature within a thunderstorm characterized by a local minimum in radar reflectivity at low levels which extends upward into, and is surrounded by, higher reflectivitys aloft.

BOUNDING ESTIMATE Estimate of exposure, dose, or risk that is higher than that incurred by the person in the population with the currently highest exposure, dose, or risk.

BOVC Base of Overcast.

BOW ECHO Radar echo which is linear but bent outward in a bow shape where damaging straight-line winds often occur near the "crest" or center of a bow echo.

BOWEN RATIO For any moist surface, the ratio of heat energy used for sensible heating (conduction and convection) to the heat energy used for latent heating (evaporation of water or sublimation of snow); ranges from about 0.1 for the ocean surface to more than 2.0 for deserts where negative values are also possible.

BOX GIRDER Hollow steel beam with a square or rectangular cross section.

BOX MODEL Computer model used to calculate air pollution concentrations; based on the assumption that pollutants are emitted into a box through which they are immediately and uniformly dispersed where the sides and bottom of the box are defined by the sidewalls and floor of the valley being studied.

BR Mist.

BRACKISH Mixed fresh and salt waters; water containing too much salt to be useful to people but less salt than ocean water.

BRACKISH ICE Ice formed from Brackish water.

BRACKISH WATER Water with a salinity intermediate between seawater and freshwater (containing from 1,000 to 10,000 milligrams per liter of dissolved solids).

BRAIDED STREAM Stream characterized by an interlacing or tangled network of several small branching and reuniting shallow channels; composed of anabranches.

BRAIN Bureau of Reclamation Acquisition Information Network.

BRAKE HORSEPOWER Actual motor input horsepower required to produce the hydraulic horsepower from a pump (flow and head) taking into account the losses incurred within the pump due to friction, leakage, etc.; ratio of hydraulic horsepower to pump efficiency.

BRANCH Addition to the main pipe in a piping system.

BRASH ICE Accumulation of floating ice made up of fragments not more than 2 meters across; the wreckage of other forms of ice.

BRC Budget Review Committee.

BRD USGS Biological Resources Division (formerly NBS).

BREACH Failed opening in a dam; a gap, rift, hole, or rupture in a dam; providing a break; allowing water stored behind a dam to flow through in an uncontrolled and unplanned manner; eroded opening through a dam which drains the reservoir; controlled breach is a constructed opening; uncontrolled breach is an unintentional opening which allows uncontrolled discharge from the reservoir.

BREACH HYDROGRAPH Flood hydrograph resulting from a dam breach.

BREAKDOWN PRODUCT Compound derived by chemical, biological, or physical action upon a pesticide; natural process that may result in a more toxic or a less toxic compound and a more persistent or less persistent compound.

BREAKERS Waves that break, displaying white water; depends on wave steepness and bottom bathymetry.

BREAKPOINT CHLORINATION Addition of chlorine to water until the chlorine demand has been satisfied.

BREAKTHROUGH Crack or break in a filter bed that allows the passage of floc or particulate matter through a filter; will cause an increase in filter effluent turbidity.

BREAKUP Time when a river whose surface has been frozen from bank to bank for a significant portion of its length begins to change to an open water flow condition; signaled by the breaking of the ice and often associated with ice jams and flooding.

BREAKUP DATE Date on which a body of water is first observed to be entirely clear of ice and remains clear thereafter.

BREAKUP JAM Ice jam that occurs as a result of the accumulation of broken ice pieces.

BREAKUP PERIOD Period of disintegration of an ice cover.

BREATHING ZONE Area of air in which an organism inhales.

BRECCIA (VOLCANIC BRECCIA) Conglomerate-like rock made up of angular pieces of volcanic rock usually bound in volcanic ash.

BRECCIATED Rock made up of highly angular, coarse fragments.

BREEDING DENSITY Density of sexually mature organisms in a given area during the breeding period.

BREEDING POTENTIAL Maximum rate of increase in numbers of individuals of a species or population under optimum conditions.

BREEDING RATE Actual rate of increase of new individuals in a given population; the breeding potential minus limiting factors.

BREEZY 15 to 25 mph winds.

BRF Brief.

BRIDGE, CLASS A Bridge located on a federal-aid highway where an examination is required every 2 years and an in-depth examination every 6 years.

BRIDGE, CLASS B Bridge not located on a federal-aid highway that is used by the public where an examination is required every 3 years.

BRIDGE, CLASS C Bridge not located on a federal-aid highway that can be designated as being a private or operating bridge and could be gated or closed to the public, if desired.

BRIGHT BAND Distinct feature observed by a radar that denotes the freezing level of the atmosphere; enhanced radar echo of snow as it melts to rain.

BRIGHT SURGE ON THE DISK (BSD) Bright gaseous stream (surge) emanating from the chromosphere.

BRIGHT SURGE ON THE LIMB (BSL) Large gaseous stream (surge) that moves outward more than 0.15 solar radius above the limb.

BRIGHTNESS Basic visual sensation describing the amount of light that appears to emanate from an object; luminance of an object.

BRINE Water that has a quantity of salt, especially sodium chloride, dissolved in it; water that contains more than 35,000 milligrams per liter of dissolved solids.

BRINE MUD Waste material, often associated with well-drilling or mining, composed of mineral salts or other inorganic compounds.

BRISK 15 to 25 mph winds.

BRISK WIND ADVISORY Small Craft Advisory issued by the National Weather Service for ice-covered waters.

BRITISH THERMAL UNITS (BTU) Standard unit for measuring the quantity of heat required to raise the temperature of 1 pound of water by 1 °F.

BRK Break.

BRM Business Reply Mail.

BROADBAND Method of signaling in which multiple signals share the bandwidth of the transmission by the subdivision of the bandwidth into channels based on frequency.

BROADCAST APPLICATION Spreading of pesticides over an entire area.

BROAD-CRESTED WEIR Overflow structure on which the nappe is supported for an appreciable length in the direction of flow.

BROCKEN SPECTER Optical phenomenon sometimes occurring at high altitudes when the image of an observer placed between the Sun and a cloud is projected on the cloud, surrounded by rings of color (glory), as a greatly magnified shadow.

BROKEN LEVEL Layer of the atmosphere with $^5/_8$ to $^7/_8$ sky cover (cloud cover).

BROWNFIELDS Abandoned, idled, or under used industrial and commercial facilities/sites where expansion or redevelopment is complicated by real or perceived environmental contamination.

BROWNOUT Partial reduction of electrical voltages that often results in lights dimming and motor-driven devices slowing down.

BS Blowing Snow.

BSC Base Safety Condition.

BSO Business Services Office.

BTR Better.

BTU British Thermal Units.

BTWN Between.

BUBBLE System under which existing emissions sources can propose alternate means to comply with a set of emissions limitations.

BUBBLE HIGH Mesoscale area of high pressure, typically associated with cooler air from the rainy downdraft area of a thunderstorm or a complex of thunderstorms.

BUBBLER GAUGE Water stage recording device that is capable of attaching to a LARC for data automation purposes.

BUCKET Part of an excavator which digs, lifts, and carries dirt.

BUDGET REVIEW COMMITTEE (BRC) Ad hoc committee of representatives from different areas that coordinates budget activities through the formulation phase.

BUDS Business Utilization Development Specialist.

BUFFER Solution or liquid whose chemical makeup is such that it minimizes changes in pH when acids or bases are added to it.

BUFFER STRIPS (FILTER STRIPS, VEGETATED FILTER STRIPS, GRASSED BUFFERS) Strips of grass or other close-growing vegetation that separates a waterway (ditch, stream, creek) from an intensive land use area (subdivision, farm).

BUFKIT Software tool used by forecasters to examine the vertical profile and other aspects of the atmosphere.

BUILDING AUTOMATION SYSTEM (BAS) System that optimizes the start-up and performance of HVAC equipment and alarm systems

BUILDING COOLING LOAD Hourly amount of heat that must be removed from a building to maintain indoor comfort (measured in British Thermal Units).

BUILDING ENVELOPE Exterior surface of a building's construction - the walls, windows, floors, roof, and floor.

BUILDING RELATED ILLNESS Diagnosable illness whose cause and symptoms can be directly attributed to a specific pollutant source within a building (e.g., Legionnaire's disease, hypersensitivity, pneumonitis).

BUILT ENVIRONMENT Human-modified environment (e.g., buildings, roads, and cities).

BULK ELECTRICAL CONDUCTIVITY Combined electrical conductivity of all material within a doughnut-shaped volume surrounding an induction probe.

Bulk Richardson Number (BRN) Non-dimensional number relating vertical stability and vertical shear (generally, stability divided by shear) where high values indicate unstable and/or weakly-sheared environments and low values indicate weak instability and/or strong vertical shear.

BULK SAMPLE Small portion (usually thumbnail size) of a suspect asbestos-containing building material collected by an asbestos inspector for laboratory analysis to determine asbestos content.

BULKHEAD Partition or structure separating compartments or to hold back water; one-piece fabricated steel unit which is lowered into guides and seals against a frame to close a water passage in a dam, conduit, spillway, etc.; object used to isolate a portion of a waterway for examination, maintenance, or repair; wall or partition erected to resist ground or water pressure.

BULKHEAD GATE Gate used either for temporary closure of a channel or conduit before dewatering it for inspection or maintenance or for closure against flowing water when the head difference is small (e.g. for diversion tunnel closure).

BULKING Increase in volume of a material due to manipulation (e.g., rock bulks upon being excavated, damp sand bulks if loosely deposited by dumping).

BULKY WASTE Large items of waste materials, such as appliances, furniture, large auto parts, trees, stumps.

BUOYANCY Tendency of a body to float or to rise when submerged in a fluid; the power of a fluid to exert an upward force on a body placed in it.

BURDEN Distance between the free face and the first row of holes or the distance between rows of holes parallel to the face.

BUREAU OF LAND MANAGEMENT (BLM) Agency within the U.S. Department of the Interior, administers 264 million acres of America's public lands, located primarily in 12 Western States that sustains the health, diversity, and productivity of the public lands for the use and enjoyment of present and future generations.

BUREAU OF RECLAMATION (USBR, RECLAMATION, BOR) Federal agency with mission to manage, develop, and protect water and related resources in an environmentally and economically sound manner in the interest of the American public.

BURIAL GROUND (GRAVEYARD) Disposal site for radioactive waste materials that uses earth or water as a shield.

BURST Transient enhancement of the solar radio emission, usually associated with an active region or flare.

BUS (BUSWORK) Conductor, or group of conductors, that serve as a common connection for two or more electrical circuits; in power plants, comprises the three rigid single-phase connectors that interconnect the generator and the step-up transformer(s).

BUSBAR Heavy metal conductor used to carry a large current.

BUST Slang for inaccurate forecast, especially one where significant weather is predicted but does not occur.

BUTT JOINT (OPEN JOINT) In pipe, flat ends that meet but do not overlap.

BUTTERFLY VALVE Valve designed for quick closure that consists of a circular leaf, slightly convex in form, mounted on a transverse shaft carried by two bearings.

BUTTRESS DAM Dam consisting of a watertight upstream part (such as a concrete sloping slab) supported at intervals on the downstream side by a series of buttresses (walls normal to the axis of the dam); comprised of reinforced masonry or stonework built against concrete; usually in the form of flat decks or multiple arches that require about 60 percent less concrete than gravity dams.

BUY-BACK CENTER Facility where individuals or groups bring recyclables in return for payment.

BW Body Weight.

BYPASS Intentional diversion of waste streams from any portion of a treatment (or pretreatment) facility.

BY-PRODUCT Material, other than the principal product, generated as a consequence of an industrial process or as a breakdown product in a living system.

C

CA Cloud-to-Air lightning.

CAA Cold Air Advection.

CADMIUM (CD) Heavy metal that accumulates in the environment.

CAIRN Pile of stones used as a marker.

CAISSON Watertight chamber or hallow floating box used in construction work under water; structure or chamber which is usually sunk or lowered by digging from the inside; used to gain access to the bottom of a stream or other body of water.

CALCAREOUS Rock or substance formed of calcium carbonate or magnesium carbonate by biological deposition or inorganic precipitation, or containing those minerals in sufficient quantities to effervesce when treated with cold hydrochloric acid.

CALCITE Light-colored mineral composed of calcium carbonate that often fills veins in igneous rocks and forms the sedimentary rock limestone.

CALDERA (CRATER) Large circular depression formed by explosion or collapse of a volcano.

CALIBRATED MODEL Model for which all residuals between calibration targets and corresponding model outputs, or statistics computed from residuals, are less than pre-set acceptable values.

CALIBRATION Derivation of a set of model parameter values that produces the "best" fit to observed data; process of refining the model representation of the hydrogeologic framework, hydraulic properties, and boundary conditions to achieve a desired degree of correspondence between the model simulations and observations of the flow system.

CALIBRATION TARGET Measured, observed, calculated, or estimated parameters that a model must reproduce, at lease approximately, to be considered calibrated.

CALM Weather condition when no air motion (wind) is detected.

CALVE To separate or break; become detached.

CAMBER Extra height added to the crest of embankment dams to ensure that the freeboard will not be diminished by foundation settlement or embankment consolidation.

CANADIAN GEODETIC VERTICAL DATUM 1928 Geodetic datum derived from a general adjustment of Canada's first order level network in 1928.

CANAL Channel, usually open, that conveys water by gravity to farms, municipalities, etc.

CANAL HEADWORKS Beginning of a canal.

CANAL PRISM Shape of the canal as seen in cross section.

CANCELLATION Refers to Section 6 (b) of the Federal Insecticide, Fungicide and Rodenticide Act (FIFRA) which authorizes cancellation of a pesticide registration if unreasonable adverse effects to the environment and public health develop when a product is used according to widespread and commonly recognized practice, or if its labeling or other material required to be submitted does not comply with FIFRA provisions.

CANCER POTENCY SLOPE FACTOR Indication of a chemical's human cancer-causing potential derived using animal studies or epidemiological data on human exposure; based on extrapolation of high-dose levels over short periods of time to low-dose levels and a lifetime exposure period through the use of a linear model.

CANDIDATE SPECIES Plant or animal species that are candidates for designation as endangered (in danger of becoming extinct) or threatened (likely to become endangered), but is undergoing status review by the U.S. Fish and Wildlife Service (FWS).

CANOPY ANGLE Measure of the openness of a stream to sunlight; angle formed by an imaginary line from the highest structure (for example, tree, shrub, or bluff) on one bank to eye level at midchannel to the highest structure on the other bank.

CANOPY-INTERCEPTION Precipitation that falls on, and is stored in the leaf or trunk of vegetation; refers to either the process or a volume.

CANYON WIND Foehn wind that is channeled through a canyon as it descends the lee side of a mountain barrier.

CAP (LID) Layer of clay, or other impermeable material installed over the top of a closed landfill to prevent entry of rainwater and minimize leachate; layer of relatively warm air aloft, usually several thousand feet above the ground, which suppresses or delays the development of thunderstorms.

CAP CLOUD Stationary cloud directly above an isolated mountain peak, with cloud base below the elevation of the peak.

CAPABILITY Maximum load that a generating unit, generating station, or other electrical apparatus can carry under specified conditions for a given period of time without exceeding approved limits of temperature and stress.

CAPABLE FAULT Active fault that is judged capable of producing macro-earthquakes and exhibits one or more of the following characteristics: Movement at or near the ground surface at least once within the past 35,000 years; macroseismicity (3.5 magnitude Richter or greater) instrumentally determined with records of sufficient precision to demonstrate a direct relationship with the fault; structural relationship to a capable fault such that movement on one fault could be reasonably expected to cause movement on the other; or (established patterns of microseismicity that define a fault, with historic macroseismicity that can reasonably be associated with that fault.

CAPACITY Load for which a generator, transmission line, or system is rated, expressed in kilowatts; amount of electric power delivered or required for which a generator, turbine, transformer, transmission circuit, station, or system is rated by the manufacturer; maximum load that a machine, station, or system can carry under existing service conditions.

CAPACITY ASSURANCE PLAN Statewide plan which supports a state's ability to manage the hazardous waste generated within its boundaries over a twenty year period.

CAPILLARITY Degree to which a material or object containing minute openings or passages, when immersed in a liquid, will draw the surface of the liquid above the hydrostatic level; phenomenon by which water is held in interstices above the normal hydrostatic level, due to attraction between water molecules.

CAPILLARY ACTION Movement of water through very small spaces due to molecular forces called capillary forces; process by which water rises through rock, sediment or soil caused by the cohesion between water molecules and an adhesion between water and other materials that pulls the water upward; property of surface tension that draws water upwards.

CAPILLARY ATTRACTION (CAPILLARY FORCE) Molecular forces which cause the movement of water through very small spaces, as between soil particles, regardless of gravity; adhesive force that holds a fluid in a capillary or a pore space; function of the properties of the fluid, and surface and dimensions of the space where if the attraction between the fluid and surface is greater than the interaction of fluid molecules, the fluid will be held in place.

CAPILLARY FRINGE See CAPILLARY ZONE.

CAPILLARY FRINGE ZONE See CAPILLARY ZONE.

CAPILLARY HEAD Potential, expressed in head of water, that causes the water to flow by capillary action.

CAPILLARY MIGRATION (CAPILLARY FLOW) Movement of water by capillary action.

CAPILLARY POTENTIAL Work required to move a unit mass of water from the reference plane to any point in the soil column.

CAPILLARY RISE Height above a free water elevation to which water will rise by capillary action.

CAPILLARY WATER Underground water held above the water table by capillary attraction; water subject to the influence of capillary action.

CAPILLARY WAVES Waves caused by the initial wind stress on the water surface causes what are known as capillary waves; wavelength of less than 1.73 cm, and the force that tries to restore them to equilibrium is the cohesion of the individual molecules; important in starting the process of energy transfer from the air to the water.

CAPILLARY ZONE Zone above the water table in which water is held by surface tension (capillary action);

porous material just above the water table which may hold water by capillarity in the smaller void spaces; water in the capillary fringe is under a pressure less than atmospheric; soil area just above the water table where water can rise up slightly through the cohesive force of capillary action; layer ranges in depth from a couple of inches, to a few feet, and it depends on the pore sizes of the materials.

CAPITAL COSTS Costs (usually long-term debt) of financing construction and equipment; usually fixed, one-time expenses which are independent of the amount of water produced; all the implements, equipment, machinery and inventory used in the production of goods and services.

CAPITAL INVESTMENT General term used to identify any money amount which is to be considered as an investment as opposed to an annual expense; can be either interest bearing or non-interest bearing.

CAPPING Region of negative buoyancy below an existing level of free convection (LFC) where energy must be supplied to the parcel to maintain its ascent; tends to inhibit the development of convection until some physical mechanism can lift a parcel to its LFC.

CAPPING INVERSION See CAP.

CAPROCK See CONFINING ZONE.

CAPS Center for Analysis and Prediction of Storms.

CAPTURE EFFICIENCY Fraction of organic vapors generated by a process that are directed to an abatement or recovery device.

CARBON ABSORBER Add-on control device that uses activated carbon to absorb volatile organic compounds from a gas stream.

CARBON ADSORPTION Treatment system that removes contaminants from groundwater or surface water by forcing it through tanks containing activated carbon treated to attract the contaminants.

CARBON CAPTURE AND STORAGE (CCS) Process of capturing carbon dioxide from an emission source, (typically) converting it to a supercritical state, transporting it to an injection site, and injecting it into deep subsurface rock formations for long-term storage.

CARBON DIOXIDE Colorless, odorless, non-poisonous gas which is the fourth most abundant constituent of dry air; product of fossil fuel combustion; greenhouse gas that traps terrestrial (i.e. infrared) radiation and contributes to the potential for global warming.

CARBON DIOXIDE EQUIVALENT Emissions of greenhouse gases are typically expressed in a common metric so that their impacts can be directly compared, as some gases are more potent (i.e. have a higher global warming potential) than others; international standard practice is to express greenhouse gases in carbon dioxide equivalents.

CARBON DIOXIDE PLUME Extent underground, in three dimensions, of an injected carbon dioxide stream.

CARBON DIOXIDE STREAM Carbon dioxide that has been captured from an emission source (e.g. power plant), plus incidental associated substances derived from the source materials and the capture process, and any substances added to the stream to enable or improve the injection process; does not apply to any carbon dioxide stream that meets the definition of a hazardous waste under 40 CFR Part 261.

CARBON MONOXIDE Colorless, odorless, poisonous gas produced by incomplete fossil fuel combustion.

CARBON TETRACHLORIDE Compound consisting of one carbon atom and four chlorine atoms, once widely used as an industrial raw material, as a solvent, and in the production of CFCs; carcinogenic.

CARBONATE ROCKS Rocks (such as limestone or dolostone) that are composed primarily of minerals (such as calcite and dolomite) containing the carbonate ion.

CARBOXYHEMOGLOBIN Hemoglobin in which the iron is bound to carbon monoxide instead of oxygen.

CARCINOGEN Substance, chemical or physical agent that can cause or aggravate cells to develop cancer.

CARNIVORE Flesh-eating or predatory organism.

CARRIER Inert liquid or solid material with non-toxic properties by itself in a pesticide product that serves as a delivery vehicle for the active ingredient; material or system that can facilitate the movement of a pollutant into the body or cells.

CARRINGTON LONGITUDE System of fixed longitudes rotating with the Sun.

CARRY OVER Quantity of water which continues past an inlet.

CARRYING CAPACITY Amount of use a recreation area can sustain without loss of quality; maximum number of animals an area can support during a given period.

CARTOGRAPHY Art and science of graphically representing the features of the Earth's surface.

CAS REGISTRATION NUMBER Number assigned by the Chemical Abstract Service to identify a chemical.

CASE STUDY Brief fact sheet providing risk, cost, and performance information on alternative methods and other pollution prevention ideas, compliance initiatives, voluntary efforts, etc.

CASING Pipe lining placed inside a drilled hole to prevent the collapse of the walls of the bore hole, to prevent pollutants from entering the well, and to house the pump and pipes.

CASK (COFFIN) Thick-walled container (usually lead) used to transport radioactive material.

CATALINA EDDY Coastal eddy forms when upper level large-scale flow off Point Conception interacts with the complex topography of the Southern California coastline; counter clockwise circulating low pressure area that forms with its center in the vicinity of Catalina Island; predominately occur between April and September with a peak in June.

CATALYST Substance that changes the speed or yield of a chemical reaction without being consumed or chemically changed by the chemical reaction.

CATALYTIC CONVERTER Air pollution abatement device that removes pollutants from motor vehicle exhaust, either by oxidizing them into carbon dioxide and water or reducing them to nitrogen.

CATALYTIC INCINERATOR Control device that oxidizes volatile organic compounds (VOCs) by using a catalyst to promote the combustion process.

CATASTROPHE Sudden and great disaster causing misfortune, destruction, or irreplaceable loss extensive enough to cripple activities in an area.

CATCH Number of fish captured, whether they are kept or released.

CATCHMENT AREA Area having a common outlet for its surface runoff.

CATCHMENT BASIN Unit watershed; area from which all the drainage water passes into one stream or other body of water.

CATEGORICAL National Weather Service precipitation descriptor for a 80, 90, or 100 percent chance of measurable precipitation (0.01 inch).

CATEGORICAL EXCLUSION Class of actions which either individually or cumulatively would not have a significant effect on the human environment and therefore would not require preparation of an environmental assessment or environmental impact statement under the National Environmental Policy Act (NEPA).

CATEGORICAL INDUSTRIAL USER (CIU) Industrial user subject to national categorical pretreatment standards.

CATEGORICAL PRETREATMENT STANDARD Technology-based effluent limitation for an industrial facility discharging into a municipal sewer system.

CATHODE Negative pole or electrode of an electrolytic cell or system; attracts positively charged particles or ions (cations).

CATHODIC PROTECTION Technique to prevent corrosion of a metal surface by making it the cathode of an electrochemical cell; electrical system for prevention of rust, corrosion, and pitting of metal surfaces which are in contact with water or soil; low-voltage current is made to flow through a liquid (water) or a soil in contact with the metal in such a manner that the external electromotive force renders the metal structure cathodic; concentrates corrosion on auxiliary anodic parts which are deliberately allowed to corrode instead of letting the structure corrode.

CATION Positively charged ion in an electrolyte solution, attracted to the cathode under the influence of a difference in electrical potential.

CAULK Material used to seal joints.

CAUTION STAGE Stage that, when reached by a rising stream, represents the level where appropriate officials (e.g., county sheriff, civil defense officials, or bypass gate operators) are notified of the threat of possible flooding.

CAVITATION Formation and collapse of gas pockets or bubbles on the blade of an impeller or the gate of a valve; collapse of pockets or bubbles that drives water with such force that it can cause pitting of the gate or valve surface and is usually accompanied by noise and vibration; formation of voids or cavities caused in a liquid due to turbulence or temperature which causes the pressure in local zones of the liquid to fall below the vapor pressure; attack on surfaces caused by subatmospheric pressures immediately downstream from an obstruction or offset.

CAVITATION DAMAGE Damage caused when partial vacuums formed in a liquid by a swiftly moving solid body (e.g. propeller) pit and wear away solid surfaces (e.g., metal or concrete); attack on surfaces caused by the implosion of bubbles of water vapor.

CAVU Clear or Scattered Clouds (visibility greater than 10 mi).

CB Cumulonimbus cloud (thunderhead), characterized by strong vertical development in the form of mountains or huge towers topped at least partially by a smooth, flat, often fibrous anvil.

CBMAM Cumulonimbus Mamma.

CC Cloud-to-Cloud Lightning.

CCITT Consultative Committee for International Telephone and Telegraph.

CD Cold.

CDB Computing Development Branch (NCEP).

CDT Central Daylight Time.

CEILING (CIG) Height of the lowest layer of clouds, when the sky is broken or overcast.

CEILOMETER Device using a laser or other light source to determine the height of a cloud base; uses triangulation to determine the height of a spot of light projected onto the base of the cloud by measuring the time required for a pulse of light to be scattered back from the cloud base.

CELL In Biology, smallest structural part of living matter capable of functioning as an independent unit; in meteorology, convection in the form of a single updraft, downdraft, or updraft/downdraft couplet, typically seen as a vertical dome or tower as in a towering cumulus cloud; in modeling, distinct one-two-or three dimensional model unit representing a discrete portion of a physical system with uniform properties assigned to it; in solid waste disposal, holes where waste is dumped, compacted, and covered with layers of dirt on a daily basis.

CELL VOLUME (BIOVOLUME) DETERMINATION Used to estimate biomass of algae in aquatic systems; used frequently in aquatic surveys as an indicator of algal production; number of cells of any organism that is counted by using a microscope and grid or counting cell.

CELSIUS (°C) Unit of temperature; standard scale where the freezing point of water is 0 °F and the boiling point is 100 °F; to convert a Fahrenheit temperature to Celsius, multiply it by $5/9$ and then subtract 32.

°C See CELSIUS.

CEMENT Material used to support and seal the well casing to the rock formations exposed in the borehole, protects the casing from corrosion and prevents movement of injectate up the borehole.

CEMENTITIOUS Densely packed and non-fibrous friable materials.

CENTER Vertical axis of a tropical cyclone, usually defined by the location of minimum wind or minimum pressure.

CENTER PIVOT IRRIGATION Automated sprinkler system involving a rotating pipe or boom that supplies water to a circular area of an agricultural field through sprinkler heads or nozzles.

CENTIMETER BURST Solar radio burst in the centimeter wavelength range.

CENTRAL COLLECTION POINT Location were a generator of regulated medical waste consolidates wastes originally generated at various locations in his facility.

CENTRAL FLYWAY Important international migration route for many birds.

CENTRAL MERIDIAN PASSAGE (CMP) Passage of an Active Region or other feature across the longitude meridian that passes through the apparent center of the solar disk.

CENTRAL NERVOUS Toxicants that deaden the central nervous system (CNS).

CENTRALIZED AUTOMATED DATA ACQUISITION SYSTEM (CADAS) System of two minicomputers in NWSH that interrogates LARCs and DARDCs by telephone every 6 hours and transmits the data to AFOS via HADS.

CENTRALIZED WASTEWATER TREATMENT SYSTEM Managed system consisting of collection sewers and a single treatment plant used to collect and treat wastewater from an entire service area.

CENTRIFUGAL COLLECTOR Mechanical system using centrifugal force to remove aerosols from a gas stream or to remove water from sludge.

CENTRIFUGAL PUMP Pump that moves water by an impeller fixed on a rotating shaft that whirls the water around, building up enough pressure to force the water through the discharge outlet and imparting energy to the water.

CENTROID Center of mass.

CERCLIS Federal Comprehensive Environmental Response, Compensation, and Liability Information System is a database that includes all sites which have been nominated for investigation by the Superfund program.

CERTIFICATION SIGNATURE Signatures of those who co-facilitated the risk analysis signifying that methodology, processes, and requirements were followed.

CERTIFIED WATER RIGHT State-issued document that serves as legal evidence that an approved application has been physically developed and the water put to beneficial use; establishes: priority date, type of beneficial use, and the maximum amount of water that can be used.

CFC Chlorofluorocarbon.

CFP Cold Front Passage.

CFS See CUBIC FEET PER SECOND.

CFS-DAY See CUBIC FEET PER SECOND-DAY.

CG Cloud-to-Ground Lightning.

CHAMFER To bevel or slope an edge or corner.

CHANCE National Weather Service precipitation descriptor for 30, 40, or 50 percent chance of measurable precipitation (0.01 inch); when the precipitation is convective in nature, the term scattered is used.

CHANNEL General term for any naturally or artificially created open conduit that may convey water, with a definite bed and banks to confine and conduct continuously or periodically flowing water; forms a connecting link between two bodies of water; may be single or braided; a natural channels is termed a river, creek, run, branch, anabranch, and tributary and an artificial channels is termed a canal or floodway.

CHANNEL BARS Lowest prominent geomorphic features higher than the channel bed.

CHANNEL INFLOW Water, which at any instant, is flowing into the channel system form surface flow, subsurface flow, base flow, and rainfall that has directly fallen onto the channel.

CHANNEL INVERT Lowest point in the channel.

CHANNEL LEAD Elongated opening in the ice cover caused by a water current.

CHANNEL MARGIN DEPOSITS Narrow sand deposits which line channel banks.

CHANNEL ROUTING Process of determining progressively timing and shape of the flood wave at successive points along a river.

CHANNEL SCOUR Erosion by flowing water and sediment on a stream channel; results in removal of mud, silt, and sand on the outside curve of a stream bend and the bed material of a stream channel.

CHANNEL STABILIZATION Stable channel is neither progressively aggrading nor degrading, or changing its cross-sectional area through time; stabilization techniques consist of bank protection and other measures that work to transform an unstable channel into a stable one.

CHANNEL STORAGE Volume of water at a given time in the channel, river reach or over the flood plain of streams in a drainage basin.

CHANNELED HIGH WINDS Channeled air through constricted passages producing high winds.

CHANNELIZATION Modification of a natural river channel that may include deepening, widening, or straightening to permit water to move faster or to drain a wet area for farming; often done for flood control or for improved agricultural drainage or irrigation.

CHARACTERISTIC Any one of the four categories used in defining hazardous waste: ignitability, corrosivity, reactivity, and toxicity.

CHARACTERIZATION OF ECOLOGICAL EFFECTS Part of ecological risk assessment that evaluates ability of a stressor to cause adverse effects under given circumstances.

CHARACTERIZATION OF EXPOSURE Portion of an ecological risk assessment that evaluates interaction of a stressor with one or more ecological entities.

CHC Chance.

CHECK DAM Small dam designed to retard the flow of water and sediment in a channel, used especially to control soil erosion; small barrier constructed in a gully or other small watercourse to decrease flow velocity, minimize channel scour, and promote deposition of sediment.

CHECK STRUCTURE Structure used to regulate the upstream water surface and control the downstream flow in a canal.

CHECK VALVE Device that allows fluid or air to pass through it in only one direction; valve with a hinged disc or flap that opens in the direction of normal flow and is forced shut when flows attempt to go in the reverse or opposite (backflow) direction of normal flow.

CHECKED SIGNATURE Checked signatures verify that all probability estimates, inputs and outputs and their distributions, were entered correctly into event trees, and that any other calculations, figures, or tables have been checked.

CHECK-VALVE TUBING PUMP Water sampling tool.

CHEMICAL CASE Grouping of chemically similar pesticide active ingredients (e.g., salts and esters of the same chemical) into chemical cases.

CHEMICAL COMPOUND Distinct and pure substance formed by the union or two or more elements in definite proportion by weight.

CHEMICAL ELEMENT Fundamental substance comprising one kind of atom; the simplest form of matter.

CHEMICAL OXYGEN DEMAND (COD) Measure of the oxygen-consuming capacity of inorganic and organic matter present in wastewater expressed as the amount of oxygen consumed in mg/l; measure of the oxygen required to oxidize all compounds, both organic and inorganic, in water and furnishes an approximation of the amount of organic and reducing material present.

CHEMICAL STRESSORS Chemicals released to the environment through industrial waste, auto emissions, pesticides, and other human activity that can cause illnesses and even death in plants and animals.

CHEMICAL TREATMENT Any one of a variety of technologies that use chemicals or a variety of chemical processes to treat waste.

CHEMISTRY MODEL Computer model used in air pollution investigations that simulates chemical and photochemical reactions of the pollutants during their transport and diffusion.

CHEMNET Mutual aid network of chemical shippers and contractors that assigns a contracted emergency response company to provide technical support if a representative of the firm whose chemicals are involved in an incident is not readily available.

CHEMOSTERILANT Chemical that controls pests by preventing reproduction.

CHEMTREC Industry-sponsored Chemical Transportation Emergency Center; provides information and other assistance to emergency responders.

CHG Change.

CHGS Changes.

CHILD RESISTANT PACKAGING (CRP) Packaging that protects children or adults from injury or illness resulting from accidental contact with or ingestion of residential pesticides that meet or exceed specific toxicity levels that is required by FIFRA regulations; term used for protective packaging of medicines.

CHILLER Device that generates a cold liquid that is circulated through an air-handling unit's cooling coil to cool the air supplied to the building.

CHILLING EFFECT Lowering of the Earth's temperature because of increased particles in the air blocking the Sun's rays.

CHIMNEY DRAIN Vertical or inclined layer of pervious material in an embankment to facilitate and control drainage of the embankment fill.

CHINOOK Region-specific term used for Foehn Winds in the lee of the Rocky Mountains in the United States that are warm, dry winds that occur in the lee of high mountain ranges.

CHINOOK ARCH Foehn cloud formation appearing as a bank of altostratus clouds east of the Rocky Mountains, heralding the approach of a chinook that forms in the rising portion of standing waves on the lee side of the mountains.

CHIPPING Loosening of shallow rock by light blasting or air hammers.

CHIRONOMID Group of two-winged flying insects who live their larval stage underwater and emerge to fly about as adults.

CHISEL PLOWING Preparing croplands by using a special implement that avoids complete inversion of the soil as in conventional plowing that can leave a protective cover or crops residues on the soil surface to help prevent erosion and improve filtration.

CHLORDANE Organochlorine insecticide no longer registered for use in the U.S. that is a mixture in which the primary components are cis- and trans-chlordane, cis- and trans-nonachlor, and heptachlor; Octachloro-4,7-methanotetrahydroindane.

CHLORINATED HYDROCARBONS Chemicals containing only chlorine, carbon, and hydrogen that linger in the environment and accumulate in the food chain (e.g., DDT, aldrin, dieldrin, heptachlor, chlordane, lindane, endrin, Mirex, hexachloride, and toxaphene); chlorinated organic compounds including chlorinated solvents (e.g., dichloromethane, trichloromethylene, chloroform).

CHLORINATED SOLVENT Organic solvent containing chlorine atoms (e.g., methylene chloride and 1,1,1-trichloromethane) with uses including aerosol spray containers, in highway paint, and dry cleaning fluids.

CHLORINATION Application of chlorine to drinking water, sewage, or industrial waste to disinfect or oxidize undesirable compounds and other biological or chemical results (aiding coagulation and controlling tastes and odors).

CHLORINATOR Device that adds chlorine, in gas or liquid form, to water or sewage to kill infectious bacteria.

CHLORINE-CONTACT CHAMBER
Part of a water treatment plant where effluent is disinfected by chlorine.

CHLOROFLUOROCARBONS (CFCS)
Class of volatile compounds consisting of carbon, chlorine, and fluorine (i.e. freons) that are inert, non-toxic, and easily liquefied to be used as coolants, packaging, insulation, computer-chip cleaners, blowing agents, and propellants; chemicals with long lifetime that are not destroyed in the lower atmosphere and drift into the upper atmosphere destroying stratospheric ozone.

CHLOROPHENOXY Class of herbicides that may be found in domestic water supplies and cause adverse health effects.

CHLOROSIS Discoloration of normally green plant parts caused by disease, lack of nutrients, or various air pollutants.

CHOLINESTERASE Enzyme found in animals that regulates nerve impulses by the inhibition of acetylcholine that causes a variety of acute symptoms including: nausea, vomiting, blurred vision, stomach cramps, and rapid heart rate.

CHROMIUM See HEAVY METALS.

CHROMOSPHERE Layer of the solar atmosphere above the photosphere and beneath the transition region and the corona.

CHROMOSPHERIC EVENTS
Flares that are just Chromospheric Events without Centimetric Bursts or Ionospheric Effects.

CHRONIC Stimulus that lingers or continues for a relatively long period of time, often one-tenth of the life span or more; considered a relative term depending on the life span of an organism; measurement of a chronic effect can be reduced growth, reduced reproduction, etc., in addition to lethality.

CHRONIC EFFECT Adverse effect on a human or animal in which symptoms recur frequently or develop slowly over a long period of time; process by which small amounts of toxic substances are taken into the body over an extended period.

CHRONIC HEALTH EFFECT Possible result of exposure over many years to a drinking water contaminant at levels above its MCL.

CHRONIC TOXICITY Capacity of a substance to cause long-term poisonous health effects in humans, animals, fish, and other organisms.

CHRYSENE See POLYCYCLIC AROMATIC HYDROCARBON (PAH).

CHUTE Portion of spillway between the gate or crest structure and the terminal structure, where open- channel flow conditions will exist; conduit for conveying free-flowing materials at high velocity to lower elevations.

CIENAGA Marshy area where the ground is wet due to the presence of seepage or springs.

CIO Chief Information Officer.

CIPOLLETTI WEIR (TRAPEZOIDAL WEIR) Contracted weir of trapezoidal shape in which the sides of the notch are given a slope of 1 horizontal to 4 vertical.

CIRCLE OF INFLUENCE Circular outer edge of a depression produced in the water table by the pumping of water from a well.

CIRCUIT Complete path of an electric current, including the generating apparatus or other source; specific segment or section of the complete path.

CIRCUIT BREAKER Safety device in an electrical circuit that automatically shuts off the circuit when it becomes overloaded and can be manually reset.

CIRCUIT MILE For single circuit electric power transmission lines, circuit miles are equal to geographic miles or pole miles.

CIRCULATION Flow, or movement, of a fluid (e.g., water or air) in or through a given area or volume.

CIRCUMNEUTRAL Water with a pH between 5.5 and 7.4; pH modifier used in the U.S. Fish and Wildlife Service wetland classification system.

CIRQUE Bowl-like depression carved into a mountaintop by ice at the head of a glacier.

CIRRIFORM High altitude ice clouds with a very thin wispy appearance.

CIRROCUMULUS Cirriform cloud characterized by thin, white patches, each of which is composed of very small granules or ripples at high altitudes (20,000-40,000 ft or 6,000-12,000 m).

CIRROSTRATUS Cloud of a class characterized by a composition of ice crystals and often by the production of halo phenomena and appearing as a whitish and usually somewhat fibrous veil, often covering the whole sky and sometimes so thin as to be hardly discernible at high altitudes (20,000-40,000 ft or 6,000-12,000 m).

CIRRUS CLOUDS (CI) High-level clouds (16,000 feet or higher), composed of ice crystals and appearing in the form of white, delicate filaments or white or mostly white patches or narrow bands.

CISTERN Small tank (usually covered) or storage facility used to store water for a home or farm; often used to store rain water.

CIVIL DAWN Time of morning at which the Sun is 6 degrees below the horizon.

CIVIL DEFENSE AGENCY State and/or local agency responsible for emergency operations, planning, mitigation, preparedness, response, and recovery for all hazards.

CIVIL DUSK Time at which the Sun is 6 degrees below the horizon in the evening.

CIVIL EMERGENCY MESSAGE (CEM) Message issued by the National Weather Service in coordination with federal, state or local government to warn the general public of a non-weather related time-critical emergency which threatens life or property (e.g., nuclear accident, toxic chemical spill, etc.).

CL Abbreviation on climate outlook maps issued by CPC to indicate areas where equal chances of experiencing below-normal, normal, and above-normal conditions are possible.

CLADOPHORA Filamentous green alga.

CLAM SHELL GATE High pressure regulating gate consisting of two curved leaves which open and close over the end of a conduit that is used for free discharge into air with minimal cavitation damage.

CLARIFICATION Clearing action that occurs during wastewater treatment when solids settle out that is often aided by centrifugal action and chemically induced coagulation in wastewater.

CLARIFIER Tank in which solids settle to the bottom and are subsequently removed as sludge.

CLASS (PIPE AND FITTINGS) Working pressure rating of a specific pipe (i.e., cast iron, ductile iron, asbestos cement, and some plastic pipe) for use in water distribution systems that includes allowances for surges.

CLASS I AREA Geographic area designated by the Clean Air Act where only a small amount or increment of air quality deterioration is permissible.

CLASS I SUBSTANCE One of several groups of chemicals with an ozone depletion potential of 0.2 or higher.

CLASS I WELLS Technologically sophisticated wells that inject hazardous waste, non-hazardous industrial waste, or municipal wastewater into deep, isolated rock formations below the lowermost underground source of drinking water (USDW).

CLASS II SUBSTANCE Substance with an ozone depletion potential of less than 0.2.

CLASS II WELLS Wells (salt water disposal, enhanced recovery, and hydrocarbon storage) that inject brines and other fluids associated with oil and gas production, or storage of hydrocarbons.

CLASS III WELLS Wells that inject fluids associated with solution mining of minerals (salt solution, in-situ leaching of uranium, and sulfur).

CLASS IV WELLS Wells that inject hazardous or radioactive wastes into or above a underground source of drinking water (USDW) and are banned unless authorized under a federal or state groundwater remediation project.

CLASS V INJECTION WELL Shallow well used to place a variety of fluids at shallow depths below the land surface, including a domestic onsite wastewater treatment system serving more than 20 people.

CLASS V WELLS Wells not included in Classes I to IV and Class VI that inject non-hazardous fluids into or above a underground source of drinking water (USDW) and are typically shallow, on-site disposal systems; however, this class also includes some deeper injection operations.

CLASS VI WELLS Wells that inject carbon dioxide for the purposes of long-term storage known as CO2 geologic sequestration.

CLASTIC Rock or sediment composed principally of broken fragments that are derived from preexisting rocks which have been transported from their place of origin (e.g. sandstone).

CLAY Fine-grained soil or the fine-grained portion of soil that passes a No. 200 (0.075 mm) United States Standard sieve and can be made to exhibit plasticity (putty-like properties) within a range of moisture contents, and that exhibits considerable strength when air-dry.

CLAY SIZE Portion of the soil finer than 0.002 mm (0.005 mm in some cases).

CLAY SOIL Soil material containing more than 40 percent clay, less than 45 percent sand, and less than 40 percent silt.

CLEAN COAL TECHNOLOGY Any technology not in widespread use prior to the Clean Air Act Amendments of 1990 that will achieve significant reductions in pollutants associated with the burning of coal.

CLEAN FUELS Blends or substitutes for gasoline fuels, including compressed natural gas, methanol, ethanol, and liquefied petroleum gas.

CLEAN WATER ACT See FEDERAL WATER POLLUTION CONTROL ACT OF 1948.

CLEANER TECHNOLOGIES SUBSTITUTES ASSESSMENT Document that systematically evaluates the relative risk, performance, and cost trade-offs of technological alternatives; serves as a repository for all the technical data (including methodology and results) developed by pollution prevention or education project.

CLEANUP Actions taken to deal with a release or threat of release of a hazardous substance that could affect humans and/or the environment.

CLEAR AIR TURBULENCE (CAT) Sudden severe turbulence occurring in cloudless regions that causes violent buffeting of aircraft.

CLEAR CUT Harvesting all the trees in one area at one time, a practice that can encourage fast rainfall or snowmelt runoff, erosion, sedimentation of streams and lakes, and flooding, and destroys vital habitat.

CLEAR ICE Thin coating that is relatively transparent of ice on terrestrial objects, caused by rain that freezes on impact.

CLEAR SLOT Local region of clearing skies or reduced cloud cover, with respect to severe thunderstorms, indicating an intrusion of drier air; often seen as a bright area with higher cloud bases on the west or southwest side of a wall cloud.

CLEAR WELL Reservoir for storing filtered water of sufficient quantity to prevent the need to vary the filtration rate with variations in demand; used to provide chlorine contact time for disinfection.

CLEARANCE Procedure used to establish, under tightly controlled discipline, a safe environment for maintenance, repair, or inspection that includes systematically isolating pertinent equipment from all sources of hazardous energy (hydraulic, electrical, mechanical, pneumatic, chemical, etc.), attaching safety tags or locks to the appropriate controls, and a written statement that documents isolation of the equipment.

CLEARING Removal of all vegetation such as trees, shrubs, brush, stumps, exposed roots, down timber, branches, grass, and weeds; removal of all rubbish and all other objectionable material.

CLIENT AGENCY Public fire service or wildlands management agency, federal or non-federal, which requires and uses National Weather Service fire and forestry meteorological services.

CLIMATE Composite or generally prevailing weather conditions of a region, throughout the year, averaged over a series of years.

CLIMATE CHANGE Non-random change in climate that is measured over several decades or longer that may be due to natural or human-induced causes.

Climate Diagnostic Center (CDC) Advance national capabilities to interpret the causes of observed climate variations, and to apply this knowledge to improve climate models and forecasts and develop new climate products that better serve the needs of the public and decision-makers.

CLIMATE DIAGNOSTICS BULLETIN (CDB) Monthly CPC Bulletin reports on the previous months' status of the ocean-atmosphere climate system and provides various seasonal ENSO-related outlooks.

CLIMATE DIAGNOSTICS CENTER (CDC) Mission of NOAA's Climate Diagnostics Center is to identify the nature and causes for climate variations on time scales ranging from a month to centuries.

CLIMATE MODEL Mathematical model for quantitatively describing, simulating, and analyzing the interactions between the atmosphere and underlying surface (e.g., ocean, land, and ice).

CLIMATE OUTLOOK Issued by the CPC gives probabilities that conditions, averaged over a specified period, will be below-normal, normal, or above-normal.

CLIMATE PREDICTION CENTER One of several centers under the National Centers for Environmental Prediction (NCEP) part of the National Weather Service (NWS) in the National Oceanic and Atmospheric Administration (NOAA) that serves the public by assessing and forecasting the impacts of short-term climate variability, emphasizing enhanced risks of weather-related extreme events, for use in mitigating losses and maximizing economic gains.

CLIMATE SYSTEM System consisting of the atmosphere (gases), hydrosphere (water), lithosphere (solid rocky part of the Earth), and biosphere (living) that determine the Earth's climate.

CLIMATIC YEAR Continuous 12-month period during which a complete climactic cycle occurs.

CLIMATOLOGICAL OUTLOOK Based upon climatological statistics for a region (abbreviated as CL on seasonal outlook maps) that indicates if climate outlook has an equal chance of being above normal, normal, or below normal.

CLIMATOLOGY Science that deals with the phenomena of climates or climatic conditions.

CLIMO Climatology/Climatological.

CLIMOMETER Instrument that measures angles of inclination; used to measure cloud ceiling heights.

CLONING Obtaining a group of genetically identical cells from a single cell; making identical copies of a gene.

CLOSED BASIN Basin draining to some depression or pond within its area, from which water is lost only by evaporation or percolation; basin without a surface outlet for precipitation.

CLOSED BASIN LAKE FLOODING Flooding that occurs on lakes with either no outlet or a relatively small one where water may stay at flood stage for weeks, months, or years; seasonal increases in rainfall cause the lake level to rise faster than it can drain.

CLOSED LOW Low pressure area with a distinct center of cyclonic circulation which can be completely encircled by one or more isobars or height contour lines.

CLOSED-LOOP RECYCLING Reclaiming or reusing wastewater for non-potable purposes in an enclosed process.

CLOSTRIDIUM PERFRINGENS (C. PERFRINGENS) Spore-forming bacterium that is common in the feces of human and other warmblooded animals; used experimentally as an indicator of past fecal contamination and the presence of microorganisms that are resistant to disinfection and environmental stresses.

CLOSURE Procedure a landfill operator must follow when a landfill reaches its legal capacity for solid ceasing acceptance of solid waste and placing a cap on the landfill site.

CLOUD (CLD) Visible aggregate of minute water droplets or ice particles in the atmosphere above the Earth's surface.

CLOUD CEILING Height of the cloud base for the lowest broken or overcast cloud layer.

CLOUD CONDENSATION NUCLEI Small particles in the air on which water vapor condenses and forms cloud droplets.

CLOUD LAYER Array of clouds whose bases are at approximately the same level.

CLOUD MOVEMENT Direction toward which a cloud is moving.

CLOUD STREETS Rows of cumulus or cumulus-type clouds aligned parallel to the low-level flow.

CLOUD TAGS Ragged, detached cloud fragments; fractus or scud.

CLOUDBURST Intense torrential downpour of rain that suggests the immediate bursting and discharge of an entire cloud.

CLOUDY 7/8ths or more of the sky is covered by clouds.

CLR Clear.

CLRG Clearing.

CLUTTER Radar echoes that interfere with observation of desired signals on the radar display.

CM Centimeter; Combined Moment.

CMP Corrugated Metal Pipe.

CMPLT Complete.

CMPLX Complex.

CNIF Calibration Network Information Files.

CNTR Center.

CNTRL Central.

CNVG Converge.

CNVTV Convective.

COAGULATION Clumping of particles in wastewater to settle out impurities often induced by chemicals such as lime, alum, and iron salts.

COAL CLEANING TECHNOLOGY Precombustion process by which coal is physically or chemically treated to remove some of its sulfur so as to reduce sulfur dioxide emissions.

COAL GASIFICATION Conversion of coal to a gaseous product by one of several available technologies.

COALESCENCE Process by which water droplets in a cloud collide and come together to form raindrops.

COARSE GRAVEL PROTECTION Gravel generally placed in a layer upon a finished surface to protect the finished surface from deterioration or erosion.

COASTAL FLOODING Flooding that occurs from storms (e.g., hurricanes, "nor'easter," or tropical storms) where water is driven onto land from an adjacent body of water over and above normal tidal action.

COASTAL WATERS Area from a line approximating the mean high water along the mainland or island as far out as 100 nautical miles including the bays, harbors and sounds.

COASTAL WATERS FORECAST (CWF) Marine forecast for areas (e.g., bays, harbors, and sounds) from a line approximating the mean high water mark (average height of high water over a 19-year period) along the mainland or near shore islands extending out to as much as 100 nautical miles (NM).

COASTAL ZONE Lands and waters adjacent to the coast that exert an influence on the uses of the sea and its ecology, or whose uses and ecology are affected by the sea.

COASTAL/LAKESHORE FLOOD ADVISORY Minor flooding is possible (i.e., over and above normal high tide levels).

COASTAL/LAKESHORE FLOOD WARNING Flooding that will pose a serious threat to life and property is occurring, imminent or highly likely.

COASTAL/LAKESHORE FLOOD WATCH Flooding with significant impacts is possible.

COASTER GATE Rectangular gate similar in construction and appearance to the tractor gate except sealing is accomplished in much the same way as a ring seal gate.

COATING Protective material applied to the outer surface of a material, frequently metalwork.

COBBLE (COBBLESTONE) Rock fragment, usually rounded or semi-rounded, with an average dimension between 3 to 12 inches; particle of rock that will pass a 12-inch (300-mm) square opening and be retained on a 3-inch (75-mm) U.S.A. Standard sieve.

CODE OF FEDERAL REGULATIONS (CFR) Compilation of all federal rules currently in effect that is updated annually; codification of the final rules published daily in the Federal Register; document that codifies all rules of the executive departments and agencies of the Federal Government.

CODE SELECTION Process of choosing the appropriate computer code, algorithm, or other analysis technique capable of simulating those characteristics of the physical system required to fulfill the modeling project's objective(s).

COE U.S. Army Corps of Engineers.

COEFFICIENT OF COMPRESSIBILITY Slope of a one-dimensional compression curve relating void ratio to effective stress.

COEFFICIENT OF CONSOLIDATION Coefficient that relates the change in excess pore pressure with time to the excess pore pressure diffusion in the soil mass in terms of soil mass and pore fluid characteristics.

COEFFICIENT OF HAZE (COH) Measurement of visibility interference in the atmosphere.

COFFERDAM Temporary structure enclosing all or part of the construction area so that construction can proceed in the dry where a diversion cofferdam diverts a stream into a pipe, channel or tunnel; temporary barrier, usually an earthen dike, constructed around a worksite in a reservoir or on a stream, so the worksite can be dewatered or the water level controlled.

CO-FIRE Burning of two fuels in the same combustion unit (e.g., coal and natural gas, or oil and coal).

COGENERATION Consecutive generation of useful thermal and electric energy from the same fuel source.

COGENERATOR Generating facility that produces electricity and another form of useful thermal energy (i.e., heat or steam), used for industrial, commercial, heating, or cooling purposes.

COHERENT RADAR Radar that utilizes both signal phase and amplitude to determine target characteristics.

COHESION Mutual attraction of soil particles due to molecular and capillary forces in the presence of water where attraction is high in clay (especially dry) and low in silt or sand; ability of a substance to stick to itself and pull itself together; molecular attraction that holds two particles together.

COHESIONLESS MATERIALS (COHESIONLESS SOIL) Soil materials that when unconfined have little or no strength when air-dried and that have little or no cohesion when submerged; soil that has little tendency to stick together whether wet or dry (e.g., sands and gravels).

COHESIVE SOIL Predominantly clay and silt soil, fine-grained particles, that sticks together whether wet or dry; soil that, when unconfined, has considerable strength when air-dried, and that has significant cohesion when submerged.

COKE One of the basic materials used in blast furnaces for the conversion of iron ore into iron.

COKE OVEN Industrial process which converts coal into coke.

COLD ADVECTION Transport of cold air into a region by horizontal winds.

COLD AIR AVALANCHE Downslope flow pulsations that occur at more or less regular intervals as cold air builds up on a peak or plateau, reaches a critical mass, and then cascades down the slopes.

COLD AIR DAM Shallow cold air mass which is carried up the slope of a mountain barrier, but with insufficient strength to surmount the barrier; cold air, trapped upwind of the barrier alters the effective terrain configuration of the barrier to larger-scale approaching flows.

COLD AIR DAMMING (CAD) Phenomenon in which a low-level cold air mass is trapped topographically.

COLD AIR FUNNEL Funnel cloud or (rarely) a small, relatively weak tornado that can develop from a small shower or thunderstorm when the air aloft is unusually cold.

COLD FRONT Zone separating two air masses, of which the cooler, denser mass is advancing and replacing the warmer.

COLD JOINT Unplanned joint resulting when a concrete surface hardens before the next batch is placed against it; fresh concrete placed on harden concrete.

COLD OCCLUSION Frontal zone formed when a cold front overtakes a warm front and, being colder than the air ahead of the warm front, slides under the warm front, lifting it aloft.

COLD POOL Region of relatively cold air, represented on a weather map analysis as a relative minimum in temperature surrounded by closed isotherms.

COLD TEMPERATURE CO Standard for automobile emissions of carbon monoxide (CO) emissions to be met at a low temperature (i.e. 20 °F).

COLD-WATER FISHERY Water or water system which has an environment suitable for salmonid fishes.

COLIFORM Organisms common to the intestinal tract of humans and animals; group of related bacteria whose presence in water indicates fecal pollution and potentially adverse contamination by pathogens.

COLIFORM INDEX Rating of the purity of water based on a count of fecal bacteria.

COLIPHAGES Viruses that infect and replicate in coliform bacteria that are indicative of sewage contamination of water and of the survival and transport of viruses in the environment.

COLLAR Open end of a drill hole.

COLLAR CLOUD Generally circular ring of cloud that may be observed on rare occasions surrounding the upper part of a wall cloud.

COLLARING Starting a drill hole.

COLLECTION EFFICIENCY Fraction of droplets approaching a surface that actually deposit on that surface.

COLLECTOR Public or private hauler that collects non-hazardous waste and recyclable materials from residential, commercial, institutional and industrial sources.

COLLECTOR SEWERS Pipes used to collect and carry wastewater from individual sources to an interceptor sewer that will carry it to a treatment facility.

COLLOIDAL PARTICLES Soil particles that are so small (less than 0.001 mm) that the surface activity has an appreciable influence on the properties of the aggregate.

COLLOIDAL SUSPENSION Method of sediment transport in which water turbulence (movement) supports the weight of the sediment particles, thereby keeping them from settling out or being deposited.

COLLOIDS Very small, finely divided solids (that do not dissolve) that remain dispersed in a liquid for a long time due to their small size and electrical charge.

COLLUVIUM General term applied to loose and incoherent deposits, usually at the foot of a slope and brought there chiefly by gravity.

COLOR UNIT Produced by 1 milligram per liter of platinum in the form of the chloroplatinate ion; color expressed in units of the platinum-cobalt scale.

COLORADO LOW Low pressure storm system that forms in winter in southeastern Colorado or northeastern New Mexico and tracks northeastward across the central plains of the U.S. over a period of several days, producing blizzards and hazardous winter weather.

COLUMNAR ICE Ice consisting of columnar shaped grain where ordinary black ice is usually columnar-grained.

COMBINED CYCLE PLANT Plant which achieves higher efficiency by employing two cycles in tandem (e.g. heat rejected from a gas-fired turbine is used to generate steam to operate a steam turbine).

COMBINED HEAT AND POWER (CHP) Generation of electricity and the capture and use of otherwise wasted heat energy byproducts.

COMBINED SEAS Combination or interaction of wind waves and swells in which the separate components are not distinguished.

COMBINED SEWER OVERFLOW (CSO) Discharge of untreated sewage and stormwater to a stream when the capacity of a combined storm/sanitary sewer system is exceeded during rainstorms by storm runoff in one piping system.

COMBINED SEWER SYSTEM (CSS) Wastewater collection system which conveys sanitary wastewaters (domestic, commercial and industrial wastewaters) and storm-water runoff through a single pipe to a publicly owned treatment works for treatment prior to discharge to surface waters.

COMBUSTION Burning, or rapid oxidation, accompanied by release of energy in the form of heat and light; controlled burning of waste, in which heat chemically alters organic compounds, converting into stable inorganics such as carbon dioxide and water.

COMBUSTION CHAMBER Actual compartment where waste is burned in an incinerator.

COMBUSTION PRODUCT Substance produced during the burning or oxidation of a material.

COMMA CLOUD Synoptic scale cloud pattern with a characteristic comma-like shape.

COMMA ECHO Thunderstorm radar echo which has a comma-like shape.

COMMAND POST Centralized base of operations established near the site of a hazardous materials incident.

COMMAND-AND-CONTROL REGULATIONS Specific requirements prescribing how to comply with specific standards defining acceptable levels of pollution.

COMMENT PERIOD Time provided for the public to review and comment on a proposed action or rulemaking after publication in the Federal Register.

COMMERCIAL OPERATION Commercial operation begins when control of the loading of the generator is turned over to the system dispatcher.

COMMERCIAL RIVER TRIP Trips organized by boating companies that conduct tours for paying passengers or customers.

COMMERCIAL USER DAY Amount of commercial use within a 24-hour period or any portion thereof.

COMMERCIAL WASTE All solid waste emanating from business establishments such as stores, markets, office buildings, restaurants, shopping centers, and theaters.

COMMERCIAL WASTE MANAGEMENT FACILITY Treatment, storage, disposal, or transfer facility which accepts waste from a variety of sources, as compared to a private facility which normally manages a limited waste stream generated by its own operations.

COMMERCIAL WATER USE Water used for motels, hotels, restaurants, office buildings, other commercial facilities, and institutions that may be obtained from a public supply or may be self-supplied.

COMMERCIAL WITHDRAWALS See COMMERCIAL WATER USE.

COMMINGLED RECYCLABLES Mixed recyclables that are collected together.

COMMINUTER Machine that shreds or pulverizes solids to make waste treatment easier.

COMMINUTION Mechanical shredding or pulverizing of waste used in both solid waste management and wastewater treatment.

COMMON EXCAVATION All materials excavated not considered as rock; boulders or detached pieces of solid rock less than 1 cubic yard in volume are classified as common excavation.

COMMON MATERIAL All earth materials which do not fall under the definition of rock.

COMMON SENSE INITIATIVE Voluntary program to simplify environmental regulation to achieve cleaner, cheaper, smarter results, starting with six major industry sectors.

COMMUNITY All members of a specified group of species present in a specific area at a specific time; a group of people that see themselves as a unit; species that interact in a common area.

COMMUNITY COHESION Ability of a community to have a unified response when facing a problem.

COMMUNITY COMPONENT Any portion of a biological community; may pertain to the taxonomic group (fish, invertebrates, algae), the taxonomic category (phylum, order, family, genus, species), the feeding strategy (herbivore, omnivore, carnivore), or organizational level (individual, population, community association) of a biological entity within the aquatic community.

COMMUNITY RELATIONS Effort to establish two-way communication with the public to create understanding of programs and related actions, to ensure public input into decision-making processes related to affected communities, and to make certain that the Agency is aware of and responsive to public concerns.

COMMUNITY WATER SYSTEM Public water system which serves at least 15 service connections used by year-round residents or regularly serves at least 25 year-round residents.

COMPACT FLUORESCENT LAMP (CFL) Small fluorescent lamps used as more efficient alternatives to incandescent lighting.

COMPACTED BACKFILL Backfill which has been reduced to bulk by rolling, tamping, or soaking.

COMPACTED EMBANKMENT Embankment which has been reduced in bulk by rolling, tapping, or soaking.

COMPACTION Reduction of the bulk of solid waste by rolling and tamping; to increase soil density by reducing the voids through mechanical manipulation.

COMPACTION CURVE (PROCTOR CURVE, MOISTURE-DENSITY CURVE) Curve showing the relationship between the dry density (dry unit weight) and the moisture content of a soil for a given compactive effort.

COMPACTION TEST Laboratory compacting procedure whereby a soil at a known moisture content is placed in a specified manner into a mold of given dimensions, subjected to a compactive effort of controlled magnitude, and the resulting dry unit weight determined.

COMPARATIVE RISK ASSESSMENT Process that generally uses the judgment of experts to predict effects and set priorities among a wide range of environmental problems.

COMPLETE TREATMENT Method of treating water that consists of the addition of coagulant chemicals, flash mixing, coagulation-flocculation, sedimentation, and filtration.

COMPLETELY MIXED CONDITION Concentration of pollutant does not change across transect of the water body.

COMPLEX GALE/STORM Area for which gale/storm force winds are forecast or are occurring but for which no single center is the principal generator of these winds.

COMPLEX TERRAIN Refers to mountainous terrain; coastal regions and heterogeneous landscapes.

COMPLIANCE Act of meeting all state and federal drinking water regulations.

COMPLIANCE COAL Any coal that emits less than 1.2 pounds of sulfur dioxide per million Btu when burned.

COMPLIANCE COATING Coating whose volatile organic compound content does not exceed that allowed by regulation.

COMPLIANCE CYCLE Nine (9) year calendar year cycle, beginning January 1, 1993, during which public water systems must monitor where each cycle consists of three 3-year compliance periods.

COMPLIANCE MONITORING Collection and evaluation of data, including self-monitoring reports, and verification to show whether pollutant concentrations and loads contained in permitted discharges are in compliance with the limits and conditions specified in the permit.

COMPLIANCE SCHEDULE Negotiated agreement between a pollution source and a government agency that specifies dates and procedures by which a source will reduce emissions and, thereby, comply with a regulation.

COMPOSITE Average that is calculated according to specific criteria.

COMPOSITE HYDROGRAPH Stream discharge hydrograph which includes base flow; stream discharge hydrograph that corresponds to a net rain storm of duration longer than one unit period.

COMPOSITE POWER VALUE Value of power including both a capacity and energy component, usually expressed in mills per kilowatt-hour.

COMPOSITE SAMPLE Series of water samples taken over a given period of time and weighted by flow rate.

COMPOST Humus or soil-like material created from aerobic, microbial decomposition of organic materials such as food scraps, yard trimmings, and manure.

COMPOSTING Controlled biological decomposition of organic material in the presence of air to form a humus-like material.

COMPOSTING FACILITIES Off-site facility where the organic component of municipal solid waste is decomposed under controlled conditions; aerobic process in which organic materials are ground or shredded and then decomposed to humus in windrow piles or in mechanical digesters, drums, or similar enclosures.

COMPREHENSIVE EAP EXERCISE In depth exercise of an EAP that involves the interaction of the dam owner with the state and local emergency management agencies in a stressful environment with time constraints.

Comprehensive Environmental Response, Compensation, and Liability Act (CERCLA) of 1980 Federal statute that authorized "Superfund," is administered by EPA and provides funding for cleanups and emergency response actions for hazardous substances at the worst hazardous waste sites in the United States.

COMPREHENSIVE FACILITY REVIEW (CFR)
Review performed on a high- or significant-hazard dam every 6 years, which includes a field examination and a state-of-the-art review of a structure's design assumptions, construction practices, and integrity under various loading conditions; detailed examination performed on dams with a senior dam engineer.

COMPREHENSIVE FLARE INDEX (CFI)
Indicative of solar flare importance.

COMPRESSED NATURAL GAS (CNG)
Alternative fuel for motor vehicles that is considered one of the cleanest because of low hydrocarbon emissions and its vapors are relatively non-ozone producing.

COMPRESSIBILITY Property of a soil describing its susceptibility to decrease in volume when subjected to load.

COMPRESSION Reduction in volume of a soil mass resulting from an increase in effective stress.

COMPRESSION CURVE Any function relating volume change to effective stress.

COMPUTATION DURATION
User-defined time window used in hydrologic modeling.

COMPUTATION INTERVAL User-defined time step used by a hydrologic model for performing mathematical computations.

COMPUTER BASE STATION
Computer equipment that is designated to receive data from the data collection platforms and translate that data into useable information for the decision makers.

COMPUTER CODE Assembly of numerical techniques, bookkeeping, and control language that represents the model from acceptance of input data and instructions to delivery of output.

CONCENTRATED FLOW PATH
Existing hypothetical avenue of concentrated seepage.

CONCENTRATION Relative amount of a substance mixed with another substance; ratio of the quantity of any substance present in a sample of a given volume or a given weight compared to the volume or weight of the sample.

CONCENTRATION GRADIENT
Rate of change in solute concentration per unit distance at a given point and in a given direction.

CONCENTRATION TIME See TIME OF CONCENTRATION.

CONCENTRIC RINGS Quasi-constant intensifying or weakening of intense hurricanes that mark the end the period of intensification.

CONCEPTUAL MODEL Interpretation of the characteristics and dynamics of an aquifer system which is based on an examination of all available hydrogeological data for a modeled area.

CONCEPTUALIZATION ERROR
Modeling error where model formulation is based on incorrect or insufficient understanding of the modeled system.

CONCORDANT FLOWS Flows at different points in a river system that have the same recurrence interval, or the same frequency of occurrence.

CONCRETE DAM Concrete dam that generally requires a sound rock foundation.

CONCRETE LIFT Vertical distance between successive horizontal construction joints.

CONCURRENT FLOODS Flood flows expected at a point on the river system below a dam at the same time a flood inflow occurs above the dam.

CONCUSSION Shock or sharp air waves caused by an explosion or heavy blow.

CONDENSATE Water created by cooling steam or water vapor; liquid formed when warm landfill gas cools as it travels through a collection system.

CONDENSATE RETURN SYSTEM System that returns the heated water condensing within steam piping to the boiler and thus saves energy.

CONDENSATION Process by which water changes from the vapor state into the liquid or solid state.

CONDENSATION FUNNEL Funnel-shaped cloud associated with rotation and consisting of condensed water droplets.

CONDITIONAL REGISTRATION Under special circumstances, the Federal Insecticide, Fungicide, and Rodenticide Act (FIFRA) permits registration of pesticide products that is "conditional" upon the submission of additional data.

CONDITIONALLY EXEMPT GENERATORS (CE) Persons or enterprises which produce less than 220 pounds of hazardous waste per month.

CONDITIONALLY UNSTABLE AIR Atmospheric condition that exists when the environmental lapse rate is less than the dry adiabatic lapse rate but greater than the moist adiabatic lapse rate.

CONDUCTANCE Measure of the ability of a solution to carry and electrical current; rapid method of estimating the dissolved solids content of water supply by determining the capacity of a water sample to carry an electrical current.

CONDUCTION Flow of heat in response to a temperature gradient within an object or between objects that are in physical contact.

CONDUCTIVITY Measure of the ability of a solution to carry an electrical current.

CONDUCTOR Substance, body, device, or wire that readily conducts or carries electrical current.

CONDUIT Closed channel to convey water through, around, or under a dam; covered portion of spillway between the gate or crest structure and the terminal structure, where open channel flow and/or pressure flow conditions may exist; portion of an outlet works between the intake structure and gate chamber and/or the control structure.

CONE OF DEPRESSION See CONE OF INFLUENCE.

CONE OF INFLUENCE Depression, roughly conical in shape, produced in a water table, or other piezometric surface, by the pumping of water from a well at a given rate; volume of the cone will vary with rate of withdrawal of water.

CONE PENTEROMETER TESTING (CPT) Direct push (DP) system used to measure lithology based on soil penetration resistance where sensors in the tip of the cone of the DP rod measure tip resistance and side-wall friction, transmitting electrical signals to digital processing equipment on the ground surface.

CONFIDENTIAL BUSINESS INFORMATION (CBI) Material that contains trade secrets or commercial or financial information that has been claimed as confidential by its source (e.g. a pesticide or new chemical formulation registrant).

CONFIDENTIAL STATEMENT OF FORMULA (CSF) List of the ingredients in a new pesticide or chemical formulation.

CONFINED AQUIFER (ARTESIAN AQUIFER) An aquifer that is completely filled with water under pressure and that is overlain by material that restricts the movement of water; aquifer that is bound above and below by dense layers of rock and contains water under pressure that is significantly greater than atmospheric pressure; aquifer containing water between two relatively impermeable boundaries where the water level can rise above the ground surface yielding a flowing well.

CONFINED GROUNDWATER Ground water held under an aquiclude or an aquifuge called artesian if the pressure is positive.

CONFINED SPACE Space that is large enough to and so configured that a person can bodily enter and perform assigned work; space that has limited or restricted means for entry and exit; space not designed for continuous occupancy.

CONFINING BED (CONFINING UNIT) Hydrogeologic unit of less permeable material bounding one or more aquifers.

CONFINING LAYER Body of impermeable or low permeability that bounds an aquifer.

CONFINING ZONE Geological formation (or group or part of a formation) capable of limiting fluid movement out of an injection zone.

CONFLUENCE point at which two or more streams converge; place where stream is formed by two joining streams; place where a tributary joins the main stream; pattern of wind flow in which air flows inward toward an axis oriented parallel to the general direction of flow.

CONFLUENT GROWTH Continuous bacterial growth covering all or part of the filtration area of a membrane filter in which the bacteria colonies are not discrete.

CONGESTUS (CUMULUS CONGESTUS) See TOWERING CUMULUS.

CONGLOMERATE Sedimentary rock composed of rounded gravel (pebbles, cobbles, and boulders) cemented together, usually found with sandstone; coarse-grained sedimentary rock composed of fragments larger than 2 millimeters in diameter.

CONGRESSIONAL ORGANIC ACT OF 1890
Act that assigned the responsibility of river and floor forecasting for the benefit of the general welfare of the Nation's people and economy to the Weather Bureau, and subsequently the National Weather Service.

CONIFERS Cone-bearing trees or shrubs, mostly evergreens such as pine, cedar, and spruce.

CONING Pattern of wildfire plume dispersion in a neutral atmosphere, in which the plume attains the form of a cone with its vertex at the top of the stack.

CONJUGATE POINTS Two points on the Earth's surface, at opposite ends of a geomagnetic field line.

CONJUNCTIVE USE Coordinated use of surface water and groundwater resources.

CONNECTION Physical connection (e.g., transmission lines, transformers, switch gear, etc.) between two electric systems permitting the transfer of electric energy in one or both directions.

CONSENT DECREE Legal document, approved by a judge, that formalizes an agreement reached between agency and potentially responsible parties (PRPs) regarding the cleanup action at a Superfund site.

CONSEQUENCES Potential loss of life or property damage downstream of a dam caused by floodwaters released at the dam or by waters released by partial or complete failure of the dam that includes effects of landslides upstream of the dam on property located around the reservoir; potential number of lives that would be lost from a dam failure and an uncontrolled reservoir release (considers load, failure modes, and estimated population distribution warning times); construction joint; interface between two successive placements or pours of concrete where bond, and not permanent separation, is intended.

CONSERVATION Increasing the efficiency of energy use, water use, production, or distribution; preserving and renewing human and natural resources.

CONSERVATION EASEMENT
Easement restricting a landowner to land uses that that are compatible with long-term conservation and environmental values.

CONSERVATION STORAGE Storage of water for later release for usual purposes such as municipal water supply, power, or irrigation in contrast with storage capacity used for flood control.

CONSISTENCY Relative ease with which a soil can be deformed.

CONSOLIDATED ICE COVER Ice cover formed by the packing and freezing together of floes, brash ice and other forms of floating ice.

CONSOLIDATION Reduction in particle spacing in a soil, and decrease in water content, resulting from an increase in external pressure.

CONSOLIDATION CURVE Any function relating volume change to time.

CONSOLIDATION GROUTING Strengthening an area of ground by injecting grout.

CONSOLIDOMETER Apparatus used for one-dimensional consolidation/compression testing.

CONSTANT ANGLE ARCH DAM Arch dam in which the angle subtended by any horizontal section is constant throughout the whole height of the dam.

CONSTANT PRESSURE CHART Weather map representing conditions on a surface of equal atmospheric pressure; alternate term for Isobaric Chart.

CONSTANT RADIUS ARCH DAM Arch dam in which every horizontal segment or slice of the dam has approximately the same radius of curvature.

CONSTANT-HEAD BOUNDARY Boundary in the discretized groundwater flow model domain (node) where the hydraulic head remains the same over the time period considered.

CONSTANT-HEAD NODE Location in the discretized groundwater flow model domain (node) where the hydraulic head remains the same over the time period considered.

CONSTITUENT Chemical or biological substance in water, sediment, or biota that can be measured by an analytical method.

CONSTITUENT OF CONCERN Specific chemicals that are identified for evaluation in the site assessment process.

CONSTRUCTION AND DEMOLITION WASTE Waste building materials, dredging materials, tree stumps, and rubble resulting from construction, remodeling, repair, and demolition of homes, commercial buildings and other structures and pavements that may contain lead, asbestos, or other hazardous substances.

CONSTRUCTION BAN If an area's planning requirements for correcting non-attainment is not approved, construction or modification of any major stationary source of the pollutant is not permitted for which the area is in non-attainment.

CONSTRUCTION JOINT Placed in concrete to facilitate construction; to reduce initial shrinkage stresses and cracks; to allow time for the installation of embedded metalwork; to allow for the subsequent placing of other concrete.

CONSUMER SURPLUS Value of a commodity, good, or opportunity above the cost to the consumer; measured using willingness to pay, as specified in federal guidelines for water resources planning.

CONSUMPTIVE WATER USE Use which lessens the amount of water available for another use; quantity of water that is not manufacturing, plant or animal tissue.

CONTACT GROUTING Filling, with cement grout, any voids existing at the contact of two zones of different materials usually carried out at low pressure.

CONTACT LOAD Sediment particles that roll or slide along in almost continuous contact with the streambed.

CONTACT PESTICIDE Chemical that kills pests when it touches them, instead of by ingestion; soil that contains the minute skeletons of certain algae that scratch and dehydrate waxy-coated insects.

CONTACT RECREATION Recreational activities, such as swimming and kayaking, in which contact with water is prolonged or intimate, and in which there is a likelihood of ingesting water.

CONTAINMENT LEVEE Dike or embankment to contain stream flow.

CONTAMINANT Any physical, chemical, biological, or radiological substance or matter that has an adverse effect on air, water, or soil; material added by humans or natural activities that may, in sufficient concentrations, render the environment unacceptable for biota.

CONTAMINANT FATE Chemical changes and reactions that change the chemical nature of the contaminant, effectively removing the contaminant from the subsurface hydrologic system.

CONTAMINANT TRANSPORT MODEL Model describing the movement of contaminants in the environment.

CONTAMINANT TRANSPORT VELOCITY Rate in which contamination moves through an aquifer.

CONTAMINATION Degradation of water quality compared to original or natural conditions due to human activity; introduction into water, air, and soil of microorganisms, chemicals, toxic substances, wastes, or wastewater in a concentration that makes the medium unfit for its next intended use.

CONTAMINATION SOURCE INVENTORY Inventory of contaminant sources within delineated State Water-Protection Areas.

CONTENTS Volume of water in a reservoir.

CONTIGUOUS Actual contact with; also, near or adjacent to.

CONTINENTAL AIR MASS Dry air mass originating over a large land area.

CONTINENTAL SHELF Zone bordering a continent and extending to a depth, usually around 100 FM, from which there is a steep descent toward greater depth.

CONTINGENCIES Used in appraisal and feasibility estimates to estimate overruns on quantities, changed site conditions, change orders, etc.

CONTINGENCY PLAN Document setting out an organized, planned, and coordinated course of action to be followed in case of a fire, explosion, or other accident that releases toxic chemicals, hazardous waste, or radioactive materials that threaten human health or the environment.

CONTINGENT VALUATION Survey method asking for the maximum values that users would pay for access to a particular activity.

CONTINUOUS DISCHARGE Routine release to the environment that occurs without interruption, except for infrequent shutdowns for maintenance, process changes, etc.

CONTINUOUS MODEL Model that tracks the periods between precipitation events, as well as the events themselves.

CONTINUOUS SAMPLE Flow of water, waste or other material from a particular place in a plant to the location where samples are collected for testing.

CONTINUOUS-FLOW IRRIGATION System of irrigation water delivery where each irrigator receives their allotted quantity of water at a continuous rate.

CONTINUOUS-RECORD STATION Site where data are collected with sufficient frequency to define daily mean values and variations within a day.

CONTINUUM STORM (CTM) General term for solar noise lasting for hours and sometimes days.

CONTOUR Line of constant elevation.

CONTOUR DITCH Irrigation ditch laid out approximately on the contour.

CONTOUR FARMING System of farming used for erosion control and moisture conservation whereby field operations are performed approximately on the contour; conservation-based method of farming in which all farming operations (e.g., tillage and planting) are performed across (rather than up and down) the slope.

CONTOUR FLOODING Method of irrigation by flooding from contour ditches.

CONTOUR MAP Topographic map that portrays relief or elevation differences by the use of lines (contours) indicating equal elevation.

CONTOUR PLOWING Soil tilling method that follows the shape of the land to discourage erosion.

CONTOUR STRIP FARMING Contour farming in which row crops are planted in strips, between alternating strips of close-growing, erosion resistant forage (i.e., grass, grain, hay) crops.

CONTRACT LABS Laboratories under contract that analyze samples taken from waste, soil, air, and water or carry out research projects.

CONTRACT RATE Repayment or water service rate set forth in a contract to be paid by a district to the United States.

CONTRACTED WEIR Crest and sides of a rectangular weir are far enough from the bottom and sides of the channel so that their effect on flow is negligible.

CONTRACTION JOINT Joints placed in concrete to provide for volumetric shrinkage of a monolithic unit or movement between monolithic units; joints have no bond between the concrete surfaces forming the joint.

CONTRIBUTING AREA Area in a drainage basin that contributes water to streamflow or recharge to an aquifer.

CONTROL Designates a feature in the channel that physically affects the water-surface elevation and thereby determines the stage-discharge relation at the gauge; natural constriction of the channel, a long reach of the channel, a stretch of rapids, or an artificial structure downstream from a gaging station that determines the stage-discharge relation at the gauge; may be complete or partial constriction of the channel, a bedrock outcrop, a gravel bar, an artificial structure, or a uniform cross section over a long reach of the channel.

CONTROL AREA Part of a power system, or a combination of systems, to which a common electrical generation allocation scheme is applied.

CONTROL HOUSE See CONTROL STRUCTURE.

CONTROL JOINT Joints placed in concrete to provide for control of initial shrinkage stresses and cracks of monolithic units; unbonded joints to provide weak areas for cracking; joints that will transfer moment but not shear unless keyed.

CONTROL POINTS Small monuments securely embedded in the surface of the dam monitored to detect horizontal and vertical movement of the dam.

CONTROL STRUCTURE (CONTROL HOUSE)
Structure on a stream or canal that is used to regulate the flow or stage of the stream or to prevent the intrusion of saltwater; concrete portion of an outlet works, located at the downstream end of the tunnel or conduit, housing the control (regulation) gates; water regulating structure; structure on a stream or canal that is used to regulate the flow or stage of the stream.

CONTROL TECHNIQUE GUIDELINES (CTG)
EPA documents designed to assist state and local pollution authorities to achieve and maintain air quality standards for certain sources through reasonably available control technologies (RACT).

CONTROLLED LOW STRENGTH MATERIAL (CLSM)
Mixture of pozzolan, Portland cement, water, coarse aggregate, and occasionally soil, typically used for pipe bedding and backfill that is also known as flowable fill or cement-slurry backfill.

CONTROLLED REACTION Chemical reaction under temperature and pressure conditions maintained within safe limits to produce a desired product or process.

CONUS Continental United States.

CONVECTION Transport of heat and moisture by the movement of a fluid; vertical transport of heat and moisture in the atmosphere.

CONVECTIVE AVAILABLE POTENTIAL ENERGY (CAPE)
Measure of the amount of energy available for convection; directly related to the maximum potential vertical speed within an updraft; thus, higher values indicate greater potential for severe weather.

CONVECTIVE BOUNDARY LAYER
Unstable boundary layer that forms at the surface and grows upward through the day as the ground is heated by the Sun and convective currents transfer heat upwards into the atmosphere.

CONVECTIVE CLOUDS Vertically developed family of clouds are cumulus and cumulonimbus with the height of bases ranging from as low as 1,000 feet to a bit more than 10,000 feet.

CONVECTIVE CONDENSATION LEVEL (CCL) Level in the atmosphere to which an air parcel, if heated from below, will rise dry adiabatically, without becoming colder than its environment just before the parcel becomes saturated.

CONVECTIVE INHIBITION (CIN OR B-) Measure of the amount of energy needed in order to initiate convection; numerical measure of the strength of "capping," typically used to assess thunderstorm potential.

CONVECTIVE OUTLOOK Forecast containing the area(s) of expected thunderstorm occurrence and expected severity over the contiguous United States, issued several times daily by the SPC.

CONVECTIVE OVERDEVELOPMENT Convection that covers the sky with clouds, thereby cutting off the sunshine that produces convection.

CONVECTIVE TEMPERATURE Approximate temperature that the air near the ground must warm to in order for surface-based convection to develop, based on analysis of a sounding.

CONVENTIONAL FILTRATION See COMPLETE TREATMENT.

CONVENTIONAL POLLUTANTS Pollutants typical of municipal sewage (e.g., organic waste, sediment, acid, BOD, TSS, fecal coliform bacteria, viruses, nutrients, oil and grease) for which municipal secondary treatment plants are typically designed.

CONVENTIONAL SEPTIC SYSTEM Wastewater treatment system consisting of a septic tank and a typical trench or bed subsurface wastewater infiltration system.

CONVENTIONAL SITE ASSESSMENT Assessment in which most of the sample analysis and interpretation of data is completed off-site.

CONVENTIONAL SYSTEMS Systems that have been traditionally used to collect municipal wastewater in gravity sewers and convey it to a central primary or secondary treatment plant prior to discharge to surface waters.

CONVENTIONAL TILLAGE Traditional method of farming in which soil is prepared for planting by completely inverting it with a moldboard plow.

CONVENTIONAL TILLING Tillage operations considered standard for a specific location and crop and that tend to bury the crop residues.

CONVERGENCE Contraction of a vector field.

CONVEYANCE LOSS (DISTRIBUTION LOSS) Loss of water from pipes, channels, conduits, and ditches due to leakage, seepage, evaporation, or evapotranspiration; water not available for further use.

CONVEYANCE SYSTEM EFFICIENCY Ratio of the volume of water delivered to users in proportion to the volume of water introduced into the conveyance system.

CONVEYOR Device that transports material by belts, cables, or chains.

COOLING DEGREE DAYS (CDD) Form of Degree Day used to estimate energy requirements for air conditioning or refrigeration, typically calculated as how much warmer the mean temperature at a location is than 65 °F on a given day.

COOLING ELECTRICITY USE Amount of electricity used to meet the building cooling load.

COOLING TOWER Structure that helps remove heat from water used as a coolant (e.g. in electric power generating plants); device which dissipates the heat from water-cooled systems by spraying the water through streams of rapidly moving air.

COOLING TOWER BLOWDOWN WATER Water released from a cooling tower to maintain proper water mineral concentration.

COOLING TOWER MAKE-UP WATER Water added to a cooling tower to replace water lost to evaporation or blowdown.

COOLING WATER Water used for cooling purposes, such as of condensers and nuclear reactors.

COOPERATIVE AGREEMENT Formal document that states the obligations to one or more other parties providing the authority to issue funding, transfers money, property, services or anything of value.

COOPERATIVE ELECTRIC UTILITY Electric utility legally established to be owned by and operated for the benefit of those using its service.

COOPERATIVE OBSERVER Individual (or institution) who takes precipitation and temperature observations-and in some cases other observations such as river stage, soil temperature, and evaporation-at or near their home, or place of business.

COOPERATIVE STATION REPORT (B-44 FORM) Weather Service form documenting station management, exposure, topography, driving instructions, payment information, hydrometeorologic equipment, and observing information.

COORDINATED OPERATION Operation of two or more interconnected electrical systems to achieve greater reliability and economy.

COORDINATED UNIVERSAL TIME (UNIVERSAL TIME COORDINATED, UTC) Under international agreement, the local time at the prime meridian that passes through Greenwich, England; known as "Z time," "Zulu Time" or prior to 1972, Greenwich Mean Time (GMT) .

COORDINATION Practice by which two or more interconnected electric power systems augment the reliability of bulk electric power supply by establishing planning and operating standards; by exchanging pertinent information regarding additions, retirements, and modifications, to the bulk electric power supply system; and by joint review of these changes to assure that they meet the predetermined standards.

COR Correction.

CORAL REEF Ridge of limestone, composed chiefly of coral, coral sands, and solid limestone resulting from organic secretion of calcium carbonate that generally occur along continents and islands where the temperature is above 18 °C.

CORDUROY Road made of logs laid crosswise on the ground or on other logs.

CORE (IMPERVIOUS CORE OR IMPERVIOUS ZONE) Zone of low permeability material in an embankment dam that is sometimes referred to as central core, inclined core, puddle clay core, rolled clay core, or impervious zone; cylindrical piece of an underground formation cut and raised by a rotary drill with a hollow bit.

CORE DRILL Rotary drill, usually a diamond drill, equipped with a hollow bit and a core lifter.

CORE PROGRAM COOPERATIVE AGREEMENT Assistance agreement whereby EPA supports states or tribal governments with funds to help defray the cost of non-item-specific administrative and training activities.

CORE PUNCH Penetration by a vehicle into the heavy precipitation core of a thunderstorm.

CORE SAMPLE Sample of rock, soil, or other material obtained by driving a hollow tube into the undisturbed medium and withdrawing it with its contained sample.

CORE WALL Wall of substantial thickness built of impervious material, usually of concrete or asphaltic concrete, in the body of an embankment dam to prevent seepage.

CORIOLIS FORCE Fictitious force used to account for the apparent deflection (caused by the rotation of the Earth) of a body in motion with respect to the Earth, as seen by an observer on the Earth.

CORN SNOW ICE Rotten granular ice.

CORNER EFFECTS Small-scale convergence effect that can be quite severe that occurs around steep islands and headlands.

CORONA Outermost layer of the solar atmosphere characterized by low densities (¡ 1.0 x 10^9/cc) and high temperatures (¿ 1.0 x 10^6 K); white or colored circle or set of concentric circles of light of small radius seen around a luminous body (e.g., Sun or Moon).

CORONAL HOLE Extended region of the corona, exceptionally low in density and associated with unipolar photospheric regions.

CORONAL RAIN (CRN) Material condensing in the corona and appearing to rain down into the chromosphere as observed at the solar limb above strong sunspots.

CORONAL TRANSIENTS General term for short-time-scale changes in the corona, but principally used to describe outward-moving plasma clouds.

CORPS OF ENGINEERS See U.S. ARMY CORPS OF ENGINEERS.

CORRECTIVE ACTION Use of Director-approved methods to assure that wells within the area of review do not serve as conduits for the movement of fluids into USDWs; EPA can require treatment, storage and disposal (TSDF) facilities handling hazardous waste to undertake corrective actions to clean up spills resulting from failure to follow hazardous waste management procedures or other mistakes.

CORRELATED SHEAR Output of the mesocyclone detection algorithm indicating a 3-dimensional shear region (i.e. vertically correlated) that is not symmetrical.

CORRELATION Relation between two or more variables; measure of similarity between variables of functions; statistical test that sets a numerical value to the amount of interdependence.

CORRELATIVE ESTIMATE Amount determined by correlation that represents a similar value of the amount for any particular period.

CORRIDOR Narrow strip of land reserved for location of transmission lines, pipelines, and service roads.

CORROSION Dissolution and wearing away of metal caused by a chemical reaction (i.e., rusting or acids) such as between water and the pipes, chemicals touching a metal surface, or contact between two metals; starts at the surface of a material and moves inward.

CORROSIVE Chemical agent that reacts with the surface of a material causing it to deteriorate or wear away; chemical that destroys or irreversibly alters living tissue by direct chemical action at the site of contact.

COSMIC RAY Extremely energetic (relativistic) charged particle.

COST ALLOCATION Procedures used to allocate joint project costs to specific purposes in a multipurpose project.

COST EFFECTIVENESS Economic efficiency, obtaining the best method for the least amount of money.

COST RECOVERY Legal process by which potentially responsible parties who contributed to contamination at a Superfund site can be required to reimburse the Trust Fund for money spent during any cleanup actions by the Federal Government.

COST SHARING Publicly financed program through which society, as a beneficiary of environmental protection, shares part of the cost of pollution control with those who must actually install the controls.

COST-BENEFIT ANALYSIS (CBA) Quantitative evaluation of the costs which would be incurred versus the overall benefits to society of a proposed action.

COST-EFFECTIVE ALTERNATIVE Alternative control or corrective method identified after analysis as being the best available in terms of reliability, performance, and cost.

COULEES Small streams or dry streambeds; deep gulch or ravine that is usually dry in the summer.

COULOMB Quantity of electricity that passes any point in an electric circuit in 1 second when the current is maintained constant at 1 ampere.

COUNTY WARNING AND FORECAST AREA (CWFA) Group of counties for which a National Weather Service Forecast Office is responsible for issuing warnings and weather forecasts.

COUNTY WARNING AREA Group of counties for which a National Weather Service Forecast Office is responsible for issuing warnings.

COUPLED ATMOSPHERE-OCEAN MODEL
See COUPLED MODEL.

COUPLED MODEL Refers to a numerical model which simulates both atmospheric and oceanic motions and temperatures and which takes into account the effects of each component on the other.

COUPLET Adjacent maxima of radial velocities of opposite signs.

COVER CROP Crop that provides temporary protection for delicate seedlings and/or provides a cover canopy for seasonal soil protection and improvement between normal crop production periods.

COVER MATERIAL Soil used to cover compacted solid waste in a sanitary landfill.

COYOTE HOLES Horizontal tunnels in which explosives are packed for blasting a high rock face.

CPC Climate Prediction Center.

CRACK Separation formed in an ice cover of floe that does not divide it into two or more pieces.

CRACK MONITORS (WHITTEMORE GAUGES, DIAL GAUGES, DEPTH MICROMETERS, AND AVONGARD CRACK MONITORS)
Measure movements transverse and along a joint or crack.

CRADLE-TO-GRAVE SYSTEM (MANIFEST SYSTEM)
Procedure in which hazardous materials are identified and followed as they are produced, treated, transported, and disposed of by a series of permanent, linkable, descriptive documents (e.g. manifests).

CRAZING Network pattern of fine cracks in concrete that do not penetrate much below the surface.

CREEK Relatively small stream of water that serves as the natural drainage course for a drainage basin of nominal (i.e. small size).

CREEP Slow movement of rock debris or soil usually imperceptible except to observations of long duration; time-dependent strain or deformation.

CREPUSCULAR RAYS Alternating bands of light and dark (rays and shadows) seen at the Earth's surface when the Sun shines through clouds.

CREST Top surface of a dam, dike, spillway, or weir, to which water must rise before passing over the structure excluding any parapet wall, railing, curb, etc.; crown of the roadway or the level of the walkway which crosses the dam; highest stage or level of a flood wave as it passes a point; highest point in a wave.

CREST GATE (SPILLWAY GATE)
Gate on the crest of a spillway to control the discharge or reservoir water level.

CREST GAUGE Gage used to obtain a record of flood crests at sites where recording gauges are installed.

CREST LENGTH (LENGTH OF DAM)
Measured length of the dam along

the axis, centerline crest (e.g. top) or roadway surface on the crest.

CREST STRUCTURE Portion of spillway between the inlet channel and the chute, tunnel or conduit, which does not contain gates.

CREST WIDTH (TOP THICKNESS) Thickness or width of a dam at the level of the crest (top) of the dam (excluding corbels or parapets) where the term thickness is used for gravity and arch dams and width is used for other dams.

CRIB DAM Gravity dam built up of boxes, cribs, crossed timbers or gabions, filled with earth or rock.

CRITERIA Numeric values and the narrative standards that represent contaminant concentrations that are not to be exceeded in the receiving environmental media (surface water, groundwater, sediment) to protect beneficial uses.

CRITERIA CONTINUOUS CONCENTRATION (CCC) EPA national water quality criteria recommendation for the highest in-stream concentration of a toxicant or an effluent to which organisms can be exposed indefinitely without causing unacceptable effect.

CRITERIA MAXIMUM CONCENTRATION (CMC) EPA national water quality criteria recommendation for the highest in-stream concentration of a toxicant or an effluent to which organisms can be exposed for a brief period of time without causing an acute effect.

CRITERIA POLLUTANTS 1970 amendments to the Clean Air Act required EPA to set National Ambient Air Quality Standards for certain pollutants known to be hazardous to human health and welfare where EPA identified and set standards for six pollutants: ozone, carbon monoxide, total suspended particulates, sulfur dioxide, lead, and nitrogen oxide.

CRITERION Standard rule or test on which a judgment or decision can be based.

CRITICAL DAMPING Minimum amount of damping that prevents free oscillatory vibration.

CRITICAL DEPTH Depth of flow at which the discharge is maximum for a given specific energy; depth of flow at which a given discharge occurs with minimum specific energy; depth of flow when the Froude number equals one.

CRITICAL DISCHARGE Maximum discharge for a given specific energy; discharge that will occur with minimum specific energy.

CRITICAL EFFECT First adverse effect, or its known precursor, that occurs as a dose rate increases.

CRITICAL FLOW Flow where Froude number is equal to one; flow at critical depth; in reference to the Reynolds' critical velocities, point at which the flow changes from streamline or non-turbulent to turbulent flow.

CRITICAL HABITAT Areas designated as critical for the survival and recovery of threatened or endangered species.

CRITICAL HEIGHT Maximum height at which a vertical or sloped bank of soil will stand unsupported under a given set of conditions.

CRITICAL LIFE STAGE Period of time in an organism's lifespan in which it is the most susceptible to adverse effects caused by exposure to toxicants, usually during early development (egg, embryo, larvae).

CRITICAL RAINFALL PROBABILITY (CRP) Probability that a given rainfall will cause a river, or stream to rise above flood stage; probability that the actual precipitation during a rainfall event has exceeded or will exceed the flash flood guidance value.

CRITICAL SLOPE Maximum angle with the horizontal at which a sloped bank of soil or rock of given height will stand unsupported; slope that will sustain a given discharge at uniform critical depth in a given channel.

CRITICAL SPECIES Species that exists at the site and is listed as threatened or endangered; species for which there is evidence that the loss of the species from the site is likely to cause an unacceptable impact on a commercially or recreationally important species, a threatened or endangered species, the abundances of a variety of other species, or the structure or function of the community.

CRITICAL VELOCITY Mean velocity when the discharge is critical.

CROCHET Sudden deviation in the sunlit geomagnetic field (H component) associated with large solar flare X-ray emission.

CROP CONSUMPTIVE USE Amount of water transpired during plant growth plus what evaporated from the soil surface and foliage in the crop area.

CROP IRRIGATION REQUIREMENT Quantity of water, exclusive of effective precipitation, that is needed for crop production.

CROP MOISTURE INDEX Index to assess short-term crop water conditions and needs across major crop-producing regions.

CROP ROOT ZONE Soil depth from which a mature crop extracts most of the water needed for evapotranspiration equal expressed in inches or feet.

CROP ROTATION Pattern of changing the crops grown in a specific field from year to year in order to control pests and maintain soil fertility.

CROP SUBSIDY Price support paid by the government to farmers.

CROP WATER REQUIREMENT Crop consumptive use plus the water required to provide the leaching requirements.

CROPPING PATTERN Acreage distribution of different crops in any one year in a given farm area such as a county, water agency, or farm.

CROSS CONTAMINATION Movement of underground contaminants from one level or area to another due to invasive subsurface activities.

CROSS-CONNECTION Actual or potential connection between a drinking water system and an unapproved water supply or other source of contamination.

CROSS-SECTION Slice of the channel and adjacent valley made perpendicular to the assumed flow; elevation view of a dam formed by passing a plane through the dam perpendicular to the axis.

CROSS-SECTIONAL AREA Area of a stream, channel or waterway, usually measured perpendicular to the flow; area perpendicular to the direction of flow.

CROSS-VALLEY WIND SYSTEM Thermally driven wind that blows during daytime across the axis of a valley toward the heated sidewall.

CROWN Highest point of the interior of a circular conduit, pipe, or tunnel (also referred to as the soffit); point in an arch dam which generally corresponds with where the height of the dam is a maximum; elevation of a road center above its sides.

CROWN FIRE Fire where flames travel from tree to tree at the level of the tree's crown or top.

CROWNING Movement of a fire from the understory into the crown of a forest canopy.

CRS Console Replacement System.

CRUMB RUBBER Ground rubber fragments the size of sand or silt used in rubber or plastic products, or processed further into reclaimed rubber or asphalt products.

CRUSHED GRAVEL Gravel which has been produced by a machine.

CRUSHED ROCK Rock which has been reduced in size by a machine.

CRUSHER Machine that reduces rocks to smaller and more uniform sizes.

CRYOLOGY Science of the physical aspects of snow, ice, hail, and sleet and other forms of water produced by temperatures below 0 °C.

CRYPTOSPORIDIUM Microorganism commonly found in lakes and rivers that are highly resistant to disinfection that can cause gastrointestinal illness.

CRYSTALLINE ROCKS Rocks (e.g., igneous or metamorphic) consisting wholly of crystals or fragments of crystals.

CSI Conditional Symmetric Instability.

CST Central Standard Time.

CSTL Coastal.

CTY City.

CUBIC FEET PER MINUTE (CFM) Measure of the volume of a substance flowing through air within a fixed period of time; with regard to indoor air, refers to the amount of air, in cubic feet, that is exchanged with outdoor air in a minute's time.

CUBIC FEET PER SECOND (CFS) Flow rate or discharge equal to one cubic foot per second; equivalent to approximately 7.48 gallons per second.

CUBIC FEET PER SECOND (CFS, FT3/S) Unit expressing rates of discharge; rate of streamflow (volume in cubic feet) of water passing a reference point in 1 second; approximately 7.48 gallons per second, 448.8 gallons per minute (gpm), 1.98 acre-feet per day or 0.02832 cubic meter per second.

CUBIC FEET PER SECOND PER SQUARE MILE (CFSM) Average number of cubic feet of water per second flowing from each square mile of area drained by a stream, assuming that the runoff is distributed uniformly in time and area.

CUBIC FEET PER SECOND-DAY (CFS-DAY) Volume of water represented by a flow of 1 cubic foot per second for 24 hours; 1 cfs-day is 24 x 60 x 60 = 86,000 ft^3, 1.983471 acre-feet, or 646,317 gallons; average flow in cubic feet per second for any time period is the volume of flow in cfs-days.

CUBIC FOOT PER SECOND PER SQUARE MILE (CFSM, (FT3/S)/MI2) Average number of cubic feet of water flowing per second from each square mile of area drained, assuming the runoff is distributed uniformly in time and area.

CUBIC FOOT PER SECOND-DAY (CFS-DAY, CFS-DAY, (FT3/S)/D) Volume of water represented by a flow of 1 cubic foot per second for 24 hours; approximately 86,400 ft^3, 1.98347 acre-feet, 646,317 gallons, or 2,446.6 m^3.

CUFRA Cumulus Fractus.

CULLET Crushed glass.

CULTURAL EUTROPHICATION Increasing rate at which water bodies "die" by pollution from human activities.

CULTURAL PATRIMONY Ancestral heritage and entitlement.

CULTURAL RESOURCE(S) Any building, site, district, structure, or object significant in history, architecture, archeology, culture, or science.

CULTURES AND STOCKS Infectious agents and associated biologicals including cultures from medical and pathological laboratories; cultures and stocks of infectious agents from research and industrial laboratories; waste from the production of biologicals; discarded live and attenuated vaccines; culture dishes and devices used to transfer, inoculate, and mix cultures.

CULVERT Pipe or small bridge for drainage under a road or structure; conduit for the free passage of surface drainage water under a highway, railroad, canal, or other embankment.

CUMULATIVE ECOLOGICAL RISK ASSESSMENT Consideration of the total ecological risk from multiple stressors to a given eco-zone.

CUMULATIVE EXPOSURE Sum of exposures of an organism to a pollutant over a period of time.

CUMULATIVE WORKING LEVEL MONTHS (CWLM) Sum of lifetime exposure to radon working levels expressed in total working level months.

CUMULIFORM Descriptive of all clouds with vertical development in the form of rising mounds, domes, or towers.

CUMULIFORM ANVIL Thunderstorm anvil with visual characteristics resembling cumulus-type clouds that arises from rapid spreading of a thunderstorm updraft.

CUMULUS BUILDUPS Clouds which develop vertically due to unstable air.

CUMULUS CLOUDS (CU) Detached clouds, generally dense and with sharp outlines, showing vertical development in the form of domes, mounds, or towers.

CUMULUS CONGESTUS Large, towering cumulus cloud with great vertical development but lacking the characteristic anvil of a cumulonimbus.

CURB STOP (CURB COCK) Water service shutoff valve located in a water service pipe near the curb and between the water main and the building that is usually operated by a wrench or valve key and is used to start or stop flows in the water service line to a building.

CURBSIDE COLLECTION Method of collecting recyclable materials at homes, community districts or businesses.

CURRENT Movement of electrons through a conductor, measured in amperes; horizontal movement of water.

CURRENT METER Device used to measure the speed of flowing water (i.e. velocity) or current in a river.

CURTAIN DRAIN Drain constructed at the upper end of the area to be drained, to intercept surface or groundwater flowing toward the protected area from higher ground, and carry it away from the area.

CURTAIN GROUTING Process of pressure grouting deep holes under a dam or in an abutment to form a watertight barrier and effectively seal seams, fissures, fault zones, or fill cavities in the foundation or abutment.

CURVED GRAVITY DAM Gravity dam that is curved in plan.

CUSEC Cubic foot per second.

CUT To lower an existing grade or surface level, or an area where this has been done.

CUTIE-PIE Instrument used to measure radiation levels.

CUTOFF Impervious construction that prevents water from passing through foundation material.

CUTOFF LOW Closed upper-level low which has become completely displaced from basic westerly current and moves independently of that current.

CUTOFF TRENCH (KEYWAY) Excavation in the foundation of a dam, usually located upstream of the dam axis or centerline crest which extends to bedrock or to an impervious stratum, that is later to be filled with impervious material to limit seepage beneath the dam.

CUTOFF WALL Wall of impervious material (e.g., concrete, asphaltic concrete, timber, steel sheet piling, or impervious grout curtain) constructed in the foundation and abutments to reduce seepage beneath and adjacent to the dam.

CUTTING Excavating; lowering a grade.

CUTTINGS Spoils left by conventional drilling with hollow stem auger or rotary drilling equipment.

CVR Cover.

CWA Clean Water Act; County Warning Area.

CWFA County Warning and Forecast Area.

CYCLE Regularly recurring succession of events (e.g. seasons).

CYCLIC MOBILITY Phenomenon in which a cohesionless soil loses shear strength during earthquake ground vibrations and acquires a degree of mobility sufficient to permit intermittent movement up to several feet, as contrasted to liquefaction where continuous movements of several hundred feet are possible.

CYCLIC STORM Thunderstorm that undergoes cycles of intensification and weakening (pulses) while maintaining its individuality.

CYCLING Power plant operation to meet the intermediated portion of the load (9 to 14 hours per day).

CYCLOGENESIS (CYCLGN) Formation or intensification of a cyclone or low-pressure storm system.

CYCLONE (CYC) Area of low pressure around which winds rotate counterclockwise in the Northern Hemisphere and clockwise in the Southern Hemisphere.

CYCLONE COLLECTOR Device that uses centrifugal force to remove large particles from polluted air.

CYCLONIC CIRCULATION Circulation (or rotation) which is in the same sense as the Earth's rotation (i.e. counterclockwise in the Northern Hemisphere).

CYCLOPEAN DAM Gravity dam in which the mass masonry consists primarily of large one-man or derrick stone embedded in concrete.

CYLINDER GATE Gate that resembles a large barrel, reinforced to withstand external pressure, with no top or bottom; outlet gate with a vertical hollow cylinder being raised and lowered to expose openings in the outer (usually concrete) wall through which water enters the central area of the gate.

CYPRESS DOME Small, isolated, circular, depressional, forested wetlands, in which cypress predominates, that have convex silhouettes when viewed from a distance.

D

D REGION Daytime layer of the Earth's ionosphere approximately 50 to 90 km in altitude.

DAILY DISCHARGE Discharge of a pollutant measured during any 24-hour period that reasonably represents a calendar day for purposes of sampling.

DAILY FLOOD PEAK Maximum mean daily discharge occurring in a stream during a given flood event.

DAILY MAXIMUM LIMIT Maximum allowable discharge of pollutant during a calendar day.

DAILY MEAN SUSPENDED-SEDIMENT CONCENTRATION Time-weighted mean concentration of suspended sediment passing a stream cross section during a 24-hour day.

DAILY RECORD STATION Site where data are collected with sufficient frequency to develop a record of one or more data values per day.

DALR Dry Adiabatic Lapse Rate.

DAM Barrier built to impound or divert water for the purpose of storage or control of water; barrier (e.g. structure) that obstructs, directs, holds back, or stores the flow of water.

DAM FAILURE Catastrophic type of failure characterized by the sudden, rapid, and uncontrolled release of impounded water or the likelihood of such an uncontrolled release.

DAM FAILURE HYDROGRAPH Flood hydrograph resulting from a dam breach prepared for a specific location downstream from the dam.

DAM FOUNDATION Excavated surface or undisturbed material upon which a dam is placed.

DAM SAFETY Art and science of ensuring the integrity and viability of dams such that they do not present unacceptable risks to the public, property, and the environment.

DAM SAFETY DEFICIENCY Physical condition capable of causing the sudden uncontrollable release of reservoir water by partial or complete failure of a dam, appurtenant structure, or facility.

DAM SAFETY ISSUE Dam safety related issues and concerns that if not adequately addressed could/would lead to a failure or malfunction resulting in an uncontrolled release of stored water that would place the downstream population potentially at risk or compromise the agency's ability to detect developing adverse dam performance and prudently respond to that performance.

DAM TENDER Person responsible for the daily or routine operation and maintenance activities of a dam and its appurtenant structures.

DAMBRK Dam Break Forecasting Model.

DAMPING Resistance which reduces vibration by energy absorption.

DAMPING RATIO Ratio of the actual damping to the critical damping, critical damping being the minimum amount of damping which prevents free oscillatory vibration; design water level; maximum water elevation, including the flood surcharge, that a dam is designed to withstand.

DAPM Data Acquisition Program Manager.

DARK SURGE ON DISK (DSD) Dark gaseous ejections visible in H-alpha.

DART LEADER Faint, negatively charged channel that travels more or less directly and continuously from cloud to ground.

DATA CALL-IN Part of the Office of Pesticide Programs (OPP) process of developing key required test data, especially on the long-term, chronic effects of existing pesticides, in advance of scheduled Registration Standard reviews.

DATA COLLECTION PLATFORM (DCP) Electronic equipment device that gathers digital and analog data from monitoring devices and remotely relays that data to a central computer for interpretation and archiving.

DATA COMPARABILITY Characteristics that allow information from many sources to be of definable or equivalent quality so that this information can be used to address program objectives not necessarily related to those for which the data were collected.

DATA LOGGER Microprocessor-based data acquisition system designed specifically to acquire, process, and store data.

DATA POINT Location on a river/stream for which observed data is input to hydrologic forecast procedures or included in public hydrologic products.

DATA QUALITY OBJECTIVES (DQOS) Characteristics or goals that are determined by a monitoring or interpretive program to be essential to the usefulness of the data; qualitative and quantitative statements of the overall level of uncertainty that a decision-maker will accept in results or decisions based on environmental data.

DATACOL Software System that supports RFC gateway functions.

DATANET Hydrologic Data Network Analysis Software.

DATAWEB Electronic presentation of the Bureau of Reclamation's Project Data book and contains historical, statistical, and technical information on the projects of the Bureau of Reclamation.

DATUM Surface or point relative to which measurements of height and/or horizontal position are reported; vertical datum is a horizontal surface used as the zero point for measurements of gauge height, stage, or elevation; horizontal datum is a reference for positions given in terms of latitude-longitude, State Plane coordinates, or Universal Transverse Mercator (UTM) coordinates.

DATUM PLANE Horizontal plane to which ground elevations or water surface elevations are referenced.

DAWN Time of morning at which the Sun is 6 degrees below the horizon; See CIVIL DAWN.

DAY LENGTH Duration of the period from sunrise to sunset.

DAY TANK Deaerating tank.

DBE Design Basis Earthquake.

DBZ Non-dimensional "unit" of radar reflectivity which represents a logarithmic power ratio (in decibels, or dB) with respect to radar reflectivity factor, Z.

DCP (Data Collection Platform) Electronic device that connects to a river or rainfall gauge that records data from the gauge and at predetermined times transmits that data through a satellite to a remote computer.

DDS Data Distribution System.

DDT (DICHLORO-DIPHENYL-TRICHLOROETHANE) First chlorinated hydrocarbon insecticide chemical name: has a half-life of 15 years and can collect in fatty tissues of certain animals.

DEAD CAPACITY (DEAD STORAGE) Reservoir capacity from which stored water cannot be evacuated by gravity.

DEAD END End of a water main which is not connected to other parts of the distribution system.

DEAD STORAGE Volume (e.g. storage) that lies below the lowest outlet invert and cannot readily be withdrawn from the reservoir.

DEADMEN Anchors drilled or cemented into the ground to provide additional reactive mass for DP sampling rigs.

DEBRIS CLOUD Rotating "cloud" of dust or debris, near or on the ground, often appearing beneath a condensation funnel and surrounding the base of a tornado.

DEBRIS FAN Sloping mass of boulders, cobbles, gravel, sand, silt and clay formed by debris flows at the mouth of a tributary.

DEBRIS FLOW Flash flood consisting of a mixture of rocks and sediment containing less than 40 percent water, by volume; forms a debris fan.

DECADAL Occurring over a 10-year period.

DECANT To draw off the upper layer of liquid after the heaviest material (a solid or another liquid) has settled.

DECAY PRODUCTS Degraded radioactive materials, often referred to as "daughters" or "progeny".

DECHLORINATION Removal of chlorine from a substance.

DECIBEL Unit for expressing the relative intensity of sounds on a scale from zero for the average least perceptible sound to about 130 for the average level at which sound causes pain to humans.

DECIDUOUS Plants that shed foliage at the end of the growing season.

DECISION CRITERIA Set of formal criteria that is used to determine if an event is threatening and what action to implement.

DECISION MAKERS Designated personnel responsible for interpreting data that has been collected and determining if an event is threatening enough to issue an alert or warning.

DECISION MAKING Second of five Early Warning System components consisting of the processes and facilities necessary to translate incoming data about the threatening event into decisions to alert or warn the population at risk.

DECKING Separating charges of explosives by inert material which prevents passing of concussion, and placing a primer in each charge.

DECLINATION Latitude that the Sun is directly over at a given time; 23° N at the summer solstice, 23° S at the winter solstice, and 0° (over the equator) at the spring and autumn equinoxes.

DECOMPOSITION Breakdown of matter by bacteria and fungi, changing the chemical makeup and physical appearance of materials; subdividing a failure mode into discrete elements or sequential events so that the failure mode could be better understood, and probabilities can be more reasonably estimated for each step in an event tree.

DECONTAMINATION Removal of harmful substances such as noxious chemicals, harmful bacteria or other organisms, or radioactive material from exposed individuals, rooms and furnishings in buildings, or the exterior environment.

DEEP PERCOLATION Movement of water by gravity downward through the soil profile, beyond the root zone, to regions of deeper ground-water aquifers.

DEEP PERCOLATION LOSS Water that percolates downward through the soil beyond the reach of plant roots.

DEEP SEEPAGE Infiltration which reaches the water table.

DEEP WELL Well whose pumping head is too great to permit use of a suction pump.

DEEPENING Decrease in the central pressure of a surface low pressure system (i.e. storm is intensifying).

DEEP-WATER HABITATS Permanently flooded lands that lie below the deep-water boundary of wetlands.

DEEP-WELL INJECTION Deposition of raw or treated, filtered hazardous waste by pumping it into deep wells, where it is contained in the pores of permeable subsurface rock.

DEFICIENCY VERIFICATION ANALYSIS (DVA)
Analysis that verifies the existence or non-existence of Safety of Dams (SOD) deficiencies at the dam.

DEFLAGRATION To burn with sudden and startling combustion.

DEFLATION Force of wind erosion (e.g. blowouts).

DEFLECT Decrease in the vertical diameter of a flexible pipe.

DEFLECTION Upstream or downstream movement of a dam or dike.

DEFLOCCULATING AGENT Material added to a suspension to prevent settling.

DEFLUORIDATION Removal of excess fluoride in drinking water to prevent the staining of teeth.

DEFOLIANT Herbicide that removes leaves from trees and growing plants.

DEFORMATION Change in shape or size.

DEFORMATION ZONE Change in shape of a fluid mass by variations in wind, specifically by stretching and/or shearing.

DEFORMED ICE Ice which has been squeezed together and forced upwards and downwards in places.

DEGASIFICATION Water treatment that removes dissolved gases from the water.

DEGRADATION Geologic process by which stream beds, sandbars, floodplains, and the bottoms of other water bodies are lowered in elevation by the removal of material from the boundary, by the action of wind and water.

DEGRADATION CONSTANT Term used to address the decay of contaminant concentration due to factors other than dispersion.

DEGRADATION PRODUCTS Compounds resulting from transformation of an organic substance through chemical, photochemical, and/or biochemical reactions.

DEGRADED Condition of the quality of water that has been made unfit for some specified purpose.

DEGREE DAY Measure that gauges the amount of heating or cooling needed for a building using 65 °F as a baseline.

DEIONIZE (DI) Remove ions from a solution using an ion-exchange process.

DELAY Device used to obtain detonation of charges at separate times.

DELAY PATTERN Order of firing charges obtained by arranging delays to fire separate holes or series of holes at different times.

DELEGATED STATE State (or other governmental entity such as a tribal government) that has received authority to administer an environmental regulatory program in lieu of a federal counterpart.

DELIST Use of the petition process to have a facility's toxic designation rescinded.

DELIVERED PRODUCT Green power fed into the same electric transmission and distribution system that serves the end user.

DELIVERY/RELEASE Amount of water delivered to the point of use and the amount released after use; difference between amount delivered and amount released is consumptive use.

DELTA An alluvial deposit, often in the shape of the Greek letter "delta", which is formed where a stream drops its debris load on entering a body of quieter water; alluvial sediment deposit normally formed where a river or stream enters a lake or estuary; low, nearly flat tract of land at or near the mouth of a river, resulting from the accumulation of sediment supplied by the river in such quantities that it is not removed by tides, waves, or currents.

DEMAND Maximum water use under a specified condition; rate at which electric energy is used, expressed in kilowatts, whether at a given instant, or averaged over any designated period of time.

DEMAND SCHEDULING Method of irrigation scheduling whereby water is delivered to users as needed and which may vary in flow rate, frequency and duration.

DEMAND-SIDE WASTE MANAGEMENT Prices whereby consumers use purchasing decisions to communicate to product manufacturers that they prefer environmentally sound products packaged with the least amount of waste, made from recycled or recyclable materials, and containing no hazardous substances.

DEMERSAL Fish eggs or organisms that hatch on the bottom of a lake or stream.

DEMINERALIZATION Treatment process that removes dissolved minerals from water.

DEMOGRAPHICS Relating to the statistical study of human populations.

DENDRITES Thin branch-like growth of ice on the water surface.

DENDRITIC Channel pattern of streams with tributaries that branch to form a tree-like pattern.

DENITRIFICATION Biological reduction of nitrate by denitrifying bacteria in soil that usually resulting in the escape of nitrogen gas to the air.

DENSE FOG ADVISORY Issued when fog reduces visibility to 1/8 mile or less over a widespread area; marine advisory for widespread or localized fog reducing visibilities to regionally or locally defined limitations not to exceed 1 nautical mile.

DENSE NON-AQUEOUS PHASE LIQUID (DNAPL) Non-aqueous phase liquids such as chlorinated hydrocarbon solvents or petroleum fractions with a specific gravity greater than 1.0 that sink through the water column until they reach a confining layer.

DENSE SMOKE ADVISORY Advisory for widespread or localized smoke reducing visibilities to regionally or locally defined limitations not to exceed 1 nautical mile.

DENSITY Mass per unit volume; measure of how heavy a specific volume of a solid, liquid, or gas is in comparison to water; total mass (solids plus water) per total volume; weight of a substance per unit of volume of the substance.

DENSITY CURRENT Flow of water maintained by gravity through a large body of water, such as a reservoir or lake, and retaining its unmixed identity because of a difference in density.

DENSITY OF SNOW Ratio, expressed as a percentage, of the volume which a given quantity of snow would occupy if it were reduced to water, to the volume of the snow.

DENTATE See BAFFLE BLOCK.

DEPENDABLE CAPACITY Capacity that can be relied upon to carry system load for a specified time interval and period, provide assumed reserve, and/or meet firm power obligations.

DEPENDABLE YIELD Maximum annual supply of water development that is available on demand.

DEPLETION To permanently remove water from a system for a specific use; loss of water from a stream, river, or basin resulting from consumptive use; progressive withdrawal of water from surface- or ground-water reservoirs at a rate greater than that of replenishment.

DEPLETION CURVE Graphical representation of water depletion from storage-stream channels, surface soil, and groundwater that can be drawn for base flow, direct runoff, or total flow; part of the hydrograph extending from the point of termination of the Recession Curve to the subsequent rise or alternation of inflow due to additional water becoming available for stream flow.

DEPOSITION Material settling out of the water onto the streambed; mechanical or chemical processes through which sediments accumulate in a temporary resting place; raising of the stream bed by settlement of moving sediment that may be due to local changes in the flow, or during a single flood event.

DEPRESSION Region of low atmospheric pressure that is usually accompanied by low clouds and precipitation.

DEPRESSION STORAGE Volume of water contained (stored) in an elevation concavity.

DEPRESSURIZATION Condition that occurs when the air pressure inside a structure is lower that the air pressure outdoors.

DEPTH OF CUTOFF Vertical distance that the cutoff penetrates into the dam foundation.

DEPTH OF FLOW Vertical distance from the bed of a stream to the water surface.

DEPTH OF RUNOFF Total runoff from a drainage basin, divided by its area.

DERECHO Widespread and usually fast-moving windstorm associated with convection.

DERIVED PRODUCTS Processed base data on the Doppler radar.

DERMAL ABSORPTION/PENETRATION Process by which a chemical penetrates the skin and enters the body as an internal dose.

DERMAL EXPOSURE Exposure to chemical by entry though the skin.

DERMAL TOXICITY Ability of a pesticide or toxic chemical to poison people or animals by contact with the skin.

DERRICK Non-mobile tower equipped with a hoist; crane.

DESALINATION Removing salts from ocean or brackish water by using various technologies to provide freshwater; removal of salts from soil by artificial means, usually leaching.

DESALINIZATION Removal of dissolved salts from water by natural means (leaching) or by specific water treatment processes; process of removing salt from seawater or brackish water.

DESERTIFICATION Tendency toward more prominent desert conditions in a region when soil erosion is so severe that plants and animals can no longer exist.

DESICCANT To dry up; remove moisture from a substance; chemical drying agent which is capable of removing or absorbing moisture from the atmosphere.

DESICCATION Process of drying out; process used to thoroughly dry air; to remove virtually all moisture from air.

DESIGN BASIS EARTHQUAKE (DBE) Earthquake which the structure is required to safely withstand with repairable damage.

DESIGN CAPACITY Average daily flow that a treatment plant or other facility is designed to accommodate.

DESIGN CRITERIA Hypothetical flood used in the sizing of the dam and the associated structures to prevent dam failure by overtopping, especially for the spillway, outlet works, channel culvert, and inlet sizing.

DESIGN FLOW Used for steady-state waste load allocation modeling.

DESIGN RESPONSE SPECTRA Smooth, broad-banded spectra appropriate for specifying the level of seismic design force, or displacement, for earthquake-resistant design purposes.

DESIGN SUMMARY Document that summarizes the designers' development of the design that results in the specifications.

DESIGN VALUE Monitored reading used by EPA to determine an area's air quality status.

DESIGN WATER LEVEL Maximum water elevation, including the flood surcharge, that a dam is designed to be able to withstand.

DESIGN WIND Most severe wind that is reasonably possible at a particular reservoir for generating wind setup and runup.

DESIGNATED FLOODWAY Channel of a water course and those portions of the adjoining flood plain required to provide for the passage of a selected flood with a small increase in flood stage above that of natural conditions.

DESIGNATED FREQUENCY FLOOD Probability that a flood will occur in a given year; chance that such a flood will be equaled or exceeded in any one year; 100-year flood has a 0.01 annual exceedance probability (i.e. 1 chance in 100 that this flood flow level will be equaled or exceeded in any given year).

DESIGNATED POLLUTANT Air pollutant which is neither a criteria nor hazardous pollutant, as described in the Clean Air Act, but for which new source performance standards exist.

DESIGNATED USES Water uses identified in state water quality standards that must be achieved and maintained as required under the Clean Water Act.

DESIGNER BUGS Microbes developed through biotechnology that can degrade specific toxic chemicals at their source in toxic waste dumps or in groundwater.

DESIGNERS' OPERATING CRITERIA (DOC) Detailed operating criteria which stress the designer's intended use and operation of equipment and structures in the interest of safe, proper, and efficient use of the facilities.

DESORPTION Release or removal of an adsorbed material from the surface of a solid adsorbent; all processes by which chemicals move from the solid phase and concentrate them in the liquid phase (groundwater).

DESTINATION FACILITY Facility to which regulated medical waste is shipped for treatment and destruction, incineration, and/or disposal.

DESTRATIFICATION Vertical mixing within a lake or reservoir to totally or partially eliminate separate layers of temperature, plant, or animal life.

DESTROYED MEDICAL WASTE Regulated medical waste that has been ruined, torn apart, or mutilated through thermal treatment, melting, shredding, grinding, tearing, or breaking, so that it is no longer generally recognized as medical waste, but has not yet been treated (excludes compacted regulated medical waste).

DESTRUCTION AND REMOVAL EFFICIENCY (DRE) Percentage that represents the number of molecules of a compound removed or destroyed in an incinerator relative to the number of molecules entering the system (e.g. DRE equal to 99.99 percent means that 9,999 molecules are destroyed for every 10,000 that enter.).

DESTRUCTION FACILITY Facility that destroys regulated medical waste.

DESULFURIZATION Removal of sulfur from fossil fuels to reduce pollution.

DETECT To determine the presence of a compound.

DETECTABLE LEAK RATE Smallest leak (from a storage tank), expressed in terms of gallons- or liters-per-hour, that a test can reliably discern with a certain probability of detection or false alarm.

DETECTION First of five Early Warning System components consisting of the processes and equipment necessary to collect information about the threatening event and the response of the dam and reservoir, and relay that information to the decision makers.

DETECTION CRITERION Predetermined rule to ascertain whether a tank is leaking or not.

DETECTION LIMIT Lowest concentration of a chemical that can reliably be distinguished from a zero concentration.

DETENTION BASIN Storage structures (e.g. small unregulated reservoir) built upstream from populated areas that delay the conveyance of water downstream to prevent runoff and/or debris flows from causing property damage and loss of life; basins designed for detention of mud and rock debris are periodically excavated to maintain their storage capacity.

DETENTION DAM Dam built to store streamflow or surface runoff, and to control the release of such stored water.

DETENTION STORAGE Volume of water, other than depression storage, existing on the land surface as flowing water which has not yet reached the channel.

DETERGENT Synthetic washing agent that helps to remove dirt and oil.

DETERMINISTIC METHODOLOGY Method in which the chance of occurrence of the variable involved is ignored and the method or model used is considered to follow a definite law of certainty, and not probability.

DETONATING CORD Round, flexible textile, or plastic cord with a center core of high explosive.

DETONATION Practically instantaneous decomposition or combustion of an unstable compound, with tremendous increase in volume.

DETONATOR Device to start an explosion, as a fuse or cap.

DETRITUS Loose material that results directly from disintegration; heavier mineral debris moved by natural watercourses, usually in bed-load form; sand, grit, and other coarse material removed by differential sedimentation in a relatively short period of detention.

DEVELOPING GALE/STORM Indicate that gale/storm force winds are not now occurring but are expected before the end of the forecast period.

DEVELOPMENT DOCUMENT Report prepared during the development of an effluent limitation guideline by EPA that provides the data and methodology used to develop limitations guidelines and categorical pretreatment standards for an industrial category.

DEVELOPMENT EFFECTS Adverse effects such as altered growth, structural abnormality, functional deficiency, or death observed in a developing organism.

DEVIATOR STRESS Difference between the major and minor principal stresses in a triaxial test.

DEW Moisture in the air that condenses on solid surfaces when the air is saturated with water vapor and temperatures fall below dew point temperature.

DEW POINT (DWPT) Temperature to which air with a given quantity of water vapor must be cooled to cause condensation of the vapor in the air; greater dew point indicates more moisture present in the air.

DEW POINT DEPRESSION Difference in degrees between the air temperature and the dew point.

DEW POINT FRONT Narrow zone (mesoscale feature) of extremely sharp moisture gradient and little temperature gradient; separates moist air from dry air.

DEWATER Removing water by pumping, drainage, or evaporation; removal and control of groundwater from pores or other open spaces in soil or rock formations to the extent that allows construction activities to proceed; removal of groundwater and seepage from below the surface of the ground or other surfaces through the use of deep wells and wellpoints; remove or separate a portion of the water in a sludge or slurry to dry the sludge so it can be handled and disposed of; remove or drain the water from a tank or trench.

DEWP Dew point temperature taken at the same height as the air temperature measurement.

DEWPOINT TEMPERATURE Temperature at which dew begins to form or vapor begins to condense into a liquid.

DFUS Diffuse.

DHRO Diversity and Human Resources Office.

DI Durability Index.

DIABATIC Process which occurs with the addition or loss of heat.

DIAGNOSTIC MODEL Computer model used to calculate air pollution concentrations.

DIAGNOSTIC TOOL Any artifact which, because of form, shape, or function, provides chronological or manufacturing information.

DIAMOND DUST Fall of non-branched (snow crystals are branched) ice crystals in the form of needles, columns, or plates.

DIAPHRAGM See MEMBRANE.

DIAPHRAGM PUMP Pump that moves water by reciprocating motion of a diaphragm in a chamber having inlet and outlet check valves.

DIAPHRAGM WALL (MEMBRANE) Sheet, thin zone, or facing made of a relatively impervious material such as concrete, steel, wood, plastic, etc.

DIAPHRAGM-TYPE EARTHFILL Embankment dam which is constructed mostly of pervious material and having a diaphragm of impermeable material which forms a water barrier.

DIATOMACEOUS EARTH (DIATOMITE) Chalk-like material (fossilized diatoms) used to filter out solid waste in wastewater treatment plants; active ingredient in some powdered pesticides.

DIATOMS (BACILLARIOPHYTA) Single-celled (unicellular), colonial, or filamentous algae with siliceous cell walls constructed of two overlapping parts.

DIAZINON Insecticide.

DIBENZOFURANS Group of organic compounds, some of which are toxic.

DICHLORO-DIPHENYL-TRICHLOROETHANE (DDT) See DDT.

DICOFOL Pesticide used on citrus fruits.

DIEL Is of or pertaining to a 24-hour period of time; a regular daily cycle.

DIELDRIN Organochlorine insecticide no longer registered for use in the United States; degradation product of the insecticide aldrin.

DIFFERENTIAL HEAD (UNBALANCED HEAD) Condition in which the water pressure on the upstream and downstream sides of an object differ.

DIFFERENTIAL MOTION Cloud motion that appears to differ relative to other nearby cloud elements.

DIFFERENTIAL ROTATION Change in solar rotation rate with latitude where low latitudes rotate at a faster angular rate (approx. 14 degrees per day) than high latitudes (approx. 12 degrees per day).

DIFFUSE ICE Poorly defined ice edge limiting an area of dispersed ice; usually on the leeward side of an area of floating ice.

DIFFUSED AIR Type of aeration that forces oxygen into sewage by pumping air through perforated pipes inside a holding tank.

DIFFUSION Movement of suspended or dissolved particles (or molecules) from a more concentrated to a less concentrated area; dissipation of the energy associated with a flood wave that results in the attenuation of the flood wave.

DIFFUSION COEFFICIENT Constant of proportionality which relates the mass flux of a solute to the solute concentration gradient.

DIFLUENCE Pattern of wind flow in which air moves outward (i.e. "fan-out" pattern) away from a central axis that is oriented parallel to the general direction of the flow.

DIGESTER In wastewater treatment, a closed tank; in solid-waste conversion, a unit in which bacterial action is induced and accelerated in order to break down organic matter and establish the proper carbon to nitrogen ratio.

DIGESTION Biochemical decomposition of organic matter, resulting in partial gasification, liquefaction, and mineralization of pollutants.

DIGITIZER Device that converts analog information into a digital format.

DIKE Low wall that can act as a barrier to prevent a spill from spreading; low embankment for restraining a river or a stream; embankments which contain water within a given course.

DILATANCY Expansion of cohesionless soils when subject to shear deformation.

DILUENT Any liquid or solid material used to dilute or carry an active ingredient.

DILUTION Reduction of the concentration of a substance in air or water.

DILUTION RATIO Relationship between the volume of water in a stream and the volume of incoming water.

DIMICTIC Lakes and reservoirs that freeze over and normally go through two stratifications and two mixing cycles a year.

DINOCAP Fungicide used primarily by apple growers to control summer diseases.

DINOSEB Herbicide that is also used as a fungicide and insecticide.

DIOXIN Any of a family of compounds known chemically as dibenzo-p-dioxins.

DIP Angle between a planar feature, such as a sedimentary bed or a fault, and the horizontal plane; slope of layers of soil or rock.

DIR Direction; ten-minute average wind direction measurements in degrees clockwise from true North.

DIRECT ACCESS Arrangement in which customers can purchase electricity directly from any supplier in the competitive market, using the transmission and distribution lines of electric utilities to transport the electricity.

DIRECT CURRENT (DC) Electrical current flowing in one direction only and essentially free from pulsation.

DIRECT DIGITAL CONTROLS (DDC)
Application of microprocessor tech-

nology to building environmental controls.

DIRECT DISCHARGER Municipal or industrial facility which introduces pollution through a defined conveyance or system such as outlet pipes; point source.

DIRECT FILTRATION Method of treating water which consists of the addition of coagulant chemicals, flash mixing, coagulation, minimal flocculation, and filtration.

DIRECT FLOOD DAMAGE Damage done to property, structures, goods, etc., by a flood as measured by the cost of replacement and repairs.

DIRECT HIT Close approach of a tropical cyclone to a particular location.

DIRECT PUSH Technology used for performing subsurface investigations by driving, pushing, and/or vibrating small-diameter hollow steel rods into the ground.

DIRECT RUNOFF Water that flows over the ground surface or through the ground directly into streams, rivers, or lakes; runoff entering stream channels promptly after rainfall or snow melt.

DIRECT SHEAR TEST Shear test in which soil or rock under an applied normal load is stressed to failure by moving one section of the specimen container (shear box) relative to the other section.

DIRECT SOLAR RADIATION Component of solar radiation received by the Earth's surface only from the direction of the Sun's disk (i.e., it has not been reflected, refracted or scattered).

DIRECTIONAL SHEAR Component of wind shear which is due to a change in wind direction with height (e.g., southeasterly winds at the surface and southwesterly winds aloft).

DIRECTOR Person responsible for permitting, implementation, and compliance; authorized representative.

DISAPPEARING SOLAR FILAMENT (DSF) Sudden (timescale of minutes to hours) disappearance of a solar filament (prominence).

DISASTER Event that demands a crisis response beyond the scope of any single line agency or service, presents a threat and requires resources beyond what are available locally.

DISCHARGE Volume of a fluid or solid passing a cross section of a stream per unit time; volume of water that passes a given point within a given period of time; rate of fluid flow passing a given point at a given moment in time; flow of surface water in a stream or canal or the outflow of groundwater from a flowing artesian well, ditch, or spring; discharge of liquid effluent from a facility or to chemical emissions into the air through designated venting mechanisms; outflow; commonly expressed in volume per unit of time, cubic feet per second, cubic meters per second, million gallons per day, gallons per minute, or seconds per minute per day; often used interchangeably with flow and streamflow.

DISCHARGE AREA Area where subsurface water is discharged to the land surface, to surface water, or to the atmosphere.

DISCHARGE CAPACITY Maximum amount of water that can be safely released from a given waterway.

DISCHARGE CURVE (RATING CURVE and DISCHARGE RATING CURVE) Curve that expresses the relation between the discharge of a stream or open conduit at a given location and the stage or elevation of the liquid surface at or near that location.

DISCHARGE LENGTH SCALE Square root of the cross-sectional area of any discharge outlet.

DISCHARGE MONITORING REPORT (DMR) Form used (including any subsequent additions, revisions, or modifications) to report self-monitoring results by NPDES permittees.

DISCHARGE RATING CURVE See STAGE-DISCHARGE CURVE.

DISCHARGE TABLE (RATING TABLE) Table showing the relation between two mutually dependent quantities or variable over a given range of magnitude; table showing the relation between the gauge height and the discharge of a stream or conduit at a given gaging station.

DISCRETIZATION Process of subdividing the continuous model and/or time domain into discrete segments or cells.

DISDROMETER Equipment that measures and records the size distribution of raindrops.

DISINFECTANT Chemical (e.g., chlorine, chloramine, or ozone) or physical process (e.g. ultraviolet light) that kills microorganisms such as bacteria, viruses, and protozoa; chemical or physical process that kills pathogenic organisms in water, air, or on surfaces.

DISINFECTANT BY-PRODUCT Compound formed by the reaction of a disinfectant such as chlorine with organic material in the water supply; a chemical byproduct of the disinfection process.

DISINFECTANT TIME Time it takes water to move from the point of disinfectant application (or the previous point of residual disinfectant measurement) to a point before or at the point where the residual disinfectant is measured.

DISK Visible surface of the Sun (or any heavenly body) projected against the sky.

DISPERSANT Chemical agent used to break up concentrations of organic material such as spilled oil.

DISPERSING AGENT Agent used to assist in separating individual fine soil particles and to prevent them from flocculating when in suspension.

DISPERSION Process by which some of the water molecules and solute molecules travel more rapidly than the average linear velocity and some travel more slowly; spreading of the solute in the direction of the groundwater flow (longitudinal dispersion) or direction perpendicular to groundwater flow (transverse dispersion); distortion of the shape of a seismic-wave train because of variation of velocity with frequency; process of separating radiation into various wavelengths.

DISPERSION COEFFICIENT Measure of the spreading of a flowing substance due to the nature of the porous medium, with its interconnected channels distributed at random in all directions; sum of the coefficients of mechanical dispersion and molecular diffusion in a porous medium.

DISPERSIVITY Scale dependent property of an aquifer that determines the degree to which a dissolved constituent will spread in flowing groundwater that is comprised of three directional components (longitudinal, transverse and vertical).

DISPLACEMENT PUMP Reciprocating type of pump where a piston draws water into a closed chamber and then expels it under pressure.

DISPLACEMENT SAVINGS Saving realized by displacing purchases of natural gas or electricity from a local utility by using landfill gas for power and heat.

DISPOSABLES Consumer products, other items, and packaging used once or a few times and discarded.

DISPOSAL Final placement or destruction of toxic, radioactive, or other wastes; surplus or banned pesticides or other chemicals; polluted soils; and drums containing hazardous materials from removal actions or accidental releases.

DISPOSAL FACILITIES Repositories for solid waste, including landfills and combustors intended for permanent containment or destruction of waste materials, excluding transfer stations and composting facilities.

DISSECTED Cut by erosion into valleys, hills, and upland plains.

DISSOLVE To enter into a solution.

DISSOLVED Material in a representative water sample that passes through a 0.45-micrometer membrane filter.

DISSOLVED CONSTITUENT Operationally defined as a constituent that passes through a 0.45-micrometer filter.

DISSOLVED OXYGEN (DO) Amount of free oxygen (oxygen gas) dissolved in water; indicator of water quality.

DISSOLVED SOLIDS Disintegrated organic and inorganic material dissolved in water; indicator of salinity or hardness.

DISSOLVED-SOLIDS CONCENTRATION Quantity of dissolved material in a sample of water.

DISTILLATION Act of purifying liquids through boiling, so that the steam or gaseous vapors condense to a pure liquid.

DISTRIBUTARIES Diverging streams which do not return to the main stream, but discharge into another stream or the ocean.

DISTRIBUTION Delivery of electricity to the retail customer's home or business through low voltage distribution lines.

DISTRIBUTION COEFFICIENT Quantity of the solute, chemical or radionuclide absorbed by the solid per unit weight of solid divided by the quantity dissolved in the water per unit volume of water.

DISTRIBUTION EFFICIENCY Measure of the uniformity of irrigation water distribution over a field.

DISTRIBUTION GRAPH (DISTRIBUTION HYDROGRAPH) Unit hydrograph of direct runoff modified to show the proportions of the volume of runoff that occurs during successive equal units of time.

DISTRIBUTION LINE Low voltage electric power line, usually 69 kilovolts or less.

DISTRIBUTION SYSTEM Network of pipes leading from a treatment plant to customers' plumbing systems; system of ditches, or conduits and their appurtenances, which conveys irrigation water from the main canal to the farm units; portion of an electric system that is dedicated to delivering electric energy to an end user.

DISTRICT Entity that has a contract for the delivery of irrigation water.

DISTURBANCE Any event or series of events that disrupt ecosystem, community, or population structure and alters the physical environment.

DITCH Long, narrow excavation; constructed open channel for conducting water.

DIURNAL Daily; related to actions which are completed in the course of a calendar day, and which typically recur every calendar day.

DIURNAL CYCLES Variations in meteorological parameters such as temperature and relative humidity over the course of a day which result from the rotation of the Earth about its axis and the resultant change in incoming and outgoing radiation.

DIURNAL TEMPERATURE RANGE Temperature difference between the minimum at night (low) and the maximum during the day (high).

DIVERGENCE Expansion or spreading out of a vector field.

DIVERSION Taking of water from a stream or other body of water into a canal, pipe, or other conduit; process which, having return flow and consumptive use elements, turns water from a given path; removal of water from its natural channel for human use; use of part of a stream flow as a water supply; channel constructed across the slope for the purpose of intercepting surface runoff, changing the accustomed course of all or part of a stream; structural conveyance (or ditch) constructed across a slope to intercept runoff flowing down a hillside, and divert it to some convenient discharge point; turning aside or alteration of the natural course of a flow of water, normally considered physically to leave the natural channel.

DIVERSION CAPACITY Flow which can be passed through the canal headwork at a dam under normal head.

DIVERSION CHANNEL (CANAL OR TUNNEL) Waterway used to divert water from its natural course.

DIVERSION DAM Dam built to divert water from a waterway or stream into a different watercourse.

DIVERSION INLET Conduit or tunnel upstream from an intake structure.

DIVERSION RATE Percentage of waste materials diverted from traditional disposal such as landfilling or incineration to be recycled, composted, or re-used.

DIVERSITY Distribution and abundance of different kinds of plant and animal species (biological taxa) and communities in a specified location.

DIVERSITY INDEX (H) (SHANNON INDEX) Numerical expression of evenness of distribution of aquatic organisms.

DIVERT To direct a flow away from its natural course.

DIVIDE Ridge or high area of land that separates one drainage basin from another; high ground that forms the boundary of a watershed.

DIVIDING STREAMLINE In the blocked flow region upwind of a mountain barrier, the streamline that separates the blocked flow region near the ground from the streamlines above which go over the barrier.

DIVIDING STREAMLINE HEIGHT Height above ground of the dividing streamline, as measured far upwind of a mountain barrier.

DMNT Dominant.

DMSH Diminish.

DNA HYBRIDIZATION Use of a segment of DNA, called a DNA probe, to identify its complementary DNA; used to detect specific genes.

DNR Department of Natural Resources.

DNS Dense.

DNSTRM Downstream.

DO Dissolved oxygen.

DOBSON UNIT (DU) Unit used to measure the abundance of ozone in the atmosphere; equivalent to 2.69 x 10^{16} molecules of ozone per centimeter squared.

DOC Department of Commerce.

DOH Development and Operations Hydrologist.

DOLDRUMS Regions on either side of the equator where air pressure is low and winds are light.

DOLOMITE Sedimentary rock consisting chiefly of magnesium carbonate.

DOMAIN Geographical area over which a simulation is performed.

DOME Structure curved both in the lengthwise and crosswise direction.

DOMESTIC APPLICATION Pesticide application in and around houses, office buildings, motels, and other living or working areas.

DOMESTIC CONSUMPTION Quantity, or quantity per capita, of water consumed in a municipality or district for domestic uses or purposes during a given period, generally one day.

DOMESTIC WATER USE (WITHDRAWALS) Water for household purposes, such as watering of livestock, drinking, food preparation, bathing, washing clothes and dishes, flushing toilets, and watering lawns and gardens.

DOMESTIC WITHDRAWALS See DOMESTIC WATER USE.

DOMINANT DISCHARGE Particular magnitude of flow which is sometimes referred to as the "channel forming" discharge.

DOMINANT PLANT Plant species controlling the environment.

DOPPLER RADAR Radar that can measure radial velocity, the instantaneous component of motion parallel to the radar beam (i.e., toward or away from the radar antenna).

DOSAGE/DOSE Actual quantity of a chemical administered to an organism or to which it is exposed; amount of a substance that reaches a specific tissue (e.g. liver); amount of a substance available for interaction with metabolic processes after crossing the outer boundary of an organism.

DOSE EQUIVALENT Product of the absorbed dose from ionizing radiation and such factors as account for biological differences due to the type of radiation and its distribution in the body in the body.

DOSE RATE (DOSAGE) Dose per time unit (e.g. mg/day).

DOSE RESPONSE Shifts in toxicological responses of an individual (such as alterations in severity) or populations (such as alterations in incidence) that are related to changes in the dose of any given substance.

DOSE RESPONSE CURVE Graphical representation of the relationship between the dose of a stressor and the biological response thereto.

DOSE-RESPONSE ASSESSMENT Estimating the potency of a chemical; process of determining the relationship between the dose of a stressor and a specific biological response; evaluating the quantitative relationship between dose and toxicological responses.

DOSE-RESPONSE RELATIONSHIP Quantitative relationship between the amount of exposure to a substance and the extent of toxic injury or disease produced.

DOSIMETER Instrument to measure dosage.

DOT REPORTABLE QUANTITY Quantity of a substance specified in a U.S. Department of Transportation regulation that triggers labeling, packaging and other requirements related to shipping such substances.

DOUBLE CURVATURE ARCH DAM Arch dam which is curved in plan and elevation, with undercutting of the heel and in most instances, a downstream overhang near the crest; arch dam which is curved vertically as well as horizontally.

DOUBLE-MASS CURVE Plot on arithmetic cross-section paper of the cumulated values of one variable against the cumulated values of another or against the computed values of the same variable for a concurrent period of time.

DOWNBURST Strong downdraft current of air from a cumulonimbus cloud, often associated with intense thunderstorms.

DOWNDRAFT (DWNDFT) Small-scale column of air that rapidly sinks toward the ground, usually accompanied by precipitation as in a shower or thunderstorm.

DOWNGRADIENT Direction that groundwater flows; similar to "downstream" for surface water.

DOWNSLOPE FLOW Thermally driven wind directed down a mountain slope and usually occurring at night; part of the along-slope wind system.

DOWNSTREAM In the same direction as a stream or other flow, or toward the direction in which the flow is moving.

DOWNSTREAM FACE Inclined surface of a dam away from the reservoir.

DOWNSTREAM PROCESSORS Industries dependent on crop production (e.g., canneries and food processors).

DOWNSTREAM SLOPE Slope or face of the dam away from the reservoir water.

DOWN-VALLEY WIND Thermally driven wind directed down a valley's axis, usually occurring during nighttime; part of the along-valley wind system.

Downward Vertical Velocity (DVV) Sinking air.

DOWNWASH Deflection of air downward relative to an object that causes the deflection.

DOWNWELLING RADIATION Component of radiation directed toward the Earth's surface from the Sun or the atmosphere, opposite of upwelling radiation.

DP Deep; Dew Point.

DP HOLE Hole in the ground made with DP equipment.

DPD Dominant wave period (seconds) is the period with the maximum wave energy.

DPNG Deepening.

DPTH Depth.

DPTR Departure.

DPVA Differential Positive Vorticity Advection.

DR Direction.

DRAFT Act of drawing or removing water from a reservoir; water which is drawn or removed.

DRAFT DEPTH Depth measured perpendicularly from the water surface to the bottom of a boat, ship, etc.

DRAFT PERMIT Document prepared indicating tentative decision to issue, deny, modify, revoke and reissue, terminate, or reissue a permit; preliminary permit drafted and published that is subject to public review and comment before final action on the application.

DRAINAGE Process of removing surface or subsurface water from a soil or area; technique to improve the productivity of some agricultural land by removing excess water from the soil where surface drainage is accomplished with open ditches and subsurface drainage uses porous conduits (drain tile) buried beneath the soil surface.

DRAINAGE AREA Area that drains to a particular point on a river or stream; area having a common outlet for its surface runoff.

DRAINAGE AREA OF A STREAM At a specific location, area upstream from the location, measured in a horizontal plane, that has a common outlet at the site for its surface runoff from precipitation that normally drains by gravity into a stream.

DRAINAGE BASIN All of the land area drained by a river or stream system; land area where precipitation runs off into streams, rivers, lakes, and reservoirs; area of land that drains water, overland run-off, including tributaries and impoundments, sediment, and dissolved materials to a common outlet at some point along a stream channel; part of the surface of the earth that contains a drainage system with a common outlet for its surface runoff.

DRAINAGE BLANKET Layer of pervious material placed directly over the foundation material to facilitate drainage of the foundation and/or embankment.

DRAINAGE CURTAIN Line of vertical wells or boreholes to facilitate drainage of the foundation and abutments and to reduce water pressure.

DRAINAGE DENSITY Length of all channels above those of a specified stream order per unit of drainage area; relative density of natural drainage channels in a given area usually expressed in terms of miles of natural drainage or stream channel per square mile of area, and obtained by dividing the total length of stream channels in the area in miles by the area in square miles.

DRAINAGE DIVIDE Rim of a drainage basin; boundary line, along a topographic ridge or along a subsurface formation, separating two adjacent drainage basins.

DRAINAGE LAYER Layer of pervious material in an earthfill dam to relieve pore pressures or to facilitate drainage of the fill.

DRAINAGE SYSTEM Collection of surface and/or subsurface drains, together with structures and pumps, used to remove surface or groundwater.

DRAINAGE WELLS (RELIEF WELLS)
Vertical wells downstream of or in the downstream shell of an embankment dam to collect and control seepage through and under the dam where a line of such wells forms a "drainage curtain"; well drilled to carry excess water off agricultural fields.

DRAINER Valley or basin from which air drains continuously during nighttime rather than becoming trapped or pooled.

DRAINS (RELIEF WELLS) See DRAINAGE WELLS.

DRAW Small valley or gully.

DRAWDOWN Lowering of the surface elevation of a body of water, the water surface of a well, the water table, or the piezometric surface adjacent to the well, resulting from the withdrawal of water therefrom; process of depleting a reservoir or groundwater storage; drop in the water table or level of water in the ground when water is being pumped from a well; vertical distance the free water surface elevation is lowered or the reduction of the pressure head due to the removal of free water; difference between a water level and a lower water level in a reservoir within a particular time; amount of water used from a tank or reservoir; drop in the water level of a tank or reservoir.

DRCTN Direction.

DREDGE To dig under water; machine that digs under water.

DREDGING Scooping, or suction of underwater material from a harbor, or waterway; removal of mud from the bottom of water bodies.

DRFT Drift.

DRIFT Food organisms, including algae, plankton, and even larval fish, dislodged and moved by river current; small, nearly horizontal tunnel.

DRIFTING ICE Pieces of floating ice moving under the action of wind and/or currents.

DRIFTING SNOW Uneven distribution of snowfall/snow depth caused by strong surface winds.

DRILL EXERCISE Activity designed to evaluate a single emergency response function.

DRILLING FLUID Fluid used to lubricate the bit and convey drill cuttings to the surface with rotary drilling equipment.

DRINKING WATER EQUIVALENT LEVEL Protective level of exposure related to potentially non-carcinogenic effects of chemicals that are also known to cause cancer.

DRINKING WATER STATE REVOLVING FUND Provides capitalization grants to states to develop drinking water revolving loan funds to help finance system infrastructure improvements, assure source-water protection, enhance operation and management of drinking-water systems, and otherwise promote local water-system compliance and protection of public health.

DRINKING-WATER STANDARD OR GUIDELINE Threshold concentration for a constituent or compound in a public drinking-water supply, designed to protect human health.

DRIP IRRIGATION (TRICKLE) Irrigation system in which water is applied directly to the root zone of plants by means of applicators (orifices, emitters, porous tubing, perforated pipe, and so forth) operated under low pressure.

DRIVE CASING Heavy duty steel casing driven along with the sampling tool in cased DP systems that keeps the hole open between sampling runs and is not removed until last sample has been collected.

DRIVE POINT PROFILER Exposed groundwater DP (direct-push) system used to collect multiple depth-discrete groundwater samples.

DRIZZLE Precipitation consisting of numerous minute droplets of water less than 0.5 mm (500 micrometers) in diameter.

DROP Structure in an open conduit or canal installed for the purpose of dropping the water to a lower level and dissipating its energy.

DROP STRUCTURE Structure that conveys water to a lower elevation and dissipates the excess energy resulting from the drop.

DROP-OFF Recyclable materials collection method in which individuals bring them to a designated collection site.

DROP-SIZE DISTRIBUTION Distribution of rain drops or cloud droplets of specified sizes.

DROUGHT Prolonged period of less-than-normal precipitation such that the lack of water causes a serious hydrologic imbalance; climatic condition in which there is insufficient soil moisture available for normal vegetative growth; deficiency of moisture that results in adverse impacts on people, animals, or vegetation over a sizeable area.

DROUGHT ASSESSMENTS At the end of each month, CPC issued long-term seasonal drought assessment.

DROUGHT INDEX Computed value which is related to some of the cumulative effects of a prolonged and abnormal moisture deficiency; index of hydrological drought corresponding to levels below the mean in streams, lakes, and reservoirs.

DRUM GATE Movable crest gate in the form of a sector of a cylinder hinged at the centerline; type of spillway gate consisting of a long hollow drum that may be held in its raised position by the water pressure in a flotation chamber beneath the drum.

DRY ADIABAT Line of constant potential temperature on a thermodynamic chart.

DRY ADIABATIC LAPSE RATE (DALR)
Rate at which the temperature of a parcel of dry air decreases as the parcel is lifted in the atmosphere equal to 5.5 °F per 1000 ft or 9.8 °C per km.

DRY CRACK Crack visible at the surface but not going right through the ice cover, and therefore it is dry.

DRY FLOOD PROOFING Sealed against floodwaters where all areas below the flood protection level are made watertight.

DRY LINE Boundary separating moist and dry air masses.

DRY LINE BULGE Bulge in the dry line, representing the area where dry air is advancing most strongly at lower levels.

DRY LINE STORM Any thunderstorm that develops on or near a dry line.

DRY MASS Mass of residue present after drying in an oven at 105 °C, until the mass remains unchanged that represents the total organic matter, ash and sediment, in the sample.

DRY MICROBURST Microburst with little or no precipitation reaching the ground; most common in semi-arid regions.

DRY PUNCH Surge of drier air; normally a synoptic-scale or mesoscale process.

DRY SLOT Zone of dry (and relatively cloud-free) air which wraps east- or northeastward into the southern and eastern parts of a synoptic scale or mesoscale low pressure system.

DRY THUNDERSTORM High-based thunderstorm when lightning is observed, but little if any precipitation reaches the ground.

DRY UNIT WEIGHT Weight of solid particles per unit of total volume.

DRY WEATHER FLOW Streamflow which results from precipitation that infiltrates into the soil and eventually moves through the soil to the stream channel.

DRY WEIGHT Weight of animal tissue after it has been dried in an oven at 65 °C until a constant weight is achieved that represents total organic and inorganic matter in the tissue.

DRY WELL Deep hole, covered, and usually lined or filled with rocks, that holds drainage water until it soaks into the ground.

DRY-ADIABATIC Adiabatic process in a hypothetical atmosphere in which no moisture is present; adiabatic process in which no condensation of its water vapor occurs and no liquid water is present.

DSA Special Tropical Disturbance Statement.

DSIPT Dissipate.

DST INDEX Geomagnetic index describing variations in the equatorial ring current.

DTRT Deteriorate.

DUAL-PHASE EXTRACTION Active withdrawal of both liquid and gas phases from a well usually involving the use of a vacuum pump.

DUCTILITY Property of a material that allows it to be formed into thin sections without breaking.

DUMP Site used to dispose of solid waste without environmental controls.

DUMPED Method of compacting soil by dumping the soil into place with no compactive effort.

DUPLICATE Second aliquot or sample that is treated the same as the original sample in order to determine the precision of the analytical method.

DURATION CURVE Cumulative frequency curve that shows the percent of time during which specified units of items (e.g., discharge, head, power, etc.) were equaled or exceeded in a given period; integral of the frequency diagram.

DURATION OF ICE COVER Time from freeze-up to break-up of an ice cover.

DURATION OF STRONG GROUND MOTION "Bracketed duration" or the time interval between the first and last acceleration peaks that are equal to or greater than 0.05g.

DURATION OF SUNSHINE Amount of time sunlight detected at a given point.

DURG During.

DURN Duration.

DUSK Time at which the Sun is 6 degrees below the horizon in the evening.

DUST DEVIL Small, rapidly rotating wind that is made visible by the dust, dirt or debris it picks up.

DUST PLUME Non-rotating "cloud" of dust raised by straight-line winds.

DUST STORM Severe weather condition characterized by strong winds and dust-filled air over an extensive area.

DUST WHIRL Rotating column of air rendered visible by dust.

DUSTFALL JAR Open container used to collect large particles from the air for measurement and analysis.

DVA Deficiency Verification Analysis.

DVLP Develop.

DWNSLP Downslope.

DYNAMIC COMPACTION Method of compacting soil by dropping a heavy weight onto loose soil.

DYNAMIC EQUILIBRIUM Condition achieved when the average sand load transported by flowing water is in balance with the sand load being supplied by tributaries.

DYNAMIC ICE Pressure due to a moving ice cover or drifting ice.

DYNAMIC LIFTING Forced uplifting of air from various atmospheric processes, such as weather fronts, and cyclones.

DYNAMIC PRESSURE When a pump is operating, the vertical distance from a reference point (such as a pump centerline) to the hydraulic grade line.

DYNAMIC ROUTING Hydraulic flow routing based on the solution of the St. Venant Equation(s) to compute the changes of discharge and stage with respect to time at various locations along a stream.

DYNAMIC WAVE ROUTING MODEL (DWOPER) Computerized hydraulic routing program whose algorithms incorporate the complete one-dimensional equations of unsteady flow.

DYNAMICS Any forces that produce motion or effect change.

DYNAMOMETER Device used to place a load on an engine and measure its performance.

DYSTROPHIC LAKES Acidic, shallow bodies of water that contain much humus and/or other organic matter; contain many plants but few fish.

DZ Drizzle.

E

E East.

E REGION Daytime layer of the Earth's ionosphere roughly between the altitudes of 85 and 140 km.

E-19, REPORT ON RIVER GAUGE STATION
Report to be completed every 5 - 10 years providing a complete history of a river station and all gauges that have been used for public forecasts since the establishment of the station. Should be updated anytime a significant change occurs at a forecast point.

E-19A, ABRIDGED REPORT ON RIVER GAUGE STATION
Abridged version of an E-19, an E-19a used to be used to update the E-19 as additional information, or changes occur at the station during the intervening five year period.

E-3, FLOOD STAGE REPORT
Form that a Service Hydrologist/Hydrology Focal Point completes to document the dates in which forecast points are above flood stage, as well as the crest dates and stages.

E-5, MONTHLY REPORT OF RIVER AND FLOOD CONDITIONS
Monthly narrative report covering flooding which occurred over the past month.

E-7, FLOOD DAMAGE REPORT
Report to be completed anytime there is reported flood damage or loss of life as a direct result of flooding.

EA Environmental Assessment.

EARLY WARNING SYSTEM (EWS)
Designed system that will ensure timely recognition of a threatening event and provide a reliable and timely warning and evacuation of the population at risk from dangerous flooding associated with large operational releases or dam failure that must address five components: detection, decision making, notification, warning, and evacuation.

EARLY WARNING SYSTEM FEASIBILITY
Study of the feasibility of installing some level of Early Warning System to meet the target times for alerts and warnings.

EARTH DAM (EARTHFILL DAM)
Embankment dam in which more than 50 percent of the total volume is formed of compacted earth material generally smaller than 3-inch size where seepage through the dam is

controlled by the designed use of upstream blankets and/or internal cores constructed using compacted soil of very low permeability.

EARTH LINING Compacted layer of earth on surface of canal or other excavation.

EARTH PRESSURE Pressure or force exerted by soil on any boundary.

EARTHQUAKE Sudden motion or trembling in the Earth caused by the abrupt release of accumulated stress along a fault.

EARTHQUAKE, MAXIMUM CREDIBLE (MCE) Earthquake(s) associated with specific seismotectonic structures, source areas, or provinces that would cause the most severe vibratory ground motion or foundation dislocation capable of being produced at the site under the currently known tectonic framework.

EARTHQUAKE, MAXIMUM DESIGN (MDE) Postulated seismic event, specified in terms of specific bedrock motion parameters at a given site, which is used to evaluate the seismic resistance of manmade structures or other features at the site.

EARTHQUAKE, OPERATING BASIS (OBE) Earthquakes for which the structure is designed to resist and remain operational.

EARTHQUAKE, SAFETY EVALUATION (SEE) Earthquake, expressed in terms of magnitude and closest distance from the dam site or in terms of the characteristics of the time history of free-field ground motions, for which the safety of the dam and critical structures associated with the dam are to be evaluated.

EARTHQUAKE, SYNTHETIC Earthquake time history records developed from mathematical models that use white noise, filtered white noise, and stationary and nonstationary filtered white noise, or theoretical seismic source models of failure in the fault zone.

EARTHWORK Any one or combination of the operations involved in altering or movement of earth.

EASEMENT Right to use land owned by another for some specific purpose.

EASTERLIES Any winds with components from the east.

EASTERLY WAVE Low level disturbance of tropical origins that can develop into tropical cyclones.

EBB CURRENT Movement of a tidal current away from the coast or down an estuary.

EBND Eastbound.

EBS Emergency Broadcast System.

ECCENTRICITY Dimensionless quantity describing the elliptical shape of a planet's orbit.

ECHO Energy back scattered from a target (e.g., precipitation, clouds) and received by and displayed on a radar screen.

ECHO TOPS Height above ground of the center of the radar beam using the tilt, or scan, that contains the highest elevation where reflectivitys greater than 18 dBZ can be detected.

ECMF European Center for Meteorology Forecast model.

ECMWF European Center for Medium-Range Weather Forecasts.

ECOLOGICAL ENTITY Term referring to a species, a group of species, an ecosystem function or characteristic, or a specific habitat or biome.

ECOLOGICAL EXPOSURE Exposure of a non-human organism to a stressor.

ECOLOGICAL IMPACT Effect that a man-caused or natural activity has on living organisms and their non-living (abiotic) environment.

ECOLOGICAL INDICATOR Characteristic of an ecosystem that is related to, or derived from, a measure of biotic or abiotic variable, that can provide quantitative information on ecological structure and function; indicator can contribute to a measure of integrity and sustainability; plant or animal species, communities, or special habitats with a narrow range of ecological tolerance.

ECOLOGICAL INTEGRITY Living system exhibits integrity if, when subjected to disturbance, it sustains and organizes self-correcting ability to recover toward a biomass end-state that is normal for that system.

ECOLOGICAL RISK ASSESSMENT Application of a formal framework, analytical process, or model to estimate the effects of human actions(s) on a natural resource and to interpret the significance of those effects in light of the uncertainties identified in each component of the assessment process.

ECOLOGICAL STUDIES Studies of biological communities and habitat characteristics in NAWQA Study Units to evaluate the effects of physical and chemical characteristics of water and hydrologic conditions on aquatic biota and to determine how biological and habitat characteristics differ among environmental settings.

ECOLOGICAL/ENVIRONMENTAL SUSTAINABILITY Maintenance of ecosystem components and functions for future generations.

ECOLOGY Branch of biological science which deals with relationships between living organisms and their environments.

ECONOMIC ANALYSIS Procedure that includes both tangible and intangible factors to evaluate various alternatives.

ECONOMIC POISONS Chemicals used to control pests and to defoliate cash crops such as cotton.

ECOREGION Area of similar climate, landform, soil, potential natural vegetation, hydrology, or other ecologically relevant variables.

ECOSPHERE "Bio-bubble" that contains life on Earth, in surface waters, and in the air.

ECOSYSTEM Complex system composed of a community of people, animals, plants and bacteria as well as the chemical and physical environments; interacting system of a biological community and its non-living environmental surroundings.

ECOSYSTEM STRUCTURE Attributes related to the instantaneous physical state of an ecosystem (e.g., species population density, species richness or evenness, and standing crop biomass).

ECOTONE Habitat created by the juxtaposition of distinctly different habitats; an edge habitat; or an ecological zone or boundary where two or more ecosystems meet.

EDDY Circular current of water moving against the main current; swirling currents of air at variance with the main current.

EDT Eastern Daylight Time.

EEO Equal Employment Opportunity.

EFCT Effect.

EFFECTIVE CONCENTRATION (EC)
Point estimate of the toxicant concentration that would cause an observable adverse effect (e.g., death, immobilization, or serious incapacitation) in a given percentage of the test organisms.

EFFECTIVE DIAMETER Particle diameter corresponding to 10 percent finer on the accumulative gradation curve.

EFFECTIVE FORCE Force transmitted through a soil or rock mass by intergranular pressures.

EFFECTIVE GRAIN SIZE Diameter of the particles in an assumed rock or soil that would transmit water at the same rate as the rock or soil under consideration, and that is composed of spherical particles of equal size and arranged in a specific manner; single particle diameter that best depicts the bed material properties.

EFFECTIVE PEAK GROUND ACCELERATION
Acceleration that is most closely related to structural response and to damage potential of an earthquake.

EFFECTIVE POROSITY Ratio, usually expressed as a percentage, of the volume of water or other liquid which a given saturated volume of rock or soil will yield under any specified hydraulic condition, to the given volume of soil or rock; ratio of the volume of a soil or rock mass that can be drained by gravity to the total volume of the mass.

EFFECTIVE PRECIPITATION (RAINFALL)
Part of the precipitation that produces runoff; weighted average of current and antecedent precipitation that is "effective" in correlating with runoff; part of the precipitation falling on an irrigated area that is effective in meeting the consumptive use requirements.

EFFECTIVE TERRESTRIAL RADIATION
Difference between upwelling infrared or terrestrial radiation emitted from the Earth and the downwelling infrared radiation from the atmosphere.

EFFECTIVE TOPOGRAPHY Topography as seen by an approaching flow, which may include not only the actual terrain but also cold air masses trapped within or adjacent to the actual topography.

EFFECTIVENESS MONITORING
Documents how well the management practices meet intended objectives for the riparian area.

EFFICIENCY Ratio of useful energy output to total energy input, usually expressed as a percent; effective operation as measured by a comparison of production with cost.

EFFLUENT Wastewater, treated or untreated, that flows out of a treatment plant, sewer, or industrial outfall; outflow from a particular source (i.e., stream that flows from a lake, liquid waste that flows from a factory or sewage-treatment plant).

EFFLUENT GUIDELINES Technical documents which set effluent limitations for given industries and pollutants.

EFFLUENT LIMITATION Maximum amount of quantities, rates, and concentrations a specific substance or characteristic that can be present in effluent discharge without violating water quality standards in receiving waters.

EFFLUENT LIMITATIONS GUIDELINES (ELG) Regulation that establishes national technology-based effluent requirements for a specific industrial category.

EFFLUENT SEEPAGE Diffuse discharge of groundwater to the ground surface.

EFFLUENT STANDARD See EFFLUENT LIMITATION.

EFFLUENT STREAM Any watercourse in which all, or a portion of the water volume came from the Phreatic zone, or zone of saturation by way of groundwater flow, or baseflow.

EGL Energy Grade Line.

EH (REDOX POTENTIAL) Numerical measure of the intensity of oxidation or reducing conditions where a positive potential indicates oxidizing conditions and a negative potential indicates reducing conditions.

EIF Enhanced IFLOWS Format.

EIGHTY EIGHT D (88D) See WSR-88D.

EIS Environmental Impact Statement.

EJECTOR Device used to disperse a chemical solution into water being treated.

EL NIÑO Warming of the ocean current along the coasts of Peru and Ecuador that is generally associated with dramatic changes in the weather patterns of the region where a major El Niño event generally occurs every 3 to 7 years and is associated with changes in the weather patterns worldwide.

ELBOW Pipe fitting having two openings which causes a run of pipe to change direction 90 degrees.

ELECTRIC PLANT (PHYSICAL) Facility containing prime movers, electric generators, and auxiliary equipment for converting mechanical, chemical, and/or fission energy into electric energy.

ELECTRIC POWER INDUSTRY Public, private, and cooperative electric utility systems of the United States taken as a whole.

ELECTRIC POWER SYSTEM Physically connected electric generating, transmission, and distribution facilities operated as a unit under one control.

ELECTRIC UTILITY Corporation, person, agency, authority, or other legal entity or instrumentality that owns and/or operates facilities for the generation, transmission, distribution, or sale of electric energy primarily for use by the public.

ELECTRICAL CONDUCTIVITY Measure of the salt content of water.

ELECTRICAL DEMAND Energy requirement placed upon a utility's generation at a given instant or averaged over any designated period of time that is usually expressed in kilowatts.

ELECTRODIALYSIS Process that uses electrical current applied to permeable membranes to remove minerals from water that is often used to desalinize salty or brackish water.

ELECTROMAGNETIC GEOPHYSICAL METHODS Ways to measure subsurface conductivity via low-frequency electromagnetic induction.

ELECTROMOTIVE FORCE (EMF) Electrical pressure available to cause a flow of current when an electrical circuit is closed.

ELECTROSTATIC PRECIPITATOR (ESP) Device that removes particles from a gas stream (smoke) after combustion occurs.

ELEMENT One of the basic conditions of the atmosphere (e.g., wind, visibility, runway visual range, weather, obscurations, sky condition, temperature and dew point, and pressure).

ELEV Elevation.

ELEVATED CONVECTION Convection occurring within an elevated layer (i.e. a layer in which the lowest portion is based above the Earth's surface.).

ELEVATION Height of a point above a plane of reference.

ELEVATION HEAD Part of hydraulic head which is attributable to the elevation of a measuring point above a given datum.

ELIGIBLE COSTS Construction costs for wastewater treatment works upon which EPA grants are based.

ELONGATE Increase of the vertical diameter of a flexible pipe.

ELSW Elsewhere.

ELY Easterly.

EMAP DATA Environmental monitoring data collected under the auspices of the Environmental Monitoring and Assessment Program.

EMBANKMENT Earth structure the top of which is higher than the adjoining surface; shaped earth or rockfill dam; fill material, usually earth or rock, placed with sloping sides and with a length greater than its height.

EMBANKMENT DAM (FILL DAM) Any dam constructed of excavated natural materials (e.g., earthfill dam, rockfill dam, tailings dam, hydraulic fill dam, rolled fill dam).

EMBDD Embedded.

EMBEDDEDNESS Degree to which gravel-sized and larger particles are surrounded or enclosed by finer-sized particles.

EMC Environmental Modeling Center.

EMERGENCY Condition of a serious nature which develops unexpectedly and endangers the structural integrity of a dam or endangers downstream property and human life and requires immediate action; event that demands a crisis response beyond the scope of any single line agency or service and that presents a threat to a community or larger area; event that can be controlled within the scope of local capabilities.

EMERGENCY (CHEMICAL) Situation created by an accidental release or spill of hazardous chemicals that poses a threat to the safety of workers, residents, the environment, or property.

EMERGENCY ACTION PLAN (EAP)
Predetermined plan of action to be taken to reduce the potential for property damage and loss of life in an area affected by a dam break, excessive spillway release or large flood.

EMERGENCY ACTION PLAN (EAP) EXERCISE
Activity designed to promote emergency preparedness; test or evaluate EAPs, procedures, or facilities; train personnel in emergency management duties; and demonstrate operational capability.

EMERGENCY AND HAZARDOUS CHEMICAL INVENTORY
Annual report by facilities having one or more extremely hazardous substances or hazardous chemicals above certain weight limits.

EMERGENCY BROADCAST SYSTEM
Federally established network of commercial radio stations that voluntarily provide official emergency instructions or directions to the public during an emergency.

EMERGENCY CLASSIFICATION LEVELS
Phased system in which dam operating organizations classify dam safety emergency incidents into response levels according to how severe they are at the time of observation and as to time of occurrence.

EMERGENCY EVACUATION ZONE I
Emergency evacuation zone immediately below a dam and located on both sides of the river or stream.

EMERGENCY EVACUATION ZONE II
Second emergency evacuation zone, beyond emergency evacuation zone I and also located on both sides of the river or stream.

EMERGENCY EVACUATION ZONE III
Outermost emergency evacuation zone, extending beyond emergency evacuation zone II and also located on both sides of the river or stream.

EMERGENCY EVACUATION ZONES
Geographical areas delineated in inundation areas downstream from a dam (hazard generator) that define the potential area of impact and allow prioritizing evacuation activities based on proximity of the populations at risk to the hazard in terms of distance and flood wave travel times.

EMERGENCY EXEMPTION Provision that can grant temporary exemption to a state or another federal agency to allow the use of a pesticide product not registered for that particular use.

EMERGENCY EXERCISE Activity designed to promote emergency preparedness; evaluate emergency operations, policies, plans, procedures, and facilities; train personnel in emergency management and response duties; and demonstrate operational capability.

EMERGENCY GATE Standby or auxiliary gate used when the normal means of water control is not available; first gate in a series of flow controls, remaining open while downstream gates or valves are operating.

EMERGENCY MANAGEMENT
System by which mitigation, preparedness, response, and recovery activities are undertaken to save lives and protect property from all hazards.

EMERGENCY MANAGEMENT AGENCY
Any state or local agency responsible for emergency operations, planning, mitigation, preparedness, response, and recovery for all hazards.

EMERGENCY ON-SCENE COORDINATOR
One employee (at least) either on the premises or on call with the responsibility for coordinating all emergency response measures.

EMERGENCY OPERATIONS CENTER (EOC)
Location of facility where responsible officials gather during an emergency to direct and coordinate emergency operations, to communicate with other jurisdictions and with field emergency forces, and to formulate protective action decisions and recommendations during an emergency.

EMERGENCY OPERATIONS PLAN (EOP)
Plan, usually developed in accord with guidance contained in the Guide for the Development of State and Local Emergency Operations Plans, Civil Preparedness Guide 1-8, and other similar guides.

EMERGENCY PREPAREDNESS PLAN (EPP)
Predecessor to Emergency Action Plan (EAP).

EMERGENCY PROGRAM MANAGER
Individual responsible on a day-to-day basis for a jurisdiction's effort to develop a capability for coordinated response to and recovery from the effects of emergencies and large-scale disasters.

EMERGENCY REMOVAL ACTION
Steps taken to remove contaminated materials that pose imminent threats to local residents (e.g., removal of leaking drums or the excavation of explosive waste); state record of such removals.

EMERGENCY RESERVE FUND
Money reserved or required by contract to be reserved by an operating entity for use in emergency situations involving facilities under the entity's jurisdiction.

EMERGENCY RESPONSE VALUES
Concentrations of chemicals, published by various groups, defining acceptable levels for short-term exposures in emergencies.

EMERGENCY SERVICES Services provided in order to minimize the impact of a flood that is already happening.

EMERGENCY SPILLWAY Spillway which provides for additional safety should emergencies not contemplated by normal design assumptions be encountered (i.e., inoperable outlet works, spillway gates, or spillway structure problems).

EMERGENCY SUSPENSION Suspension of a pesticide product registration due to an imminent hazard.

EMERGENT PLANTS Erect, rooted, herbaceous plants that may be temporarily or permanently flooded at the base but do not tolerate prolonged inundation of the entire plant.

EMERGING ENVIRONMENTAL PROBLEMS Problems that may be new and (or) are becoming known because of better monitoring and use of indicators.

EMERGING FLUX REGION (EFR) Area on the Sun where new magnetic flux is erupting.

EMISSION Pollution discharged into the atmosphere from smokestacks, other vents, and surface areas of commercial or industrial facilities; from residential chimneys, motor vehicle, locomotive, or aircraft exhausts.

EMISSION CAP Limit designed to prevent projected growth in emissions from existing and future stationary sources from eroding any mandated reductions.

EMISSION FACTOR Relationship between the amount of pollution produced and the amount of raw material processed.

EMISSION INVENTORY Listing, by source, of the amount of air pollutants discharged into the atmosphere of a community; used to establish emission standards.

EMISSION STANDARD Maximum amount of air polluting discharge legally allowed from a single source, mobile or stationary.

EMISSIONS TRADING Creation of surplus emission reductions at certain stacks, vents or similar emissions sources and the use of this surplus to meet or redefine pollution requirements applicable to other emissions sources.

EMISSIVITY Ability of a surface to emit radiant energy compared to that of a black body at the same temperature and with the same area.

EML Elevated Mixed Layer.

EMULSIFIER Chemical that aids in suspending one liquid in another that is usually an organic chemical in an aqueous solution.

ENCAPSULATION Treatment of asbestos-containing material with a liquid that covers the surface with a protective coating or embeds fibers in an adhesive matrix to prevent their release into the air.

ENCLOSURE Putting an airtight, impermeable, permanent barrier around asbestos-containing materials to prevent the release of asbestos fibers into the air.

END MORAINE (TERMINAL MORAINE) Ridge of sediment piled at the front edge of a glacier.

END USER Consumer of products for the purpose of recycling that excludes products for re-use or combustion for energy recovery.

ENDANGERED SPECIES Species (animals, birds, fish, plants, or other living organisms) that is in imminent danger of becoming extinct; species or subspecies whose survival is in danger of extinction throughout all or a significant portion of its range.

ENDANGERED SPECIES ACT (ESA) Act that provides a framework for the protection of endangered and threatened species.

ENDANGERMENT Construction, operation, maintenance, conversion, plugging, or abandonment of an injection well, or the performance of other injection activities, by an owner or operator in a manner that allows the movement of fluid containing any contaminant into a USDW, if the presence of that contaminant may cause a violation of any primary drinking water regulations or may adversely affect the health of persons.

ENDANGERMENT ASSESSMENT Study to determine the nature and extent of contamination at a site on the National Priorities List and the risks posed to public health or the environment.

ENDG Ending.

ENDOCRINE SYSTEM Collection of ductless glands in animals that secrete hormones, which influence growth, gender and sexual maturity.

END-OF-THE-PIPE Technologies such as scrubbers on smokestacks and catalytic convertors on automobile tailpipes that reduce emissions of pollutants after they have formed.

ENDRIN Pesticide toxic to freshwater and marine aquatic life that produces adverse health effects in domestic water supplies.

END-USE PRODUCT Pesticide formulation for field or other end use that excludes products used to formulate other pesticide products.

ENERGY Force or action of doing work; measured in terms of the work it is capable of doing.

ENERGY DISSIPATER Device constructed in a waterway to reduce the kinetic energy of fast flowing water; structure which slows fast-moving spillway flows in order to prevent erosion of the stream channel.

ENERGY GRADE LINE (EGL, ENERGY LINE, ENERGY GRADIENT) Line showing the total energy at any point in a pipe; total energy in the flow of the section with reference to a datum line is the sum of the elevation of the pipe centerline, the piezometric height (or pressure head), and the velocity head; slopes (drops) in the direction of flow except where energy is added by mechanical devices; line representing the elevation of the total head of flow is the energy line; slope of the line is known as the energy gradient.

ENERGY HELICITY INDEX Index that incorporates vertical shear and instability, designed for the purpose of forecasting supercell thunderstorms.

ENERGY MANAGEMENT SYSTEM Control system capable of monitoring environmental and system loads and adjusting HVAC operations accordingly in order to conserve energy while maintaining comfort.

ENERGY POLICY ACT (EPACT) Comprehensive federal legislation enacted in 1992 that is resulting in fundamental changes in the electric utility industry by promoting competition in wholesale electricity markets.

ENERGY RECOVERY Obtaining energy from waste through a variety of processes (e.g. combustion).

ENERGY SOURCE Primary source that provides the power that is converted to electricity through chemical, mechanical, or other means where sources include coal, petroleum and petroleum products, gas, water, uranium, wind, sunlight, geothermal, etc.

ENFORCEABLE REQUIREMENTS Conditions or limitations in permits issued that, if violated, could result in the issuance of a compliance order or initiation of a civil or criminal action under federal or applicable state laws.

ENFORCEMENT Federal, state, or local legal actions to obtain compliance with laws, rules, regulations, or agreements and/or obtain penalties or criminal sanctions for violations.

ENFORCEMENT DECISION DOCUMENT (EDD) Document that provides an explanation to the public of EPA's selection of the cleanup alternative at enforcement sites on the National Priorities List.

ENGINEERED CONTROLS Method of managing environmental and health risks by placing a barrier between the contamination and the rest of the site, thus limiting exposure pathways.

ENGINEER'S LEVEL Telescope which is attached to a spirit-tube level, all revolving around a vertical axis and is mounted on a tripod.

ENHANCED INSPECTION AND MAINTENANCE (I&M) Improved automobile inspection and maintenance program - aimed at reducing automobile emissions that contains, at a minimum, more vehicle types and model years, tighter inspection, and better management practices.

ENHANCED V Pattern seen on satellite infrared photographs of thunderstorms, in which a thunderstorm anvil exhibits a V-shaped region of colder cloud tops extending downwind from the thunderstorm core.

ENHANCED WORDING Option used by the SPC in tornado and severe thunderstorm watches when the potential for strong/violent tornadoes, or unusually widespread damaging straight-line winds, is high; strong wording or emphasis used in a zone forecast issued by a National Weather Service Forecast Office highlighting a potential condition (e.g. "some thunderstorms may be severe").

ENHANCEMENT Improvement of a facility beyond its originally designed purpose or condition.

ENHANCEMENT FLOW Improved flows that result in better stream conditions for aquatic, terrestrial, and other resources.

ENHNCD Enhanced.

ENRICHMENT Addition of nutrients (e.g., nitrogen, phosphorus, carbon compounds) from sewage effluent or agricultural runoff to surface water, greatly increases the growth potential for algae and other aquatic plants.

ENSEMBLE Collection of numerical model results that show slightly different possible outcomes.

ENSEMBLE FORECAST Multiple predictions from an ensemble of slightly different initial conditions and/or various versions of models.

ENSEMBLE HYDROLOGIC FORECASTING
Process whereby a continuous hydrologic model is successively executed several times for the same forecast period by use of varied data input scenarios, or a perturbation of a key variable state for each model run.

ENSEMBLES Reference to a set of computer models run under the concept of Ensemble Forecasting: multiple predictions from an ensemble of models with slightly different initial conditions used as input and/or slightly different versions of models.

ENSIGN VALVE Earliest type of Bureau of Reclamation needle valve, operated by water pressure from the forebay, and mounted on the upstream side of the dam.

ENSO Abbreviation for El Niño-Southern Oscillation, a reference to the state of the Southern Oscillation.

ENSO DIAGNOSTIC DISCUSSION
Addresses the current oceanic and atmospheric conditions in the Pacific and the seasonal climate outlook for the following one to three seasons.

ENTEROCOCCUS BACTERIA
Commonly found in the feces of humans and other warmblooded animals; presence in water is an indication of fecal pollution and the possible presence of enteric pathogens (Streptococcus feacalis, Streptococcus "Bacteria.")

ENTR Entire.

ENTRAIN To trap bubbles in water either mechanically through turbulence or chemically through a reaction.

ENTRAINMENT Process by which aquatic organisms, suspended in water, are pulled through a pump or other device; carrying away of bed material produced by erosive action of moving water.

ENTRAINMENT ZONE Shallow region at the top of a convective boundary layer where fluid is entrained into the growing boundary layer from the overlying fluid by the collapse of rising convective plumes or bubbles.

ENTRANCE REGION Region upstream from a wind speed maximum in a jet stream (jet max), in which air is approaching (entering) the region of maximum winds, and therefore is accelerating; results in a vertical circulation that creates divergence in the upper-level winds in the right half of the entrance region (as would be viewed looking along the direction of flow); results in upward motion of air in the right rear quadrant (or right entrance region) of the jet max.

ENTROPY Amount of energy that is not available for work during a certain process.

ENVIRONMENT All biological, chemical, social, and physical factors to which organisms are exposed; sum of all external conditions (surroundings) affecting the life, growth, development and survival of an organism.

ENVIRONMENT CANADA Canadian Federal Government department responsible for issuing weather forecasts and weather warnings in Canada.

ENVIRONMENTAL ANALYSIS
Systematic process for consideration of environment factors in land management actions.

ENVIRONMENTAL ASSESSMENT (EA)
Environmental analysis prepared pursuant to the National Environmental Policy Act (NEPA) to determine whether a federal action would significantly affect the environment and thus require a more detailed environmental impact statement where if not, a finding of no significant impact (FONSI) is written and if so, an environmental impact statement (EIS) is written.

ENVIRONMENTAL AUDIT Independent assessment of the current status of a party's compliance with applicable environmental requirements or of a party's environmental compliance policies, practices, and controls.

ENVIRONMENTAL EQUITY/JUSTICE
Equal protection from environmental hazards for individuals, groups, or communities regardless of race, ethnicity, or economic status.

ENVIRONMENTAL EXPOSURE
Human exposure to pollutants originating from facility emissions; threshold levels are not necessarily surpassed, but low-level chronic pollutant exposure is one of the most common forms of environmental exposure.

ENVIRONMENTAL FATE Destiny of a chemical or biological pollutant after release into the environment.

ENVIRONMENTAL FATE DATA
Data that characterize a pesticide's fate in the ecosystem, considering factors that foster its degradation (light, water, microbes), pathways and resultant products.

ENVIRONMENTAL FRAMEWORK
Natural and human-related features of the land and hydrologic system, such as geology, land use, and habitat, that provide a unifying framework for making comparative assessments of the factors that govern water-quality conditions within and among Study Units.

ENVIRONMENTAL IMPACT STATEMENT (EIS)
Document required of federal agencies by the National Environmental Policy Act (NEPA) for major projects or legislative proposals significantly affecting the environment used to evaluate a range of alternatives when solving the problem would have a significant effect on the human environment; tool for decision making, it describes the positive and negative effects of the undertaking and cites alternative actions.

ENVIRONMENTAL INDICATOR
Measurement, statistic or value that provides a proximate gauge or managerially and scientifically useful evidence of the effects of environmental management programs or of the state or condition of the environment.

ENVIRONMENTAL JUSTICE Fair treatment of people of all races, cultures, incomes, and educational levels with respect to the development and enforcement of environmental laws, regulations, and policies.

ENVIRONMENTAL LAPSE RATE
Rate of decrease of air temperature

with height, usually measured with a radiosonde.

ENVIRONMENTAL LIEN Charge, security, or encumbrance on a property's title to secure payment of cost or debt arising from response actions, cleanup, or other remediation of hazardous substances or petroleum products.

ENVIRONMENTAL MEDIUM Major environmental category that surrounds or contacts humans, animals, plants, and other organisms (e.g., surface water, groundwater, soil or air) and through which chemicals or pollutants move.

ENVIRONMENTAL MONITORING FOR PUBLIC ACCESS AND COMMUNITY TRACKING
Joint federal (EPA, NOAA, and USGS) program to provide timely and effective communication of environmental data and information through improved and updated technology solutions that support timely environmental monitoring reporting, interpreting, and use of the information for the benefit of the public.

ENVIRONMENTAL PROTECTION AGENCY (EPA)
Environmental Protection Agency's mission is to protect human health and to safeguard the natural environment.

ENVIRONMENTAL RESPONSE TEAM
Federal (EPA) experts who provide around-the-clock technical assistance to EPA regional offices and states during all types of hazardous waste site emergencies and spills of hazardous substances.

ENVIRONMENTAL SAMPLE Water sample collected from an aquifer or stream for the purpose of chemical, physical, or biological characterization of the sampled resource.

ENVIRONMENTAL SETTING
Land area characterized by a unique combination of natural and human-related factors, such as row-crop cultivation or glacial-till soils.

ENVIRONMENTAL SITE ASSESSMENT
Process of determining whether contamination is present on a parcel of real property.

ENVIRONMENTAL SUSTAINABILITY
Long-term maintenance of ecosystem components and functions for future generations.

ENVIRONMENTAL TEMPERATURE SOUNDING
Instantaneous or near-instantaneous sounding of temperature as a function of height; sounding or vertical profile is usually obtained by a balloon-borne instrument, but can also be measured using remote sensing equipment.

ENVIRONMENTAL TOBACCO SMOKE
Mixture of smoke from the burning end of a cigarette, pipe, or cigar and smoke exhaled by the smoker.

ENVIRONMENTAL/ECOLOGICAL RISK
Potential for adverse effects on living organisms associated with pollution of the environment by effluents, emissions, wastes, or accidental chemical releases; energy use; or the depletion of natural resources.

EOLIAN (AEOLIAN) Windblown.

EPA Environmental Protection Agency.

EPACT Energy Policy Act.

EPCTG Expecting.

EPHEMERAL STREAM (CREEK) Stream (creek) that flows briefly only in direct response to precipitation; it receives little or no water from springs, melting snow, or other sources; its channel is at all times above the water table.

EPICENTER Focal point on the Earth's surface directly above the origin of a seismic disturbance; point on the Earth's surface located vertically above the point where the first rupture and the first earthquake motion occur.

EPIDEMIOLOGY Investigation of factors contributing to disease or toxic effects in studies of the general population; study of the distribution of disease, or other health-related states and events in human populations, as related to age, sex, occupation, ethnicity, and economic status in order to identify and alleviate health problems and promote better health.

EPIFAUNA Animals which live on the benthos.

EPILIMNION Upper, or top, layer of a lake or reservoir with essentially uniform warmer temperatures; upper layer of water in a thermally stratified lake or reservoir; layer consists of the warmest water and has a fairly constant temperature; layer is readily mixed by wind action.

EPISODE (POLLUTION) Air pollution incident in a given area caused by a concentration of atmospheric pollutants under meteorological conditions that may result in a significant increase in illnesses or deaths; may also describe water pollution events or hazardous material spills.

EPT (RICHNESS) INDEX Total number of distinct taxa within the insect orders Ephemeroptera (mayflies), Plecoptera (stoneflies), and Trichoptera (caddisflies), that are composed primarily of species considered to be relatively intolerant to environmental alterations; index usually decreases with pollution.

EPV Equivalent Potential Vorticity.

EQUAL-WIDTH INCREMENT (EWI) SAMPLE Composite sample of water collected across a section of stream with equal spacing between verticals and equal transit rates within each vertical that yields a representative sample of stream conditions.

EQUILIBRIUM Balanced; state at which the radioactivity of consecutive elements within a radioactive series is neither increasing nor decreasing.

EQUILIBRIUM DEPTH Minimum water depth for the condition of no sediment transport.

EQUILIBRIUM DRAWDOWN Ultimate, constant drawdown for a steady rate of pumped discharge.

EQUILIBRIUM LEVEL (EL) On a sounding, the level above the level of free convection (LFC) at which the temperature of a rising air parcel again equals the temperature of the environment; height of the EL is the height at which thunderstorm updrafts no longer accelerate upward.

EQUILIBRIUM LOAD Amount of sediment that a system can carry for a given discharge without an overall accumulation (deposit) or scour (degradation).

EQUILIBRIUM SURFACE DISCHARGE Steady rate of surface discharge which results from a long-continued, steady rate of net rainfall, with discharge rate equal to net rainfall rate.

EQUILIBRIUM TIME Time when flow conditions become substantially equal to those corresponding to equilibrium discharge or equilibrium drawdown.

EQUINOX Time when the Sun crosses the Earth's equator, making night and day of approximately equal length all over the Earth and occurring about March 21 (the spring or vernal equinox) and September 22 (autumnal equinox).

EQUI-POTENTIAL LINE Line connecting points of equal hydraulic head (potential); line, in a field of flow, such that the total head is the same for all points on the line, and therefore the direction of flow is perpendicular to the line at all points; set of such lines provides a contour map of a potentiometric surface.

EQUIVALENCY Any body of procedures and techniques of sample collection and (or) analysis for a parameter of interest that has been demonstrated in specific cases to produce results not statistically different to those obtained from a reference method.

EQUIVALENT DIAMETER (EQUIVALENT SIZE) Diameter of a hypothetical sphere composed of material having the same specific gravity as that of the actual soil particle and of such size that it will settle in a given liquid at the same terminal velocity as the actual soil particle.

EQUIVALENT FLUID Hypothetical fluid having a unit weight such that it will produce a pressure against a lateral support presumed to be equivalent to that produced by the actual soil.

EQUIVALENT METHOD Any method of sampling and analyzing for air pollution which has been demonstrated to be, under specific conditions, an acceptable alternative to normally used reference methods.

EQUIVALENT POTENTIAL TEMPERATURE Temperature a parcel at a specific pressure level and temperature would have if it were raised to 0 mb, condensing all moisture from the parcel, and then lowered to 1000 mb.

ERLY Early.

ERN Eastern.

ERODE To wear away or remove the land surface by wind, water, or other agents.

EROSION Gradual wearing away of surface (soil, rock, bank, streambed, embankment, or other surface) by running water, glaciers, winds, and waves can be subdivided into three process: Corrasion, Corrosion, and Transportation; concrete surface disturbance caused by cavitation, abrasion from moving particles in water, impact of pedestrian or vehicular traffic, or impact of ice floes; surface displacement of soil caused by weathering, dissolution, abrasion, or other transporting; gradual wearing away of material as a result of abrasive action; process whereby materials of the Earth's crust are loosened, dissolved, or worn away and simultaneously moved from one place to another; wearing away of land surface by wind or water, intensified by land-clearing practices related to farming, residential or industrial development, road building, or logging.

ERRATIC Boulder transported by a glacier and left behind when the ice melted.

ERUPTIVE Solar activity levels with at least one radio event (10 cm) and several chromospheric events per day (Class C Flares).

ERUPTIVE PROMINENCE ON LIMB (EPL) Solar prominence that becomes activated and is seen to ascend from the Sun.

ESA Endangered Species Act (of 1973).

ESCAPEMENT Unharvested spawning stocks that return to the streams; cliff or steep slope that separates two level or gently sloping areas; cliff or steep slope edging higher land.

ESCHERICHIA COLI (E. COLI) Bacteria present in the intestine and feces of warmblooded animals; member species of the fecal coliform group of indicator bacteria.; concentrations are expressed as number of colonies per 100 mL of sample.

ESP Extended Streamflow Prediction.

ESPINIT ESP Initialization Program.

ESSENTIAL ELEMENT Structural or geologic feature or an equipment item whose failure under the particular loading condition or set of circumstances being considered would create a dam safety deficiency; equipment item or procedure required for safe operation of the dam or reservoir.

EST Eastern Standard Time.

ESTABLISHED TREATMENT TECHNOLOGIES Technologies for which cost and performance data are readily available.

ESTHETICS Emotional judgment about what is beautiful or pleasing.

ESTIMATED (E) VALUE OF A CONCENTRATION Reported when an analyte is detected and all criteria for a positive result are met.

ESTIMATED ENVIRONMENTAL CONCENTRATION Estimated pesticide concentration in an ecosystem.

ESTUARINE HABITAT Tidal habitats and adjacent tidal wetlands that are usually semi-enclosed by land but have open, partly obstructed, or sporadic access to the open ocean and in which ocean water is at least occasionally diluted by freshwater runoff from the land .

ESTUARINE WETLANDS Tidal wetlands in low-wave-energy environments where the salinity of the water is greater than 0.5 part per thousand and is variable owing to evaporation and the mixing of seawater and freshwater; tidal wetlands of coastal rivers and embayments, salty tidal marshes, mangrove swamps, and tidal flats.

ESTUARY Area of mixing of fresh water and salt water where a river flows into an ocean and where tidal effects are evident; thin zone along a coastline where freshwater systems and rivers meet and mix with a salty ocean (such as a bay, mouth of a river, salt marsh, lagoon); an arm of the ocean at the lower end of a river.

ESTURINE WATERS Deepwater tidal habitats and tidal wetlands that are usually enclosed by land but have access to the ocean and are at least occasionally diluted by freshwater runoff from the land (such as bays, mouths of rivers, salt marshes, lagoons).

ESTURINE ZONE Area near the coastline that consists of estuaries and coastal saltwater wetlands.

ET See EVAPOTRANSPIRATION.

ETA Estimated Time of Arrival.

ETA MODEL Now referred to as North American Meso (NAM) is one of the operational numerical forecast models run at NCEP where Eta is run four times daily, with forecast output out to 84 hours.

ETENTION TIME Theoretical calculated time required for a small amount of water to pass through a tank at a given rate of flow; actual time that a small amount of water is in a settling basin, flocculating basin, or rapid-mix chamber; length of time water will be held in a storage reservoir before being used.

ETHANOL Alternative automotive fuel derived from grain and corn; usually blended with gasoline to form gasohol.

ETHYLENE DIBROMIDE (EDB) Chemical used as an agricultural fumigant and in certain industrial processes; extremely toxic and found to be a carcinogen in laboratory animals, EDB has been banned for most agricultural uses in the United States.

EUGLENOIDS (EUGLENOPHYTA) Group of algae that usually are free-swimming and rarely creeping; have the ability to grow either photosynthetically in the light or heterotrophically in the dark.

EUTROPHIC Nutrient enrichment of a body of water that contains more organic matter than existing biological oxidization processes can consume; body of water that has become, either naturally or by pollution, rich in nutrients and often seasonally deficient in dissolved oxygen; reservoirs and lakes that are rich in nutrients and very productive in terms of aquatic animal and plant life.

EUTROPHIC LAKES Shallow, murky bodies of water with concentrations of plant nutrients causing excessive production of algae.

EUTROPHICATION Process by which water becomes enriched with plant nutrients, most commonly phosphorus and nitrogen; process where more organic matter is produced than existing biological oxidization processes can consume; increase in the nutrient levels of a lake or other body of water; slow aging process during which a lake, estuary, or bay evolves into a bog or marsh and eventually disappears.

EVACUATION Fifth of five Early Warning System components consisting of the plans, personnel, equipment, and facilities needed to move the population at risk to safety; involves taking protective actions to leave an area of risk until the hazard has passed and the area is safe for return.

EVACUATION WARNING Public warning message that local officials would issue following declaration of Response Level III by personnel of the dam operating organization; intended to notify the population at risk to evacuate flood inundation areas.

EVALUATION REPORT (REPORT FOR EXAMINATION NO. 1, FIRST FORMAL EXAMINATION) Report of the initial SEED onsite examination.

EVAPORATION Process by which water is changed from the liquid or the solid state into the vapor state; water vapor losses from water surfaces, sprinkler irrigation, and other related factors; loss of water to the atmosphere; water from land areas, bodies of water, and all other moist surfaces is absorbed into the atmosphere as a vapor; vaporization and sublimation that takes place at a temperature below the boiling point.

EVAPORATION DEMAND Maximum potential evaporation generally determined using an evaporation pan.

EVAPORATION OPPORTUNITY (RELATIVE EVAPORATION) Ratio of the rate of evaporation from a land or water surface in contact with the atmosphere, to the evaporability under existing atmospheric conditions.

EVAPORATION PAN Pan used to hold water during observations for the determination of the quantity of evaporation at a given location; US National Weather Service class A pan is 4 feet in diameter, 10 inches deep, initially filled with 8 inches of water, set up on a timber grillage so that the top rim is about 16 inches from the ground.

EVAPORATION PONDS Areas where sewage sludge is dumped and dried.

EVAPORATION RATE Quantity of water, expressed in terms of depth of liquid water, which is evaporated from a given surface per unit of time; rate at which a chemical changes into a vapor.

EVAPORATION, TOTAL Sum of water lost from a given land area during any specific time by transpiration from vegetation and building of plant tissue; by evaporation from water surfaces, moist soil, and snow; and by interception.

EVAPORATION-MIXING FOG Fog that forms when the evaporation of water raises the dew point of the adjacent air.

EVAPORATIVITY Potential rate of evaporation.

EVAPORIMETER Instrument which measures the evaporation rate of water into the atmosphere.

EVAPORITE MINERALS (DEPOSITS) Minerals or deposits of minerals formed by evaporation of water containing salts that are common in arid climates.

EVAPOTRANSPIRATION (ET) Sum of evaporation and transpiration; collective term that includes water lost (discharged to the atmosphere) through evaporation from the soil and surface-water bodies and by plant transpiration in a specific time period.

EVE Evening.

EVENT-BASED MODEL Model that simulates some hydrologic response to a precipitation event.

EWD Eastward.

EWS Early Warning System.

EXAMINATION REPORT Report that documents the condition of the facility during the examination, operation and maintenance activities accomplished since the last examination, and recommendations necessary for the continued safe and efficient operation of the facility.

EXCAVATION Action or process of excavating (to dig or remove earth).

EXCEEDANCE Violation of the pollutant levels permitted by environmental protection standards.

EXCEEDANCE PROBABILITY Probability that an event selected at random will exceed a specified magnitude.

EXCESS (EFFECTIVE) PRECIPITATION Precipitation in excess of infiltration capacity, evaporation, transpiration, and other losses.

EXCESS CAPACITY Power generation capacity available on a short-term basis that exceeds the firm energy on a long-term contract offered to an electricity customer.

EXCESS LAND Irrigable land, other than exempt land, owned by any landowner in excess of the maximum ownership entitlement under applicable provision of Reclamation law.

EXCESS RAIN Effective rainfall in excess of infiltration capacity.

EXCESSIVE HEAT Combination of high temperatures (significantly above normal) and high humidities.

EXCESSIVE HEAT OUTLOOK Combination of temperature and humidity over a certain number of days, is designed to provide an indication of areas of the country where people and animals may need to take precautions against the heat during May to November.

EXCESSIVE HEAT WARNING Issued within 12 hours of the onset of the following criteria: heat index of at least 105 °F for more than 3 hours per day for 2 consecutive days, or heat index more than 115 °F for any period of time.

EXCESSIVE HEAT WATCH Issued when heat indices in excess of 105 °F (41 °C) during the day combined with nighttime low temperatures of 80 °F (27 °C) or higher are forecast to occur for two consecutive days.

EXCHANGE CAPACITY Capacity to exchange ions as measured by the quantity of exchangeable ions in a soil or rock.

EXCLD Exclude.

EXCLUSION One of several situations that permit a Local Education Agency (LEA) to delete one or more of the items required by the Asbestos Hazard Emergency Response Act (AHERA).

EXCLUSIONARY ORDINANCE Zoning that excludes classes of persons or businesses from a particular neighborhood or area.

EXCLUSIVE FLOOD CONTROL CAPACITY Reservoir capacity assigned to the sole purpose of regulating flood inflows to reduce flood damage downstream.

EXCLUSIVE FLOOD CONTROL STORAGE CAPACITY Space in a reservoir reserved for the sole purpose of regulating flood inflows to abate flood damage.

EXEMPT LAND Irrigation land in a district to which the acreage limitation and pricing provisions do not apply.

EXEMPT SOLVENT Specific organic compounds not subject to requirements of regulation because they are deemed to be of negligible photochemical reactivity.

EXEMPTED AQUIFER Underground bodies of water defined in the Underground Injection Control (UIC) program as aquifers that are potential sources of drinking water though not being used as such, and thus exempted from regulations barring underground injection activities.

EXEMPTION Federal or state permission for a water system not to meet a certain drinking water standard involving Maximum Contaminant Level (MCL), treatment technique, or both, if the system cannot comply due to compelling economic or other factors, or because the system was in operation before the requirement or MCL was instituted and the exemption will not create a public health risk.

EXERCISE Simulated emergency condition carried out for the purpose of testing and evaluating the readiness of a community or organization to handle a particular type of emergency.

EXISTENCE VALUE Value people place on simply knowing an area or feature continues to exist in a particular condition.

EXISTING GROUND Earth's surface as it is prior to any work.

EXISTING USES Uses actually attained in the water body on or after November 28, 1975, whether or not they are included in the water quality standards.

EXIT CHANNEL Outlet channel.

EXIT CONFERENCE Discussion following a facility review examination involving examination team members and interested representatives of the water uses, project, and region.

EXIT REGION Region downstream from a wind speed maximum in a jet stream (jet max), in which air is moving away from the region of maximum winds, and therefore is decelerating; results in divergence in the upper-level winds in the left half of the exit region (as would be viewed looking along the direction of flow); results in upward motion of air in the left front quadrant (or left exit region) of the jet max.

EXOSPHERE Upper most layer of the Earth's atmosphere; the only layer where atmospheric gases can escape into outer space.

EXOTIC SPECIES Species (plants or animals) that is not indigenous (non-native) to a region.

EXPANSION Increase in volume of a soil mass.

EXPANSION JOINT Separation between adjoining parts of a concrete structure which is provided to allow small relative movements, such as those caused by temperature changes, to occur independently; flexible filler is provided in the joint, reinforcement does not pass through the joint; joint that permits pipe to move as a result of expansion.

EXPERIMENTAL PRODUCT Experimental product is in the final stages of testing and evaluation.

EXPERIMENTAL USE PERMIT Obtained by manufacturers for testing new pesticides or uses thereof whenever they conduct experimental field studies to support registration on 10 acres or more of land or one acre or more of water.

EXPERIMENTAL WELLS Class V wells that are used to test new technologies.

EXPLOIT Excavate in such a manner as to utilize material in a particular vein or layer, and waste or avoid surrounding material.

EXPLOSIVE Chemical that causes a sudden, almost instantaneous release of pressure, gas, and heat when subjected to sudden shock, pressure, or high temperatures.

EXPLOSIVE DEEPENING Decrease in the minimum sea-level pressure of a tropical cyclone of 2.5 mb/hr for at least 12 hours or 5 mb/hr for at least six hours.

EXPLOSIVE LIMITS Amounts of vapor in the air that form explosive mixtures; expressed as lower and upper limits and give the range of vapor concentrations in air that will explode if an ignition source is present.

EXPORTS Municipal solid waste and recyclables transported outside the state or locality where they originated.

EXPOSURE Amount of radiation or pollutant present in a given environment that represents a potential health threat to living organisms.

EXPOSURE ASSESSMENT Identifying the pathways by which toxicants may reach individuals, estimating how much of a chemical an individual is likely to be exposed to, and estimating the number likely to be exposed.

EXPOSURE CONCENTRATION Concentration of a chemical or other pollutant representing a health threat in a given environment.

EXPOSURE INDICATOR Characteristic of the environment measured to provide evidence of the occurrence or magnitude of a response indicator's exposure to a chemical or biological stress.

EXPOSURE LEVEL Amount (concentration) of a chemical at the absorptive surfaces of an organism.

EXPOSURE PATHWAY Path from sources of pollutants via, soil, water, or food to man and other species or settings.

EXPOSURE ROUTE Way a chemical or pollutant enters an organism after contact (i.e., ingestion, inhalation, or dermal absorption).

EXPOSURE-RESPONSE RELATIONSHIP Relationship between exposure level and the incidence of adverse effects.

EXTD Extend/Extended.

EXTENDED FORECAST DISCUSSION Issued once a day around 2 PM EST

(3 PM EDT) and is primarily intended to provide insight into guidance forecasts for the 3- to 5-day forecast period.

EXTENSIVE GARDEN Gardens that have thinner soil depths and require less management and less structural support than intensive gardens; do not require artificial irrigation; plants are low-maintenance, hardy species that do not have demanding habitat requirements.

EXTIRPATED SPECIES Species which has become extinct in a given area.

EXTRACTABLE ORGANIC HALIDES (EOX) Organic compounds that contain halogen atoms such as chlorine; semi-volatile and extractable by ethyl acetate from air-dried streambed sediment; concentration is reported as micrograms of chlorine per gram of the dry weight of the streambed sediment.

EXTRACTION PROCEDURE (EP TOXIC) Determining toxicity by a procedure which simulates leaching; if a certain concentration of a toxic substance can be leached from a waste, that waste is considered hazardous.

EXTRACTION WELL Discharge well used to remove groundwater or air.

EXTRADOS Curved upstream surface of horizontal arch elements in an arch dam.

EXTRAPOLATION Estimation of unknown values by extending or projecting from known values.

EXTRATERRESTRIAL RADIATION Theoretically-calculated radiation flux from the Sun at the top of the atmosphere, before losses by atmospheric absorption.

EXTRATROPICAL Term used in advisories and tropical summaries to indicate that a cyclone has lost its "tropical" characteristics.

EXTRATROPICAL CYCLONE Cyclone in the middle and high latitudes often being 2000 kilometers in diameter and usually containing a cold front that extends toward the equator for hundreds of kilometers.

EXTRATROPICAL LOW Low pressure center which refers to a migratory frontal cyclone of middle and higher latitudes.

EXTREME ULTRAVIOLET (EUV) Portion of the electromagnetic spectrum from approximately 100 to 1000 angstroms.

EXTREME WIND WARNING (EWW) Informs the public of the need to take immediate shelter in an interior portion of a well-built structure due to the onset of extreme tropical cyclone winds.

EXTREMELY HAZARDOUS SUBSTANCE (EHS) Hundreds of hazardous chemicals identified as toxic to provide a focus for state and local emergency planning activities.

EXTREMELY LOW FREQUENCY (ELF) Portion of the radio frequency spectrum from 30 to 3000 hertz.

EXTRM Extreme.

EXTSV Extensive.

EYE Relatively calm center in a hurricane that is more than one half surrounded by wall cloud.

EYE WALL Organized band of cumuliform clouds that immediately surrounds the center (eye) of a hurricane; fiercest winds and most intense rainfall typically occur near the eye wall.

F

F (FUJITA) SCALE System of rating the intensity of tornadoes.

F CORONA Of the white-light corona (that is, the corona seen by the eye at a total solar eclipse), that portion which is caused by sunlight scattered or reflected by solid particles (dust) in inter-planetary space.

F REGION Upper layer of the ionosphere, approximately 120 to 1500 km in altitude; F region is subdivided into the F1 and F2 regions where F2 region is the most dense and peaks at altitudes between 200 and 600 km and F1 region is a smaller peak in electron density, which forms at lower altitudes in the daytime.

FA Forecast Area.

FAA Federal Aviation Administration.

FABRIC FILTER Cloth device that catches dust particles from industrial emissions.

FACE Exposed surface of dam materials (earth, rockfill, or concrete), upstream and downstream; external surface of a structure, such as the surface of a dam; vertical surface of rock exposed by blasting or excavating; cutting end of a drill hole.

FACILITIES Structures associated with irrigation projects, municipal and industrial water systems, power generation facilities, including all storage, conveyance, distribution, and drainage systems.

FACILITIES PLANS Plans and studies related to the construction of treatment works necessary to comply with the Clean Water Act or RCRA.

FACILITY EMERGENCY COORDINATOR Representative of a facility covered by environmental law (e.g., a chemical plant) who participates in the emergency reporting process with the Local Emergency Planning Committee (LEPC).

FACING With reference to a wall or concrete dam, a coating of a different material, masonry or brick, for architectural or protection purposes (e.g., stonework facing, brickwork facing); with reference to an embankment dam, an impervious coating or face on the upstream slope of the dam.

FACT SHEET Document summarizes the principal facts and the significant factual, legal, methodological and policy questions considered in preparing the draft permit and tells how the public may comment; document that must be prepared for all draft individual permits for NPDES major dischargers, NPDES general permits, NPDES permits that contain variances, NPDES permits that contain sewage sludge land application plans and several other classes of permittees.

FACTOR OF SAFETY Ratio of the ultimate strength of the material to the allowable or working stress.

FACULA Bright region of the photosphere seen in white light, seldom visible except near the solar limb.

FACULTATIVE BACTERIA Bacteria that can live under aerobic or anaerobic conditions.

FAHRENHEIT (°F) Standard scale used to measure temperature in the United States where the freezing point of water is 32 °F and the boiling point is 212 °F; to convert a Celsius temperature to Fahrenheit, multiply it by $9/5$ and then add 32.

°F See FAHRENHEIT.

FAILURE Incident resulting in the uncontrolled release of water from a dam; destroyed and made useless, ceases to function as a dam; more severe and hazardous than a breach.

FAILURE MODE Physically plausible process for dam failure resulting from an existing inadequacy or defect related to a natural foundation condition, the dam or appurtenant structures design, the construction, the materials incorporated, the operations and maintenance, or aging process, which can lead to an uncontrolled release of the reservoir.

FAILURE POTENTIAL ASSESSMENT Judgment of the potential for failure of an essential element within the expected life of the project where five terms are used to describe the assessment: negligible, low, moderate, high, and urgent.

FAIR Usually used at night to describe less than $3/8$ opaque clouds, no precipitation, no extremes of visibility, temperature or winds; describes generally pleasant weather conditions.

FALL Amount of slope given to horizontal runs of pipe; season of the year which is the transition period from summer to winter occurring as the Sun approaches the winter solstice.

FALL LINE Imaginary line marking the boundary between the ancient, resistant crystalline rocks of the Piedmont province of the Appalachian Mountains, and the younger, softer sediments of the Atlantic Coastal Plain province in the Eastern United States; along rivers, this line commonly is reflected by waterfalls; a skiing term, indicating the line of steepest descent of a slope.

FALL VELOCITY Falling or settling rate of a particle in a given medium.

FALL WIND Strong, cold, downslope wind.

FALLING LIMB Portion of a hydrograph where runoff is decreasing.

FALLOW Cropland, tilled or untilled, allowed to lie idle during the whole or greater part of the growing season.

FALLSTREAK (VIGRA) Streaks or wisps of precipitation falling from a cloud but evaporating before reaching the ground.

FALSE SET See HORSEHEAD.

FANNING Pattern of plume dispersion in a stable atmosphere, in which the plume fans out in the horizontal and meanders about at a fixed height.

FARM LOSS (WATER) Water delivered to a farm which is not made available to the crop to be irrigated.

FASTST Fastest.

FATAL FLAW Any problem, lack, or conflict (real or perceived) that will destroy a solution or process; negative effect that cannot be offset by any degree of benefits from other factors.

FATALITY RATE Multiplication factor based on the estimated severity of the flood, potential warning times, and the judgment related to the understanding of the flood severity.

FATHOM Unit of water depth equal to 6 feet.

FAULT Fracture or fracture zone in the Earth along which there has been displacement of the two sides relative to one another and which is parallel to the fracture; break in rocks along which movement has occurred; shear with significant continuity that can be correlated between observation points.

FAULT, ACTIVE Fault which, because of its present tectonic setting, can undergo movement from time to time in the immediate geologic future.

FAULT, CAPABLE Active fault that is judged capable of producing macroearthquakes; structural relationship to a capable fault such that movement on one fault could be reasonably expected to cause movement on the other; established patterns of microseismicity that define a fault, with historic macroseismicity that can reasonably be associated with the fault.

FAULT-BLOCK (HORST) Uplifted section of rock bounded on both sides by faults.

FAULTING Movement which produces relative displacement along a fracture in rock.

FAUNA All animals associated with a given habitat, country, area, or period.

FAWS Flight Advisory Weather Service.

FCST Forecast; NWSRFS Forecast Program to produce operational forecasts.

FDA ACTION LEVEL Regulatory level recommended by the U.S. Environmental Protection Agency (EPA) for enforcement by the Food and Drug Administration (FDA) when pesticide residues occur in food commodities for reasons other than the direct application of the pesticide.

FEASIBILITY Determination that something can be done.

FEASIBILITY ESTIMATE Estimate used for determining the economic feasibility of a project, the probable sequence and cost for construction of a project, and as a guide in the choice between alternative locations or plans.

FEASIBILITY STUDY Analysis of the practicability of a proposal; required in some planning processes to examine the situation and determine if a workable solution can be developed and implemented usually recommends selection of a cost-effective alternative; small-scale investigation of a problem to ascertain whether a proposed research approach is likely to provide useful data.

FEATHER To blend the edge of new material smoothly into the old surface.

FEATURED (OR SPECIES EMPHASIS)
Species of high public interest and demand.

FECAL BACTERIA Bacteria found in the intestinal tracts of mammals; microscopic single-celled organisms (primarily fecal coliforms and fecal streptococci) found in the wastes of warm-blooded animals; presence in water is used to assess the sanitary quality of water for body-contact recreation or for consumption; presence indicates contamination by the wastes of warm-blooded animals and the possible presence of pathogenic (disease producing) organisms.

FECAL COLIFORM See FECAL BACTERIA.

FECAL STREPTOCOCCAL BACTERIA
Present in the intestines of warm-blooded animals and are ubiquitous in the environment; characterized as gram-positive, cocci bacteria that are capable of growth in brain-heart infusion broth.

FEDERAL EMERGENCY MANAGEMENT AGENCY (FEMA)
Federal agency responsible for enforcing the legislation for disaster and emergency planning and response; responsibilities in hazard mitigation; administers the National Flood Insurance Program (FIMP).

FEDERAL ENERGY REGULATORY COMMISSION (FERC)
Primary responsibility (established in 1977, replacing the Federal Power Commission) of ensuring the Nation's consumers adequate energy supplies at just and reasonable rates and providing regulatory incentives for increased productivity, efficiency, and competition.

FEDERAL FACILITY Any buildings, installations, structures, land, public works, equipment, aircraft, vessels, and other vehicles and property, owned by, or constructed or manufactured for the purpose of leasing to, the federal government.

FEDERAL IMPLEMENTATION PLAN
Plan to achieve attainment of air quality standards, used when a state is unable to develop an adequate plan.

FEDERAL MOTOR VEHICLE CONTROL PROGRAM
All federal actions aimed at controlling pollution from motor vehicles by such efforts as establishing and enforcing tailpipe and evaporative emission standards for new vehicles, testing methods development, and guidance to states operating inspection and maintenance programs.

FEDERAL ORGANIZATIONS
Agencies, departments, or their components of the Federal Government.

FEDERAL SNOW SAMPLER
Snow sampler consisting of five or more sections of sampling tubes, one which has a steel cutter on the end and the combined snowpack measuring depth is 150 inches.

FEDERAL WATER POLLUTION CONTROL ACT OF 1948 Establishes the basic structure for regulating discharges of pollutants into the waters of the United States and regulating quality standards for surface waters.

FEEDER BANDS Lines or bands of low-level clouds that move (feed) into the updraft region of a thunderstorm, usually from the east through south (i.e. parallel to the inflow); spiral-shaped bands of convection surrounding, and moving toward, the center of a tropical cyclone.

FEEDLOT Confined area for the controlled feeding of animals; tends to concentrate large amounts of animal waste that cannot be absorbed by the soil and, hence, may be carried to nearby streams or lakes by rainfall runoff.

FELDSPAR Group of light-colored minerals often found as crystals in intrusive igneous rocks; most common rock-forming mineral.

FEMA See FEDERAL EMERGENCY MANAGEMENT AGENCY.

FEN Peat-accumulating wetland that generally receives water from surface runoff and (or) seepage from mineral soils in addition to direct precipitation; less acidic than bogs, deriving most of their water from groundwater rich in calcium and magnesium.

FERC Federal Energy Regulatory Commission.

FERREL CELL Middle latitude cell marked by sinking motion near 30 degrees and rising motion near 60 degrees latitude.

FERROUS METALS Magnetic metals derived from iron or steel.

FERTILIZER Any of a large number of natural or synthetic materials, including manure and nitrogen, phosphorus, and potassium compounds, spread on or worked into soil to increase its fertility.

FETCH Straight line distance across a body of water subject to wind forces; distance which wind passes over water; area of the sea surface over which a wind with constant direction; effective distance which waves have traversed in open water, from their point of origin to the point where they break.

FEW National Weather Service convective precipitation descriptor for a 10 percent chance of measurable precipitation (0.01 inch); used interchangeably with isolated.

FEW CLOUDS Official sky cover classification for aviation weather observations, descriptive of a sky cover of $^1/_8$ to $^2/_8$.

FFG Flash Flood Guidance.

FIBRIL Linear pattern in the H-alpha chromosphere of the Sun, as seen through an H-alpha filter, occurring near strong sunspots and plage or in filament channels.

FIELD (MOISTURE) CAPACITY Amount of water held in soil against the pull of gravity; depth of water retained in the soil after ample irrigation or heavy rain when the rate of downward movement has substantially decreased.

FIELD CHARACTERIZATION Review of historical, on-and off-site, as well as surface and subsurface data and the collection of new data to meet project objectives.

FIELD MOISTURE DEFICIENCY Amount of water required to restore the moisture in soil to field moisture capacity.

FIELD TILE Short lengths of clay pipe that are installed as subsurface drains.

FIFRA PESTICIDE INGREDIENT Ingredient of a pesticide that must be registered with EPA under the Federal Insecticide, Fungicide, and Rodenticide Act(FIFRA).

FILAMENT Mass of gas suspended over the photosphere by magnetic fields and seen as dark lines threaded over the solar disk.

FILAMENT CHANNEL Broad pattern of fibrils in the chromosphere, marking where a filament may soon form or where a filament recently disappeared.

FILE TRANSFER PROTOCOL (FTP) Method of data transfer.

FILL Manmade deposits of natural soils or the process of the depositing; manmade deposits of natural soils or rock products and waste materials designed and installed in such a manner as to provide drainage, yet prevent the movement of soil particles due to flowing water; earth or broken rock structure or embankment; soil or loose rock used to raise a grade; soil that has no value except as bulk.

FILL DAM Any dam constructed of excavated natural materials or of industrial wastes.

FILLING Depositing dirt, mud or other materials into aquatic areas to create more dry land, usually for agricultural or commercial development purposes, often with ruinous ecological consequences; opposite of deepening; general increase in the central pressure of a low pressure system.

FILTER (FILTER ZONE) One or more layers of granular material which is incorporated in an embankment dam and is graded (either naturally or by selection) to allow seepage through or within the layers while preventing the migration of material from adjacent zones; layer or combination of layers of pervious materials designed and installed in such a manner as to provide drainage, yet prevent the movement of soil particles due to flowing water.

FILTER CAKE (MUD CAKE) Deposit of mud on the walls of a drill hole.

FILTER STRIP Strip or area of vegetation used for removing sediment, organic matter, and other pollutants from runoff and wastewater.

FILTERED Pertains to constituents in a water sample passed through a filter of specified pore diameter, most commonly 0.45 micrometer or less for inorganic analytes and 0.7 micrometer for organic analytes.

FILTERED, RECOVERABLE Amount of a given constituent that is in solution after the part of a representative water suspended sediment sample that has passed through a filter has been extracted.

FILTRATE Liquid that has been passed through a filter.

FILTRATION Treatment process, under the control of qualified operators, for removing solid (particulate) matter from water by means of porous media such as sand or a man-made filter; often used to remove particles that contain pathogens.

FINAL ACUTE VALUE (FAV) Estimate of the concentration of the toxicant corresponding to a cumulative probability of 0.05 in the acute toxicity values for all genera for which acceptable acute tests have been conducted on the toxicant.

FINANCIAL ANALYSIS Procedure that considers only tangible factors when evaluating various alternatives.

FINANCIAL ASSURANCE FOR CLOSURE Documentation or proof that an owner or operator of a facility such as a landfill or other waste repository is capable of paying the projected costs of closing the facility and monitoring it afterwards.

FINDING OF NO SIGNIFICANT IMPACT (FNSI) Document prepared by a federal agency showing why a proposed action would not have a significant impact on the environment and thus would not require preparation of an Environmental Impact Statement.

FINES Portion of a soil finer that a No. 200 U.S. Standard sieve; clay or silt particles in soil.

FINGER DRAINS Series of parallel drains of narrow width (instead of a continuous drainage blanket) draining to the downstream toe of the embankment dam.

FINISHED GRADE Elevation or surface of the earth after all earthwork has been completed (also finish grade); final grade required by specifications.

FINISHED WATER Water that has passed through all the processes in a water treatment plant and is ready to be delivered to consumers.

FINITE DIFFERENCE METHOD (FDM) Numerical model that uses a mathematical technique to obtain an approximate solution to the governing partial differential equation (in space and time) on a regular spaced mesh-or grid-points.

FINITE ELEMENT METHOD (FEM) Numerical model that uses a mathematical technique to obtain an approximate solution to the governing partial differential equation (in space and time) on an irregular spaced mesh-or grid-points.

FIRE ALGAE (PYRRHOPHYTA) Free-swimming unicells characterized by a red pigment spot.

FIRE WIND Thermally driven wind blowing radially inward toward a fire, produced by horizontal temperature differences between the heated air above the fire and the surrounding cooler free atmosphere.

FIREBRAND Any source of heat, natural or man-made, capable of igniting wildland fuels; flaming or glowing fuel particles that can be carried naturally by wind, convection currents, or gravity into unburned fuels.

FIRM ENERGY (POWER) Non-interruptible energy and power guaranteed by the supplier to be available at all times, except for uncontrollable circumstances.

FIRM YIELD Maximum quantity of water that can be guaranteed with some specified degree of confidence during a specific critical period.

FIRN (FIRN SNOW) Old snow on the top of glaciers, granular and compact but not yet converted into ice; transitional stage between snow and ice.

FIRN LINE Line separating the accumulation area from the ablation area; during the melting season, the uppermost level where fresh snow on a glacier's surface retreats.

FIRST DRAW Water that comes out when a tap is first opened.

FIRST LAW OF THERMODYNAMICS
Law of physics that states that the heat absorbed by a system either raises the internal energy of the system or does work on the environment.

FISCHER & PORTER PUNCHED TAPE RECORDER GAUGE
Precipitation gauge which converts weight into a code disk position.

FISH & WILDLIFE SERVICE (FWS)
Principal federal agency responsible for conserving, protecting and enhancing fish, wildlife and plants and their habitats for the continuing benefit of the American people; agency within the U.S. Department of the Interior.

FISH AND WILDLIFE Any nondomesticated member of the animal kingdom that includes, without limitation, any mammal, fish, bird, amphibian, reptile, mollusk, crustacean, arthropod, or other invertebrate and that includes any part, product, egg, or offspring thereof or the dead body or parts thereof.

FISH COMMUNITY See COMMUNITY.

FISH LADDER (FISHWAY) Inclined trough which carries water from above to below a dam so that fish can easily swim upstream.

FISH WEIR Type of fish ladder.

FISHERY Aquatic region in which a certain species of fish lives.

FISHING Operation of recovering an object left or dropped in a drill hole.

FIVE HUNDRED HPA (500 HPA) See 500 MB.

FIVE HUNDRED MB (500 MB)
Pressure surface (geopotential height) in the troposphere equivalent to about 18,000 feet above sea level; level of the atmosphere at which half the mass of the atmosphere lies above and half below, as measured in pressure units.

FIX A SAMPLE Adding chemicals in the field that prevent water quality indicators of interest in the sample from changing before laboratory measurements are made.

FIXED AMOUNT-FREQUENCY SCHEDULING
Method of irrigation scheduling that involves water delivery at a fixed rate or a fixed volume and at constant intervals.

FIXED BED MODEL Model in which the bed and side materials are non-erodible and deposition does not occur.

FIXED CONE VALVE (HOWELL BUNGER VALVE)
Cylinder gate mounted with the axis horizontal.

FIXED SITES Monitoring sites in NAWQA Study Units at which the most comprehensive suites of data are collected.

FIXED WHEEL GATE (FIXED ROLLER GATE, FIXED AXLE GATE) Gate consisting of a rectangular leaf mounted on wheels, particularly suited for high head situations; gate having wheels or rollers mounted on the end posts of the gate.

FIXED-LOCATION MONITORING Sampling of an environmental or ambient medium for pollutant concentration at one location continuously or repeatedly.

FIXED-STATION MONITORING Repeated long-term sampling or measurement of parameters at representative points for the purpose of determining environmental quality characteristics and trends.

FLAIL Hammer hinged to an axle so that is can be used to break or crush material.

FLAMMABLE Any material that ignites easily and will burn rapidly.

FLANGE Ridge that prevents a sliding motion; rib or rim for strength or for attachments; rim or collar attached to one end of a pipe to give support.

FLANKING LINE Line of cumulus or towering cumulus clouds connected to and extending outward from the most active part of a supercell, normally on the southwest side.

FLAP GATE Gate hinged along one edge, usually either the top or bottom edge.

FLARE Control device that burns hazardous materials to prevent their release into the environment that may operate continuously or intermittently, usually on top of a stack; sudden eruption of energy on the solar disk lasting minutes to hours, from which radiation and particles are emitted.

FLASH Sudden, brief illumination of a conductive channel associated with lightning, which may contain multiple strokes with their associated stepped leaders, dart leaders and return strokes.

FLASH FLOOD Flood which follows within a few hours (usually less than 6 hours) of heavy or excessive rainfall, dam or levee failure, or the sudden release of water impounded by an ice jam; flood of short duration with a relatively high peak rate of flow, usually resulting from a high intensity rainfall over a small area; rapid and extreme flow of high water into a normally dry are.

FLASH FLOOD GUIDANCE (FFG) Forecast guidance produced by the River Forecast Centers containing rainfall threshold values which must be exceeded in order to produce a flash flood.

FLASH FLOOD STATEMENT (FFS) Statement by the National Weather Service which provides follow-up information on flash flood watches and warnings.

FLASH FLOOD TABLE Pre-computed forecast crest stage values for small streams for a variety of antecedent moisture conditions and rain amounts.

FLASH FLOOD WARNING (FFW) Issued to inform the public, emergency management, and other cooperating agencies that flash flooding is in progress, imminent, or highly likely.

FLASH FLOOD WATCH (FFA) Issued to indicate current or developing hydrologic conditions that are favorable for flash flooding in and close to the watch area, but the occurrence is neither certain or imminent.

FLASH MULTIPLICITY Number of return strokes in a lightning flash.

FLASH POINT Lowest temperature at which evaporation of a substance produces sufficient vapor to form an ignitable mixture with air.

FLASHBOARDS Temporary structural barriers, consisting of either timber, concrete or steel, anchored to the crest of a spillway as a means of increasing the reservoir storage; length of timber, concrete, or steel placed on the crest of a spillway to raise the retention water level but which may be quickly removed in the event of a flood by a tripping device, or by deliberately designed failure of the flashboard or its supports.

FLATIRONS Triangular-shaped landforms along mountain ranges formed by erosion of steeply inclined rock layers or hogbacks.

FLEET ANGLE Angle between the position of a rope or cable at the extreme end wrap on a drum and a line drawn perpendicular to the axis of the drum; used to indicate how effective or efficient the rope or cable is for raising a load.

FLEXIBLE PIPE Pipe designed to transmit the backfill load to the soil at the sides of the pipe.

FLG Falling.

FLING Near field long period pulse from a strong ground motion resulting in a unidirectional ground heave after rupture.

FLIP BUCKET Energy dissipater located at the downstream end of a spillway and shaped so that water flowing at a high velocity is deflected upwards in a trajectory away from the foundation of the spillway.

FLOAT RECORDING PRECIPITATION GAUGE Rain gauge where the rise of a float within the instrument with increasing rainfall is recorded.

FLOATABLE DAYS Number of days during the recreation season on which it is safe to allow floating activities on recreation facilities.

FLOATABLE FLOWS River flows which make rafting and other floating recreation possible.

FLOC Loose, open-structured mass formed in a suspension by the aggregation of minute particles; clumps of bacteria and particulate impurities that have come together and formed a cluster; clumps of impurities removed from water during the purification process; formed when alum is added to impure water; cluster of frazil particles; clump of solids formed in sewage by biological or chemical action.

FLOCCULATION Process of forming flocs; process by which clumps of solids in water or sewage aggregate through biological or chemical action so they can be separated from water or sewage; a step in water filtration in which alum is added to cause particles to clump together.

FLOE Accumulation of frazil flocs (also known as a "pan") or a single piece of broken ice.

FLOOD Temporary rise in water levels resulting in inundation of areas not normally covered by water; overflow of water onto lands that are not normally covered by water; overflow of water in an established watercourse (i.e. river, stream, or drainage ditch); ponding of water at or near the point where the rain fell.

FLOOD ATTENUATION Weakening or reduction in the force or intensity of a flood.

FLOOD BOUNDARY Line drawn or outer edge of colored (inundation) area on an inundation map to show the limit of flooding.

FLOOD CATEGORIES Terms defined for each forecast point which describe or categorize the severity of flood impacts in the corresponding river/stream reach where each flood category is bounded by an upper and lower stage where the flood categories used in the National Weather Service are: Minor Flooding - minimal or no property damage, but possibly some public threat, Moderate Flooding - some inundation of structures and roads near stream with some evacuations of people and/or transfer of property to higher elevations, Major Flooding - extensive inundation of structures and roads with significant evacuations of people and/or transfer of property to higher elevations, Record Flooding - flooding which equals or exceeds the highest stage or discharge at a given site during the period of record keeping.

FLOOD CONTROL CAPACITY Reservoir capacity assigned to the sole purpose of regulating flood inflows to reduce flood damage downstream.

FLOOD CONTROL POOL (FLOOD POOL) Reservoir volume above active conservation capacity and joint use capacity that is reserved for flood runoff and then evacuated as soon as possible to keep that volume in readiness for the next flood.

FLOOD CONTROL STORAGE Storage of water in reservoirs to abate flood damage.

FLOOD CREST Maximum height of a flood wave as it passes a certain location.

FLOOD EVENT See FLOOD WAVE.

FLOOD FREQUENCY Probability (in percent) that a flood will occur in a given year.

FLOOD FREQUENCY CURVE Graph showing the number of times per year on the average, plotted as abscissa, that floods of magnitude, indicated by the ordinate, are equaled or exceeded; graph with recurrence intervals of floods plotted as abscissa.

FLOOD GATE Gate to control flood release from a reservoir.

FLOOD HYDROGRAPH Graph showing, for a given point on a stream, the discharge, height, or other characteristic of a flood with respect to time.

FLOOD IRRIGATION Method of irrigating where water is applied from field ditches onto land which has no guide preparation such as furrows, borders or corrugations; application of irrigation water whereby the entire surface of the soil is covered by ponded water.

FLOOD LOSS REDUCTION MEASURES Four basic strategy for reducing flood losses: prevention, property protection, emergency services, and structural projects.

FLOOD OF RECORD Highest observed river stage or discharge at a given location during the period of record keeping.

FLOOD PEAK Highest value of the stage or discharge attained by a flood; thus, peak stage or peak discharge.

FLOOD PLANE Position occupied by the water surface of a stream during a particular flood; elevation of the water surface at various points along the stream during a particular flood.

FLOOD POOL INDEX Ratio of the flood control pool depth to the depth below the pool, multiplied by the percent of time the reservoir water surface will be within the flood control pool.

FLOOD POTENTIAL OUTLOOK (FPO) Issued to alert the public of potentially heavy rainfall that could send area rivers and streams into flood or aggravate an existing flood.

FLOOD PREVENTION Measures that are taken in order to keep flood problems from getting worse.

FLOOD PROBLEMS Problems and damages that occur during a flood as a result of human development and actions.

FLOOD PROFILE Graph of elevation of the water surface of a river in flood, plotted as ordinate, against distance, measured in the downstream direction, plotted as abscissa; may be drawn to show elevation at a given time, crests during a particular flood, or to show stages of concordant flows.

FLOOD ROUTING Process of determining progressively the timing, shape, and amplitude of a flood wave as it moves past a dam or downstream to successive points along the river.

FLOOD SEVERITY Qualitative description of how severe a possible flood could be (High, Medium, Low) depending on failure modes (including rate of failure), flood velocity, channel width, magnitude of damage potential, rate of rise for flood waters, etc.

FLOOD SEVERITY UNDERSTANDING Understanding as to what degree flooding might affect the downstream population.

FLOOD STAGE Established gauge height within a given river reach above which a rise in water surface level is defined as a flood and begins to create a hazard to lives, property, or commerce.

FLOOD STATEMENT (FLS) Statement issued to inform the public of flooding along major streams in which there is not a serious threat to life or property.

FLOOD STORAGE Retention of water or delay of runoff either by planned operation, as in a reservoir, or by temporary filling of overflow areas, as in the progression of a flood wave through a natural stream channel.

FLOOD SURCHARGE Storage volume between the top of the active storage and the design water level.

FLOOD WARNING (FLW) Statement issued to inform the public of flooding along larger streams in which there is a serious threat to life or property.

FLOOD WATCH Statement issued to inform the public and cooperating agencies that current and developing hydrometeorological conditions are such that there is a threat of flooding, but the occurrence is neither certain nor imminent.

FLOOD WAVE Rise in streamflow to a crest and its subsequent recession caused by precipitation, snow melt, dam failure, or reservoir releases; distinct rise in stage culminating in a crest and followed by recession to lower stages.

FLOOD ZONE Area that borders a stream and is subject to equal frequency floods.

FLOOD, 100-YEAR Flood level with a 1 percent chance of being equaled or exceeded in any given year (not refer to a flood that occurs once every 100 years).

FLOOD, SAFETY EVALUATION (SEP)
Largest flood for which the safety of a dam and appurtenant structure is to be evaluated.

FLOOD-CONTROL STORAGE
Storage of water in reservoirs to abate flood damage.

FLOODED ICE Ice which has been flooded by melt water or river water and is heavily loaded by water and wet snow.

FLOODPLAIN Portion of a river valley that has been inundated by the river during historic floods; relatively level area of land bordering a stream channel and inundated during moderate to severe floods; downstream area that would be inundated or otherwise affected by the failure of a dam or by large flood flows.

FLOODPLAIN INFORMATION STUDIES
Reports usually prepared by the U.S. Army Corps of Engineers (USACE) following a survey of a flood-impacted community.

FLOODPROOFING Process of protecting a building from flood damage on site.

FLOODWALL Long, narrow concrete, or masonry embankment usually built to protect land from flooding.

FLOODWAY Part of the flood plain that facilitates the passage of flood water discharge.

FLOOR SWEEP Capture of heavier-than-air gases that collect at floor level.

FLORA All plant life (and bacteria) associated with a given habitat, country, or period.

FLOW Volume of water that passes a given point within a given period of time; qualitative reference of an air parcel(s) with respect to its direction of movement.

FLOW AUGMENTATION Release of water stored in a reservoir or other impoundment to increase the natural flow of a stream.

FLOW CHANNEL Portion of a flow net bounded by two adjacent flow lines.

FLOW CURVE Locus of points obtained from a standard liquid limit test and plotted on a graph representing moisture content as ordinate on an arithmetic scale and the number of blows as abscissa on a logarithmic scale.

FLOW DURATION CURVE Cumulative frequency curve that shows the percentage of time that specified discharges are equaled or exceeded at a given location; measure of the range and variability of a stream's flow;

FLOW DURATION PERCENTILES Values on a scale of 100 that indicate the percentage of time for which a flow is exceeded.

FLOW FAILURE Failure in which a soil mass moves over relatively long distances in a fluidlike manner.

FLOW LINE Idealized path followed by particles of water; path that a particle of water follows in its course of seepage under laminar flow conditions.

FLOW MEASUREMENT DEVICES Instruments (e.g., flowmeters, weirs, and calibrated bucket and stopwatch) that measure leakage quantities.

FLOW NET Graphical representation of flow lines and equipotential (piezometric) lines used in the study of seepage phenomena.

FLOW PATH Subsurface course a water molecule or solute would follow in a given groundwater velocity field.

FLOW RATE Rate at which a fluid escapes from a hole or fissure in a tank.

FLOW SEPARATION Process by which a separation eddy forms on the windward or leeward sides of bluff objects or steeply rising hillsides.

FLOW SLIDE Failure of a sloped bank of soil in which the movement of the soil mass does not take place along a well-defined surface of sliding.

FLOW SPLITTING Splitting of a stable airflow around a mountain barrier, with branches going around the left and right edges of the barrier, often at accelerated speeds.

FLOWABLE Pesticide and other formulations in which the active ingredients are finely ground insoluble solids suspended in a liquid.

FLOWAGE Water that floods onto adjacent land.

FLOWAGE EASEMENT Right or easement to overflow, submerge, or flood certain lands; right to prohibit building on certain floodways.

FLOWING ARTESIAN WELL Well drilled into a confined aquifer with enough hydraulic pressure for the water to flow to the surface without pumping.

FLOWING WELL Well drilled into a confined aquifer with enough hydraulic pressure for the water to flow to the surface without pumping; well or spring that taps groundwater under pressure so that if the water rises above the surface, it is known as a Artesian well.

FLOWMETER Gauge indicating the velocity of wastewater moving through a treatment plant or of any liquid moving through various industrial processes.

FLOWPATH Underground route for ground-water movement, extending from a recharge (intake) zone to a discharge (output) zone such as a shallow stream.

FLOWPATH STUDY Network of clustered wells located along a flowpath extending from a recharge zone to a discharge zone.

FLRY Flurry.

FLS River Flood Statement.

FLUCTUATING FLOWS Water released from a dam that varies in volume with time.

FLUCTUATING ZONE Area of a sandbar or vegetation zone that is within the range of fluctuating flow.

FLUCTUATION Variation in water level, up or down, as a result of project operation.

FLUE GAS Air coming out of a chimney after combustion in the burner it is venting that can include nitrogen oxides, carbon oxides, water vapor, sulfur oxides, particles and many chemical pollutants.

FLUE GAS DESULFURIZATION Technology that employs a sorbent, usually lime or limestone, to remove sulfur dioxide from the gases produced by burning fossil fuels.

FLUENCE Time integrated flux.

FLUID Matter which flows; gas or liquid.

FLUIDIZED Mass of solid particles that is made to flow like a liquid by injection of water or gas is said to have been fluidized.

FLUIDIZED BED INCINERATOR Incinerator that uses a bed of hot sand or other granular material to transfer heat directly to waste.

FLUME Natural or man-made channel that diverts water; shaped, open-channel flow sections that force flow to accelerate for purposes of measurement; artificial channel, often elevated above ground, used to carry fast flowing water; open channel constructed with masonry, concrete or steel of rectangular or U shaped cross section and designed for medium or high velocity flow.

FLUORIDATION Addition of a chemical to increase the concentration of fluoride ions in drinking water to reduce the incidence of tooth decay.

FLUORIDES Gaseous, solid, or dissolved compounds containing fluorine that result from industrial processes; excessive amounts in food can lead to fluorosis.

FLUOROCARBONS (FCS) Any of a number of organic compounds analogous to hydrocarbons in which one or more hydrogen atoms are replaced by fluorine.

FLURRIES Intermittent light snowfall of short duration (generally light snow showers) with no measurable accumulation (trace category).

FLUSH To force large amounts of water through a system to clean out piping or tubing, and storage or process tanks.

FLUSHING Method used to clean water distribution lines by passing a large amount of water through the system.

FLUVIAL Pertains to streams, stream processes, rivers; growing or living in a stream or river; produced by the action of a stream or river.

FLUVIAL DEPOSIT Sedimentary deposit consisting of material transported by suspension or laid down by a river or stream.

FLUX Flowing or flow; substance used to help metals fuse together; volume of fluid crossing a unit cross-sectional surface are per unit time; rate of transfer of fluids, particles or energy per unit area across a given surface (amount of flow per unit of time).

FLW Follow; Flow-Wind.

FLY ASH Non-combustible residual particles expelled by flue gas; by-product of coal-fired power plants which reacts with water and the free lime in cement while generating only half the heat of an equal amount of cement; finely divided residue resulting from the combustion of ground or powdered coal and which is transported from the firebox through the boiler by flue gases.

FLYWAY Specific air route taken by birds during migration.

FM From; fathom.

FMIN Lowest radiowave frequency that can be reflected from the ionosphere.

FNT Front.

FNTGNS Frontogenesis.

FNTLYS Frontolysis.

FOEHN Warm, dry wind on the lee side of a mountain range, the warmth and dryness of the air being due to adiabatic compression as the air descends the mountain slopes.

FOEHN PAUSE Temporary cessation of the foehn at the ground due to the formation or intrusion of a cold air layer which lifts the foehn off the ground.

FOES Maximum ordinary mode radiowave frequency capable of reflection from the sporadic E region of the ionosphere.

FOF2 Maximum ordinary mode radiowave frequency capable of reflection from the F2 region of the ionosphere.

FOG Water droplets suspended in the air at the Earth's surface.

FOG (FG OR F) Water droplets suspended in the air at the Earth's surface.

FOGBOW Rainbow that has a white band that appears in fog, and is fringed with red on the outside and blue on the inside.

FOGGING Applying a pesticide by rapidly heating the liquid chemical so that it forms very fine droplets that resemble smoke or fog; used to destroy mosquitoes, black flies, and similar pests.

FONSI Finding Of No Significant Impact.

FOOD CHAIN Sequence of organisms, each of which uses the next, lower member of the sequence as a food source.

FOOD PROCESSING WASTE Food residues produced during agricultural and industrial operations.

FOOD WASTE Uneaten food and food preparation wastes from residences and commercial establishments such as grocery stores, restaurants, and produce stands, institutional cafeterias and kitchens, and industrial sources like employee lunchrooms.

FOOD WEB Feeding relationships by which energy and nutrients are transferred from one species to another.

FOOT Twelve inches; one of a number of projections on a cylindrical drum of a tamping roller.

FOOTING Sill under a foundation; ground, in relation to its load bearing and friction qualities; portion of the foundation of a structure that transmits loads directly to the soil.

FOOT-POUND Unit of work equal to the force in pounds multiplied by the distance in feet through which it acts.

FORAGE Vegetation used for animal consumption.

FORAGE FISH Small fish that produce prolifically and are consumed by predators.

FORB Weed or a broad-leafed plant.

FORCED CHANNELING Channeling of upper winds along a valley's axis when upper winds are diverted by the underlying topography.

FORCED OUTAGE Unscheduled shut down of a generating unit or other facility for emergency or other unforeseen reasons.

FORD Place where a road crosses a stream under water.

FOREBAY (HEADRACE) Water behind (upstream) of the dam; impoundment immediately upstream from a dam or hydroelectric plant intake structure.

FORECAST Statement of prediction.

FORECAST CREST Highest elevation of river level, or stage, expected during a specified storm event.

FORECAST GUIDANCE Computer-generated forecast materials used to assist the preparation of a forecast, such as numerical forecast models.

FORECAST ISSUANCE STAGE Stage which, when reached by a rising stream, represents the level where RFCs need to begin issuing forecasts for a non-routine (flood-only) forecast point.

FORECAST PERIODS Official definitions for National Weather Service products: Today (sunrise to sunset), This afternoon (noon till 6 p.m.), This evening (6 p.m. till sunset), Tonight (sunset till sunrise), Tomorrow (sunrise to sunset of the following day).

FORECAST POINT Location that represents an area (reach of a river), where a forecast warning services are made available to the public.

FORECAST VALID FOR Period of time the forecast is in effect beginning at a given day, date and time, and ending at a given day, date and time.

FOREPOLE Plank driven ahead of a tunnel face to support the roof or wall during excavation.

FORESHORE Part of the shore between the ordinary high-and low-watermarks and generally crossed by the tide each day.

FORESIGHT Sighting on a point of unknown elevation from an instrument of known elevation

FOREST SERVICE (FS) Mission of the USDA Forest Service is to sustain the health, diversity, and productivity of the Nation's forests and grasslands to meet the needs of present and future generations; agency within the U.S. Department of Agriculture.

FORMALDEHYDE Colorless, pungent, and irritating gas used chiefly as a disinfectant and preservative and in synthesizing other compounds like resins.

FORMATION Any sedimentary, igneous, or metamorphic material represented as a unit in geology; generally called rock, but not necessarily meeting the definition of rock.

FORMATION OR GEOLOGICAL FORMATION
Layer of rock that is made up of a certain type of rock or a combination of types.

FORMULATION Substances comprising all active and inert ingredients in a pesticide.

FORMULATION PROCESS First phase performed by the Early Warning System design team, which includes an Early Warning System reliability, local capabilities assessment, and conceptual level designs for an Early Warning System.

FORWARD FLANK DOWNDRAFT
Main region of downdraft in the forward, or leading, part of a supercell, where most of the heavy precipitation is.

FOSSIL FUEL Fuel derived from ancient organic remains (e.g., peat, coal, crude oil, and natural gas).

FOSSORIAL INSECTS Insects that live in the soil.

FOUNDATION Lower part of a structure that transmits loads directly to the soil; excavated surface upon which a dam is placed; portion of the valley floor that underlies and supports the dam structure.

FOUNDATION DRAINS Tile or pipe for collecting seepage within a foundation.

FOUNDATION MATERIAL (SURFACE, SOIL)
Surface of the upper part of the earth mass carrying the load of the structure.

FOUNDATION TRENCH Trench built at and into the foundation of a dam and filled with clay or other impermeable substances to prevent water from seeping beneath the dam.

FOUNTAINHEAD Upper end of a confined-aquifer conduit, where it intersects the land surface.

FOUR-ZERO-ONE (401) (A) CERTIFICATION
Requirement of Section 401(a) of the Clean Water Act that all federally issued permits be certified by the state in which the discharge occurs.

FOUS Forecast Output United States.

FPO Flood Potential Outlook.

FPS Fujita-Pearson Scale.

FQT Frequent.

FRACTOCUMULUS Cumulus cloud presenting a ragged, shredded appearance, as if torn.

FRACTURE (JOINT) Any break or rupture formed in an ice cover or floe due to deformation; break in a rock formation due to structural stresses (e.g., faults, shears, joints, and planes of fracture cleavage); crack or break in rocks along which no movement has occurred.

FRACTURE ZONE Area that has a great number of fractures.

FRACTURING Deformation process whereby ice is permanently deformed, and fracture occurs.

FRACTUS Ragged, detached cloud fragments; same as scud.

FRAZIL (FRAZIL ICE) Fine spicules, plates, or discoids of ice suspended in water.

FRAZIL SLUSH Agglomerate of loosely packed frazil which floats or accumulates under the ice cover.

FREAK WAVE Wave of much greater height and steepness than other waves in the prevailing sea or swell system.

FREE ATMOSPHERE Part of the atmosphere that lies above the frictional influence of the Earth's surface.

FREE GROUNDWATER Unconfined groundwater whose upper boundary is a free water table.

FREE PRODUCT Petroleum hydrocarbon in the liquid free or nonaqueous phase.

FREEBOARD Vertical distance from the normal water surface to the top of a confining wall; difference in elevation between the maximum water surface in the reservoir and the dam crest; vertical distance between a stated water level and the top of a dam, without camber; vertical distance between the normal maximum level of the water surface in a channel, reservoir, tank, canal, etc., and the top of the sides of a levee, dam, etc., which is provided so that waves and other movements of the liquid will not overtop the confining structure; vertical distance from the sand surface to the underside of a trough in a sand filter.

FREEZE When the surface air temperature is expected to be 32 °F or below over a widespread area for a climatologically significant period of time.

FREEZE WARNING Issued during the growing season when surface temperatures are expected to drop below freezing over a large area for an extended period of time, regardless whether or not frost develops.

FREEZEOUT Deeply frozen over for long periods of time.

FREEZE-THAW DAMAGE Damage to concrete caused by extreme temperature variations as noted by random pattern cracking; accelerated by the presence of water and commonly more severe on the south-facing side of structures.

FREEZEUP DATE Date on which the water body was first observed to be completely frozen over.

FREEZING DRIZZLE Drizzle that falls as a liquid but freezes into glaze or rime upon contact with the cold ground or surface structures.

FREEZING DRIZZLE ADVISORY Issued when freezing rain or freezing drizzle is forecast but a significant accumulation is not expected.

FREEZING FOG Fog the droplets of which freeze upon contact with exposed objects and form a coating of rime and/or glaze.

FREEZING LEVEL Altitude at which the air temperature first drops below freezing.

FREEZING RAIN Rain that falls as a liquid but freezes into glaze upon contact with the ground.

FREEZING RAIN ADVISORY Issued when freezing rain or freezing drizzle is forecast but a significant accumulation is not expected.

FREEZING SPRAY Accumulation of freezing water droplets on a vessel caused by some appropriate combination of cold water, wind, cold air temperature, and vessel movement.

FREEZING SPRAY ADVISORY Advisory for an accumulation of freezing water droplets on a vessel at a rate of less than 2 centimeters per hour caused by some appropriate combination of cold water, wind, cold air temperature, and vessel movement.

FREEZUP JAM Ice jam formed as frazil ice accumulates and thickens.

FRENCH DRAIN Covered ditch containing a layer of fitted or loose stone or other pervious material; underground passageway for water through the interstices among stones placed loosely in a trench.

FREQUENCY Number of repetitions of a periodic process in a certain time period; how often criteria can be exceeded without unacceptably affecting the community.

FREQUENCY CURVE Curve that expresses the relation between the frequency distribution plot, with the magnitude of the variables as abscissas and the number of occurrences of each magnitude in a given period as ordinates.

FREQUENCY DEMAND SCHEDULING Method of irrigation scheduling similar to demand scheduling, but typically involves a fixed duration of the delivery.

FRESHET Annual spring rise of streams in cold climates as a result of snow melt; flood caused by rain or melting snow.

FRESHWATER Water that contains less than 1,000 milligrams per liter (mg/L) of dissolved solids.

FRESHWATER CHRONIC CRITERIA Highest concentration of a contaminant that freshwater aquatic organisms can be exposed to for an extended period of time (4 days) without adverse effects.

FRIABLE Capable of being crumbled, pulverized, or reduced to powder by hand pressure; descriptive of a rock or mineral that crumbles naturally or is easily broken, pulverized, or reduced to powder.

FRIABLE ASBESTOS Any material containing more than one-percent asbestos, and that can be crumbled or reduced to powder by hand pressure.

FRICTION Resistance to motion when one body is sliding or tending to slide over another; mechanical resistive force of one object on another object's relative movement when in contact with the first object.

FRICTION HEAD Decrease in total head caused by friction.

FRICTION LAYER (PLANETARY BOUNDARY LAYER) Layer within the atmosphere between the earth's surface and 1 km above the surface where friction affects wind speed and wind direction.

FRMG Forming.

FRONT Boundary or transition zone between two air masses of different density, and thus (usually) of different temperature.

FRONT END LOADER Tractor loader that both digs and dumps in front.

FRONTAL INVERSION Temperature inversion that develops aloft when warm air overruns the cold air behind a front.

FRONTOGENESIS Initial formation of a front or frontal zone; increase in the horizontal gradient of an air mass property, principally density, and the development of the accompanying features of the wind field that typify a front.

FROPA Frontal Passage.

FROSFC Frontal Surface.

FROST (FRST) Formation of thin ice crystals on the ground or other surfaces in the form of scales, needles, feathers, or fans.

FROST ACTION Freezing and thawing of moisture in materials and the resultant effects on these materials and on structures of which they are a part of with which they are in contact.

FROST ADVISORY Issued during the growing season when widespread frost formation is expected over an extensive area. Surface temperatures are usually in the mid-30s Fahrenheit.

FROST HEAVE Raising of a surface due to the accumulation of ice in the underlying soil or rock.

FROST LINE Greatest depth to which ground may be expected to freeze.

FROST POINT Dew point below freezing.

FROSTBITE Human tissue damage caused by exposure to intense cold.

FROUDE NUMBER Ratio of inertial forces to gravitational forces in flow; ratio of the flow velocity to the velocity of a small gravity wave in the flow; when less than one, flow is tranquil; when greater than one, flow is rapid; when equal to one, flow is critical.

FROZEN DEW When liquid dew changes into tiny beads of ice.

FRY Life stage of fish between the egg and fingerling stages.

FRZ Freeze.

FRZN Frozen.

FS U.S. Forest Service; feasibility study; factor of safety.

FSCBG Specific aerial spray dispersion model.

FT Feet; foot.

FUEL ECONOMY STANDARD Corporate Average Fuel Economy Standard (CAFE) effective in 1978 that enhanced the national fuel conservation effort imposing a miles-per-gallon floor for motor vehicles.

FUEL EFFICIENCY Proportion of energy released by fuel combustion that is converted into useful energy.

FUEL REPLACEMENT ENERGY Electric energy generated at a hydroelectric plant as a substitute for energy which would have been generated by a thermal electric plant.

FUEL SWITCHING Precombustion process whereby a low-sulfur coal is used in place of a higher sulfur coal in a power plant to reduce sulfur dioxide emissions; illegally using leaded gasoline in a motor vehicle designed to use only unleaded.

FUGITIVE DUST Dust that is not emitted from definable point sources such as industrial smokestacks where sources include open fields, roadways, storage piles, etc.

FUGITIVE EMISSIONS Emissions not caught by a capture system.

FUJITA SCALE (F SCALE) All tornadoes, and most other severe local windstorms, are assigned a single number from this scale according to the most intense damage caused by the storm; scale of tornado intensity in which wind speeds are inferred from an analysis of wind damage.

FUJIWHARA EFFECT Binary interaction where tropical cyclones within a certain distance (300-750 nm depending on the sizes of the cyclones) of each other begin to rotate about a common midpoint.

FULCRUM Pivot for a lever.

FULL GATE Maximum gate position of a turbine for a particular head.

FULL HYDRAULIC CAPACITY Designed capacity of a pipe or conduit.

FULL IRRIGATION SERVICE LAND Irrigable land now receiving, or to receive, its sole and generally adequate water supply through facilities which have been or are to be constructed by, rehabilitated by, or replaced by the Bureau of Reclamation.

FULL POOL Volume of water in a reservoir at normal water surface; reservoir level that would be attained when the reservoir is fully utilized for all project purposes, including flood control.

FULL-PHYSICS NUMERICAL MODEL Computer model used to calculate air pollution concentrations; uses a full set of equations describing the thermodynamic and dynamic state of the atmosphere and can be used to simulate atmospheric phenomena.

FULL-SCALE EXERCISE Activity in which emergency preparedness officials respond in a coordinated manner to a timed, simulated incident but includes the mobilization of field personnel and resources and the actual movement of emergency workers, equipment, and resources required to demonstrate coordination and response capability.

FUME Tiny particles trapped in vapor in a gas stream.

FUMIGANT Substance or mixture of substances that produces gas, vapor, fume, or smoke intended to destroy insects, bacteria, or rodents.

FUMIGATION Pattern of plume dispersion produced when a convective boundary layer grows upward into a plume trapped in a stable layer and the elevated plume is suddenly brought downward to the ground, producing high surface concentrations.

FUNCTIONAL EQUIVALENT Term used to describe EPA's decision-making process and its relationship to the environmental review conducted under the National Environmental Policy Act (NEPA).

FUNCTIONAL EXERCISE Activity in which participants respond in a coordinated manner to a timed, simulated incident that parallels a real operational event as close as possible.

FUNDAMENTALLY DIFFERENT FACTORS (FDF) Those components of a petitioner's facility that are determined to be so unlike those components considered by EPA during the effluent limitation guideline and pretreatment standards rulemaking that the facility is worthy of a variance from the effluent limitations guidelines or categorical pretreatment standards.

FUNGICIDE Pesticides which are used to control, deter, or destroy fungi.

FUNGISTAT Chemical that keeps fungi from growing.

FUNGUS (FUNGI) Molds, mildews, yeasts, mushrooms, and puffballs, a group of organisms lacking in chlorophyll (i.e. are not photosynthetic) and which are usually non-mobile, filamentous, and multicellular.

FUNNEL CLOUD Condensation funnel extending from the base of a towering cumulus, associated with a rotating column of air that is not in contact with the ground (and hence different from a tornado).

FUNNELLING Process whereby wind is forced to flow through a narrow opening between adjacent land areas, resulting in increased wind speed.

FURROW Natural or man-made narrow depression in the earth's surface; narrow trenchlike plowed depression in the earth surface to keep surface water away from the slopes of cuts.

FURROW IRRIGATION Irrigation method in which water travels through the field by means of small channels between each groups of rows; type of surface irrigation whereby water is applied at the upper (higher) end of a field and flows in furrows to the lower end.

FUSE Thin core of black powder surrounded by wrappings, which when lit at one end, will burn to the other at a fixed speed; string-like core of PETN, a high explosive, contained within a waterproof reinforced sheath.

FUSE PLUG SPILLWAY Auxiliary or emergency spillway comprising a low embankment or a natural saddle designed to be overtopped and eroded away during flood flows.

FUTURE LIABILITY Refers to potentially responsible parties' obligations to pay for additional response activities beyond those specified in the Record of Decision or Consent Decree.

FUTURE WITHOUT What would occur if no action were taken; future without taking any action to solve the problem.

FVT Forecast Verification Tool.

FWD Forward.

FZRA Freezing rain.

G

G/KG Grams per Kilogram.

GABION Wire basket, filled with stones, used to stabilize banks of a water course and to enhance habitat.

GABION DAM Crib dam when built with gabions.

GAGE See GAUGE.

GAGING STATION Particular site on a watercourse (e.g., stream, canal, lake, or reservoir) where systematic observations of hydrologic data are measured.

GAINING STREAM Stream or reach that receives water from the zone of saturation.

GALE Extratropical low or an area of sustained surface winds of 34 (39 mph) to 47 knots (54 mph).

GALE WARNING Warning of sustained surface winds, or frequent gusts, in the range of 34 knots (39 mph) to 47 knots (54 mph) inclusive, either predicted or occurring, and not directly associated with a tropical cyclone.

GALE WATCH Watch for an increased risk of a gale force wind event for sustained surface winds, or frequent gusts, of 34 knots (39 mph) to 47 knots (54 mph), but its occurrence, location, and/or timing is still uncertain.

GALLERY Passageway within the body of a dam, its foundation, or abutments used for inspection, foundation grouting, and/or drainage.

GALLON Unit of measure equal to four quarts or 128 fluid ounces.

GALVANIZE To coat a metal (especially iron or steel) with zinc.

GAME FISH Species like trout, salmon, or bass, caught for sport.

GAMETES Eggs or sperm.

GAMMA Unit of magnetic field intensity equal to 1 x 10^{-5} Gauss; equal to 1 nanotelsa (nT).

GAMMA RAY Type of electromagnetic radiation with a very short wavelength and high energy level.

GANTRY Overhead structure that supports machines or operating parts, such as a gantry crane.

GANTRY CRANE Fixed or traveling, bent-supported crane for handling heavy equipment.

GAP WINDS Strong winds channeled through gaps in the Pacific coastal ranges, blowing out into the Pacific

Ocean or into the waterways of the Inside Passage.

GARBAGE Animal and vegetable waste resulting from the handling, storage, sale, preparation, cooking, and serving of foods.

GAS CHROMATOGRAPH/MASS SPECTROMETER Instrument that identifies the molecular composition and concentrations of various chemicals in water and soil samples.

GAS CHROMATOGRAPHY/FLAME IONIZATION DETECTOR (GC/FID) Laboratory analytical method used as a screening technique for semivolatile organic compounds that are extractable from water in methylene chloride.

GAS LAWS Thermodynamic laws pertaining to perfect gases, including Boyle's law, Charles' law, Dalton's law and the equation of state.

GASAHOL Mixture of gasoline and ethanol derived from fermented agricultural products containing at least nine percent ethanol; emissions contain less carbon monoxide than those from gasoline.

GASEOUS SUPERSATURATION Condition of higher levels of dissolved gases in water due to entrainment, pressure increases, or heating.

GASIFICATION Conversion of solid material such as coal into a gas for use as a fuel.

GASOLINE VOLATILITY Property of gasoline whereby it evaporates into a vapor.

GATE Movable water barrier for the control of water; device that controls the flow in a conduit, pipe, or tunnel without obstructing any portion of the waterway when in the fully open position; device in which a leaf or member is moved across the waterway from an external position to control or stop flow; structure or device for controlling the rate of flow into or from a canal or ditch; movable, watertight barrier for the control of water in a waterway.

GATE CHAMBER (VALVE CHAMBER) Chamber in which a guard gate in a pressurized outlet works or both the guard and regulating gates in a free-flow outlet works are located; room from which a gate or valve can be operated, or sometimes in which the gate is located; concrete portion of an outlet works containing gates between upstream and downstream conduits and/or tunnels.

GATE HANGER Device used to maintain a set gate opening.

GATE STRUCTURE Portion of spillway between inlet channel and chute, tunnel or conduit, which contains gates, such as radial gates.

GATE VALVE Valve with a circular-shaped closing element that fits securely over an opening through which water flows; valve that utilizes a disc moving at a right angle to the flow to regulate the rate of flow.

GATED PIPE Portable pipe with small gates installed along one side for distributing irrigation water to corrugations or furrows.

GATED SPILLWAY Overflow section of dam restricted by use of gates that can be operated to control releases from the reservoir to ensure the safety of the dam.

GATE-STRUCTURE DAM Barrier built across a river, comprising a series of gates which when fully open allow the flood to pass without appreciably increasing the flood level upstream of the barrage.

GAUGE Device for indicating the magnitude or position of a thing in specific units, when such magnitude or position undergoes change, for example: elevation of a water surface, velocity of flowing water, pressure of water, amount or intensity of precipitation, depth of snowfall, etc.; act or operation of registering or measuring the magnitude or position of a thing when these characteristics are undergoing change; operation, including both field and office work, of measuring the discharge of a stream of water in a waterway.

GAUGE (GAGE) Device for registering water level, discharge, velocity, pressure, etc.; thickness of wire or sheet metal; thickness of the sheet used to make steel pipe.

GAUGE DATUM Arbitrary zero datum elevation which all stage measurements are made from; horizontal surface used as a zero point for measurement of stage or gauge height.

GAUGE HEIGHT Elevation of water surface measured by a gauge.

GAUGE HEIGHT (GH) Water-surface elevation referred to some arbitrary gauge datum (zero point). Gage height is often used interchangeably with the more general term stage, although gauge height is more appropriate when used with a reading on a gauge.

GAUGE PRESSURE Absolute pressure minus atmospheric pressure; pressure within a closed container as measured with a gauge.

GAUGE VALUES Values that are recorded, transmitted, and/or computed from a gaging station.

GAUGE ZERO Elevation of zero stage.

GAUGING STATION Specific location on a stream where systematic observations of hydrologic data are obtained through mechanical or electrical means.

GAUSS Unit of magnetic induction in the cgs (centimeter-gram-second) system.

GAUSSIAN PLUME MODEL Computer model used to calculate air pollution concentrations.

GAUSSIAN PUFF MODEL Computer model used to calculate air pollution concentrations.

GDR Direction, in degrees clockwise from true North, of the GSP, reported at the last hourly 10-minute segment.

GEMPAK General Environmental Meteorological Package.

GEN General.

GENERAL CIRCULATION Totality of large-scale organized motion for the entire global atmosphere.

GENERAL CIRCULATION MODELS (GCMS) Computer simulations that reproduce the Earth's weather patterns to predict change in the weather and climate.

GENERAL PERMIT Permit applicable to a class or category of dischargers.

GENERAL REPORTING FACILITY
Facility having one or more hazardous chemicals above the 10,000 pound threshold for planning quantities.

GENERAL WIND Winds produced by synoptic-scale pressure systems on which smaller-scale or local convective winds are superimposed.

GENERALLY RECOGNIZED AS SAFE (GRAS)
Designation by the FDA that a chemical or substance (including certain pesticides) added to food is considered safe by experts, and so is exempted from the usual FFDCA food additive tolerance requirements.

GENERATION Energy generated in kWh (kilowatt-hours) represents gross generation that consists of the total generation minus station use; process of producing electric energy by transforming other forms of energy; amount of electric energy produced.

GENERATION DISPATCH AND CONTROL
Aggregating and dispatching (sending off to some location) generation from various generating facilities, providing backup and reliability services.

GENERATOR Machine that converts mechanical energy into electrical energy; facility or mobile source that emits pollutants into the air or releases hazardous waste into water or soil; any person, by site, whose act or process produces regulated medical waste or whose act first causes such waste to become subject to regulation.

GENERATOR NAMEPLATE CAPACITY
Full-load continuous rating of a generator, prime mover, or other electric power production equipment under specific conditions as designated by the manufacturer.

GENETIC ENGINEERING Process of inserting new genetic information into existing cells in order to modify a specific organism for the purpose of changing one of its characteristics.

GENOTOXIC Damaging to DNA; pertaining to agents known to damage DNA.

GEOGRAPHIC INFORMATION SYSTEM (GIS)
Computer system designed for storing, manipulating, combining, analyzing, and displaying data in a geographic context.

GEOHYDROLOGY Geological study of the character, source, and mode of groundwater; branch of hydrology relating to subsurface, or subterranean waters.

GEOLOGIC SEQUESTRATION (GEOSEQUESTRATION)
Process of injecting carbon dioxide, which has been compressed into a liquid state, into the deep subsurface.

GEOLOGIC SEQUESTRATION PROJECT
Injection well or wells used to emplace a carbon dioxide stream beneath the lowermost formation containing a USDW; wells used for geologic sequestration of carbon dioxide that have been granted a waiver of the injection depth requirements.

GEOLOGICAL CONTROL Local rock formation or clay layer that limits (within the engineering time frame) the vertical and/or lateral movement of a stream at a particular point.

GEOLOGICAL LOG Detailed description of all underground features (depth, thickness, type of formation) discovered during the drilling of a well.

GEOLOGY Science that deals with the physical history of the Earth, the rocks of which it is comprised, and the physical changes which the Earth has undergone or is undergoing.

GEOMAGNETIC ELEMENTS Components of the geomagnetic field at the surface of the Earth.

GEOMAGNETIC FIELD Magnetic field observed in and around the Earth.

GEOMAGNETIC STORM Worldwide disturbance of the Earth's magnetic field, distinct from regular diurnal variations.

GEOMORPHIC Pertaining to the form or general configuration of the Earth or of its surface features.

GEOMORPHIC CHANNEL UNITS Fluvial geomorphic descriptors of channel shape and stream velocity.

GEOMORPHOLOGY Geological study of the configuration, characteristics, origin, and evolution of land forms and Earth features.

GEOPHYSICAL EVENTS Flares (Importance two or larger) with Centimetric Outbursts (maximum of the flux higher than the Quiet Sun flux, duration longer 10 minutes) and/or strong SID.

GEOPHYSICAL LOG Record of the structure and composition of the earth encountered when drilling a well or similar type of test hold or boring.

GEOPHYSICS Study of the physical characteristics and properties of the Earth; including geodesy, seismology, meteorology, oceanography, atmospheric electricity, terrestrial magnetism, and tidal phenomena.

GEOPOTENTIAL HEIGHT Height above sea level of a pressure level.

GEOSTATIONARY ORBITING ENVIRONMENTAL SATELLITE (GOES) Satellites orbiting at 22,370 miles above the Equator with the same rotational velocity as the Earth.

GEOSTATIONARY ORBITING ENVIRONMENTAL SATELLITE DATA COLLECTION SYSTEM (GOES DCS) Satellite-based system collects a variety of environmental data from locations in the Western Hemisphere.

GEOSTATIONARY SATELLITE Satellite that rotates at the same rate as the Earth, remaining over the same spot above the equator.

GEOSTROPHIC WIND Wind that is affected by coriolis force, blows parallel to isobars and whose strength is related to the pressure gradient (i.e. spacing of the isobars).

GEOSYNCHRONOUS Term applied to any equatorial satellite with an orbital velocity equal to the rotational velocity of the Earth.

GEOTEXTILES Any fabric or textile (natural or synthetic) used as an engineering material in conjunction with soil, foundations, or rock.

GEOTHERMAL Relating to the Earth's internal heat.

GEOTHERMAL DIRECT HEAT RETURN FLOW WELLS Class V wells that dispose of spent geothermal fluids following the extraction of heat used directly (without conversion to electricity or passage through a heat exchanger) to heat homes, swimming pools, etc.

GEOTHERMAL ELECTRIC POWER WELLS Class V wells that dispose of spent geothermal fluids following the extraction of heat for the production of electricity.

GEOTHERMAL/GROUND SOURCE HEAT PUMP Underground coils to transfer heat from the ground to the inside of a building.

GERMICIDE Any compound that kills disease-causing microorganisms.

GEYSER Geothermal feature of the Earth where there is an opening in the surface that contains superheated water that periodically erupts in a shower of water and steam.

GIANT SALVINIA (SALVINIA MOLESTA) Aquatic fern prohibited in the United States by federal law; invasive, rapidly growing plant that covers the surfaces of lakes and streams forming floating mats that shade and crowd out important native plants.

GIARDIA LAMBLIA Microorganism (protozoan) in the feces of humans and animals frequently found in rivers and lakes, which, if not treated properly, can cause severe gastrointestinal ailments

GIARDIASIS Disease that results from an infection by the protozoan parasite Giardia Intestinalis.

GIGAWATT (GW) Unit of power equal to 1 billion watts.

GIGAWATT-HOUR (GWH) One billion watt-hours of electrical energy.

GIS See GEOGRAPHIC INFORMATION SYSTEM.

GLACIAL Of or relating to the presence and activities of ice or glaciers.

GLACIAL DRIFT Rock material transported by glaciers or icebergs and deposited directly on land or in the sea.

GLACIAL LAKE Lake that derives its water, or much of its water, from the melting of glacial ice; lake that occupies a basin produced by glacial erosion.

GLACIAL MORAINE Mass of loose rock, soil, and earth deposited by the edge of a glacier.

GLACIAL OUTWASH Stratified detritus (chiefly sand and gravel) "washed out" from a glacier by meltwater streams and deposited in front of or beyond the end moraine or the margin of an active glacier.

GLACIAL STRIATIONS Lines carved into rock by overriding ice, showing the direction of glacial movement.

GLACIAL TILL Material deposited by glaciation, usually composed of a wide range of particle sizes, which has not been subjected to the sorting action of water.

GLACIATION Transformation of cloud particles from water drops to ice crystals.

GLACIER (ICE SHEET) Huge mass of ice, formed on land by the compaction and recrystallization of old snow, that moves very slowly downslope or outward due to gravity.

GLACIER DAMMED LAKE Lake formed when a glacier flows across the mouth of an adjoining valley and forms an ice dam.

GLACIER WIND Shallow downslope wind above the surface of a glacier, caused by the temperature difference between the air in contact with the glacier and the free air at the same altitude.

GLASS CONTAINERS Containers like bottles and jars for drinks, food, cosmetics and other products.

GLAZE Ice formed by freezing precipitation covering the ground or exposed objects.

GLOBAL FORECAST SYSTEM (GFS)
One of the operational forecast models run at NCEP.

GLOBAL MARITIME DISTRESS AND SAFETY SYSTEM (GMDSS)
Intended to provide more effective and efficient emergency and safety communications and disseminate Maritime Safety Information (MSI) to all ships on the world's oceans regardless of location or atmospheric conditions.

GLOBAL POSITIONING SYSTEMS (GPS)
Space-based radio positioning systems that provide 24-hour, three-dimensional position, velocity, and time information to suitably equipped users anywhere on or near the surface of the Earth.

GLOBAL TEMPERATURE CHANGE
Net result of four primary factors including the greenhouse effect, changes in incoming solar radiation, altered patterns of ocean circulations, and changes in continental position, topography and/or vegetation.

GLOBAL WARMING Increase in the near surface temperature of the Earth.

GLOBAL WARMING POTENTIAL
Ratio of the warming caused by a substance to the warming caused by a similar mass of carbon dioxide.

GLORY Optical effect characterized by concentric rings of color (red outermost and violet innermost) surrounding the shadow of an observer's head when the shadow is cast onto a cloud deck below the observer's elevation.

GLOVEBAG Polyethylene or polyvinyl chloride bag-like enclosure affixed around an asbestos-containing source (most often thermal system insulation) permitting the material to be removed while minimizing release of airborne fibers to the surrounding atmosphere.

GMN Minute of the hour that the GSP occurred, reported at the last hourly 10-minute segment.

GMT Greenwich Mean Time; Universal Coordinated Time.

GND Ground.

GNEISS Metamorphic rock that displays distinct banding of light and dark mineral layers.

GNRL General.

GOGGLE VALVE See RING FOLLOWER GATE.

GOMOOS Gulf Of Maine Ocean Observing System.

GOOSENECK Portion of a water service connection between the distribution system water main and a meter. Sometimes called a pigtail.

GPS See GLOBAL POSITIONING SYSTEMS.

GRAB SAMPLE Single sample collected at a particular time and place that represents the composition of the water, air, or soil only at that time and place.

GRABEN Down-dropped block of rock bounded on both sides by faults.

GRADATION Proportion of material of each particle size (i.e., soil, sediment, or sedimentary rock); frequency distribution of various sizes.

GRADATION CURVE Measure of the variation in grain (particle) sizes within a mixture; graph showing grain size versus the accumulated percent of material that is finer than that grain size.

GRADE (PITCH) Elevation of a surface or a surface slope; elevation of the invert of the bottom of a pipeline, canal, culvert, or conduit; fall (slope) of a line of pipe in reference to a horizontal plane; inclination or slope of a pipeline, conduit, stream channel, or natural ground surface.

GRADE STAKE Stake indicating the amount of cut or fill required to bring the ground to a specified level.

GRADED STREAM Streams that receive and carry away equal amounts of sediment.

GRADER Machine with a centrally located blade that can be angled to cast to either side, with independent hoist control on each side.

GRADIENT General slope or rate of change in vertical elevation per unit of horizontal distance of water surface of a flowing stream; slope along a specific route (i.e., road surface, channel or pipe); rate of change with respect to distance of a variable quantity in the direction of maximum change.

GRADIENT HIGH WINDS Winds that usually cover a large area and are due to synoptic-scale, extra-tropical low pressure systems.

GRADUAL COMMENCEMENT Commencement of a geomagnetic storm that has no well-defined onset.

GRAIN LOADING Rate at which particles are emitted from a pollution source.

GRAIN SHAPE FACTOR See PARTICLE SHAPE FACTOR.

GRAIN SIZE See PARTICLE SIZE.

GRAIN SIZE DISTRIBUTION (GRADATION) See GRADATION CURVE.

GRANITE Light-colored, coarse-grained intrusive igneous rock with quartz and feldspar as dominant minerals and typically peppered with mica and hornblende.

GRANITIC ROCK See GRANITE.

GRANODIORITE Coarse-grained intrusive igneous rock with less quartz and more feldspar than true granite and typically darker.

GRANULAR ACTIVATED CARBON TREATMENT Filtering system often used in small water systems and individual homes to remove organics; used by municipal water treatment plants.

GRANULATION Cellular structure of the photosphere visible at high spatial resolution.

GRAPPLE Clamshell-type bucket having three or more jaws.

GRASSCYCLING Source reduction activities in which grass clippings are left on the lawn after mowing.

GRASSED WATERWAY Natural or constructed watercourse or outlet that is shaped or graded and established in suitable vegetation for the disposal of runoff water without erosion.

GRAUPEL Snow pellets; small hail.

GRAVEL Loose rounded fragments of rock that will pass a 3-inch sieve and be retained on a No. 4 U.S. Standard sieve ($^3/_{16}''$).

GRAVEL BLANKET Thin layer of gravel spread over an area either of natural ground, excavated surface, or embankment.

GRAVEL SURFACING Layer of gravel spread over an area intended for vehicular or personnel traffic.

GRAVELFILL Gravel used to fill holes or spaces.

GRAVITY ARCH DAM Dam designed to combine load resisting features of both a gravity and arch type dam.

GRAVITY DAM Concrete structure proportioned so that its own weight provides the major resistance to the forces exerted on it.

GRAVITY IRRIGATION Irrigation method that applies irrigation water to fields by letting it flow from a higher level supply canal through ditches or furrows to fields at a lower level.

GRAVITY WAVE Wave created by the action of gravity on density variations in the stratified atmosphere (e.g., lee waves, mountains waves, etc.).

GRAY WATER Domestic wastewater composed of wash water from kitchen, bathroom, and laundry sinks, tubs, and washing machines.

GRAYBODY Hypothetical "body" that absorbs some constant fraction of all electromagnetic radiation incident upon it.

GRDL Gradual.

GREAT CIRCLE TRACK Shortest distance between two points on a sphere, and when viewed on a 2-dimensional map the track will appear curved (e.g. swell waves).

GREAT LAKES FREEZE-UP/BREAK-UP OUTLOOK (FBO) National Weather Service product to keep mariners informed of the projected freeze-up date or break-up date of ice on the Great Lakes.

GREAT LAKES MARINE FORECAST (MAFOR) National Weather Service coded summary appended to each of the Great Lakes Open Lakes forecasts.

GREAT LAKES STORM SUMMARY (GLS) National Weather Service forecast product providing updated information whenever a storm warning is in effect on any of the Great Lakes.

GREAT LAKES STOWM SUMMARY (GLS) National Weather Service forecast product providing updated information whenever a storm warning is in effect on any of the Great Lakes.

GREAT LAKES WEATHER BROADCAST (LAWEB) National Weather Service product containing an observation summary prepared to provide Great Lakes mariners with a listing of weather observations along or on the Lakes.

GREEN ALGAE (CHLOROPHYTA) Unicellular or colonial algae with chlorophyll pigments similar to those in terrestrial green plants.

GREEN INFRASTRUCTURE Array of products, technologies, and practices that use natural systems, or engineered systems that mimic natural processes, to enhance overall environmental quality and provide utility services.

GREEN LINE One of the strongest (and first-recognized) visible coronal lines.

GREEN ROOF Planted over existing roof structures, and consist of a waterproof, root-safe membrane that is covered by a drainage system, lightweight growing medium, and plants.

GREEN TAGS See RENEWABLE ENERGY CERTIFICATES.

GREENHOUSE EFFECT Warming of the Earth's atmosphere attributed to a buildup of carbon dioxide or other gases; atmospheric heating caused by solar radiation being readily transmitted inward through the Earth's atmosphere but longwave radiation less readily transmitted outward, due to absorption by certain gases in the atmosphere.

GREENHOUSE GAS Gas (e.g., carbon dioxide or methane) that contributes to potential climate change; gas that absorbs terrestrial radiation and contributes to the greenhouse effect.

GREY WATER See GRAY WATER.

GRID System of interconnected power lines and generators that is managed so that the generators are dispatched as needed to meet the requirements of the customers connected to the grid at various points; squared off areas across the terrain used to define forecast areas; digital forecast databases for meteorological elements, including temperature, wind direction, wind speed and others.

GRID OPERATOR Entity that oversees the delivery of electricity over the grid to the customer, ensuring reliability and safety.

GRINDER PUMP Mechanical device that shreds solids and raises sewage to a higher elevation through pressure sewers.

GRIZZLY (GRIZZLIE) Coarse screen used to remove oversize pieces from earth or blasted rock; gate or closure on a chute.

GROIN Contact between the upstream or downstream face of a dam and the abutments; area along the contact (or intersection) of the face of a dam with the abutments.

GROSS ALPHA/BETA PARTICLE ACTIVITY Total radioactivity due to alpha or beta particle emissions as inferred from measurements on a dry sample.

GROSS CROP VALUE Sum of annual receipts from sale of crops produced.

GROSS GENERATION Total amount of electrical energy produced by a generating station or stations, measured at generator terminals.

GROSS POWER-GENERATION POTENTIAL Installed power generation capacity that landfill gas can support.

GROUND BLIZZARD WARNING When blizzard conditions are solely caused by blowing and drifting snow.

GROUND CLUTTER Pattern of radar echoes from fixed ground targets (buildings, hills, etc.) near the radar.

GROUND COVER Plants grown to keep soil from eroding.

GROUND FLOW SYSTEM Water saturated aggregate of aquifers and confining units in which water enters and moves and which is bounded by a basal confining unit that does not allow any vertical water movement and by zones of interaction with the Earth's surface and with surface water systems.

GROUND FOG Fog produced over the land by the cooling of the lower atmosphere as it comes in contact with the ground.

GROUND FOG (GF) Fog produced over the land by the cooling of the lower atmosphere as it comes in contact with the ground.

GROUND HEAT FLUX Rate of transfer of heat from the ground to the Earth's surface; component of the surface energy budget.

GROUND MOTION General term including all aspects of ground motion, namely particle acceleration, velocity, or displacement, from an earthquake or other energy source.

GROUND MOTION PARAMETERS Numerical values representing vibratory ground motion, such as particle acceleration, velocity, and displacement, frequency content, predominant period, spectral intensity, and duration.

GROUND RECEIVE SITES Satellite dish and associated computer which receives signals from the GOES satellite, decodes the information, and transmits it to a another site for further processing.

GROUND SOURCE HEAT PUMP (GSHP) Electrically powered systems that tap the stored energy of the Earth.

GROUND STROKE Current that propagates along the ground from the point where a direct stroke of lightning hits the ground.

GROUNDED ICE Ice that has run aground or is contact with the ground underneath it.

GROUND-FAULT CIRCUIT INTERRUPTER (GFCI) Electrical device designed to protect people (not equipment) from electrical shock.

GROUND-PENETRATING RADAR Geophysical method that uses high frequency electromagnetic waves to obtain subsurface information.

GROUNDWATER That part of the subsurface water that is in the saturated zone; water under ground, such as in wells, springs and aquifers; generally all subsurface water as distinct from surface water; water that flows or seeps downward and saturates soil or rock; water stored underground in rock crevices and in the pores of geologic materials that make up the Earth's crust; phreatic water.

GROUNDWATER BASIN Groundwater system that has defined boundaries and may include more than one aquifer of permeable materials, which are capable of furnishing a significant water supply.

GROUNDWATER DISCHARGE Water released from the zone of saturation; volume of water released; groundwater entering near coastal waters which has been contaminated by landfill leachate, deep well injection of hazardous wastes, septic tanks, etc.

GROUNDWATER DISINFECTION RULE 1996 amendment of the Safe Drinking Water Act that requires disinfection as for all public water systems, including surface waters and groundwater systems.

GROUNDWATER DIVIDE Line on a water table where on either side of which the water table slopes downward.

GROUNDWATER FLOW Movement of water in the zone of saturation; streamflow which results from precipitation that infiltrates into the soil and eventually moves through the soil to the stream channel (i.e., baseflow, dry-weather flow).

GROUNDWATER FLOW MODEL Application of mathematical model to represent a regional or site-specific groundwater flow system.

GROUNDWATER FLOW SYSTEM Underground pathway by which groundwater moves from areas of recharge to areas of discharge.

GROUNDWATER HYDROLOGY Branch of hydrology that specializes in groundwater, its occurrence and movements, its replenishment and depletion, the properties of rocks that control groundwater movement and storage, and the methods of investigation and utilization of groundwater.

GROUNDWATER MINING (OVERDRAFT) Pumping of groundwater for irrigation or other uses, at rates faster than the rate at which the groundwater is being recharged.

GROUNDWATER MODELING CODE Computer code used in groundwater modeling to represent a non-unique, simplified mathematical description of the physical framework, geometry, active processes, and boundary conditions present in a reference subsurface hydrologic system.

GROUNDWATER OUTFLOW That part of the discharge from a drainage basin that occurs through the groundwater.

GROUNDWATER OVERDRAFT Pumpage of groundwater in excess of safe yield.

GROUNDWATER RECHARGE Inflow of water to a ground-water reservoir from the surface; flow to groundwater storage from precipitation, infiltration from streams, and other sources of water.

GROUNDWATER RUNOFF That part of the runoff that has passed into the ground, has become groundwater, and has been discharged into a stream channel as spring or seepage water.

GROUNDWATER TABLE Upper boundary of groundwater where water pressure is equal to atmospheric pressure.

GROUNDWATER UNDER THE DIRECT INFLUENCE (UDI) OF SURFACE WATER Any water beneath the surface of the ground with significant occurrence of insects or other microorganisms, algae, or large-diameter pathogens; any water beneath the surface of the ground with significant and relatively rapid shifts in water characteristics such as turbidity, temperature, conductivity, or pH which closely correlate to climatological or surface water conditions.

GROUNDWATER, CONFINED Ground water under pressure significantly greater than atmospheric, with its upper limit the bottom of a bed with hydraulic conductivity distinctly lower than that of the material in which the confined water occurs.

GROUP VELOCITY Speed at which a particular wave front or swell train advances.

GROUT Fluid mixture of cement and water or sand, cement, and water used to seal joints and cracks in a rock foundation; fluid material that is injected into soil, rock, concrete, or other construction material to seal openings and to lower the permeability and/or provide additional structural strength.

GROUT BLANKET Area of the foundation systematically grouted to a uniform shallow depth.

GROUT CAP Concrete filled trench or pad encompassing all grout lines constructed to impede surface leakage and to provide anchorage for grout connections.

GROUT CURTAIN (GROUT CUTOFF) One or more zones, usually thin, in the foundation into which grout is injected to reduce seepage under or around a dam; barrier produced by injecting grout into a vertical zone, usually narrow (horizontally), and in the foundation to reduce seepage under a dam.

GROUTING Filling cracks and crevices with a cement mixture.

GROWING DEGREE DAY Number of degrees that the average temperature is above a baseline value.

GROWING SEASON Frost-free period of the year during which the climate is such that crops can be produced; period of time between the last killing frost of spring and the first killing frost of autumn.

GROWLER Similar to a bergy bit, but smaller, extending less than 1 meter above the sea surface and occupying an area of 20 m^2 or less.

GRT Great.

GRTST Greatest.

GRUBBING Removal of stumps, roots, and vegetable matter from the ground surface after clearing and prior to excavation.

GSP Maximum 5-second peak gust during the measurement hour, reported at the last hourly 10-minute segment.

GST Peak 5 or 8 second gust speed (m/s) measured during the eight-minute or two-minute period.

GSTY Gusty.

GTR Greater.

GUARD GATE (EMERGENCY) First gate in a series of flow controls, remaining open while downstream gates or valves are operating; gate, usually located between the emergency and regulating gates, used in the closed position to permit servicing of the downstream regulating gate(s) or valve(s) or the downstream conduit.

GULF STREAM Warm water current extending from the Gulf of Mexico and Florida up the U.S. east coast then east northeast to Iceland and Norway.

GULLY EROSION Severe erosion in which trenches are cut to a depth greater than 30 centimeters.

GULLYING Small-scale stream erosion.

GUNGE Anything in the atmosphere that restricts visibility for storm spotting (e.g., fog, haze, precipitation, widespread low clouds, etc.).

GUST Rapid fluctuation of wind speed with variations of 10 knots or more between peaks and lulls.

GUST FRONT Leading edge of gusty surface winds from thunderstorm downdrafts.

GUSTNADO (OR GUSTINADO) Small, whirlwind that forms as an eddy in thunderstorm outflows.

GUTTATION Removal of liquid water through stomata from the healthy leaf or plant stem.

GYRES Oceanic current systems of planetary scale driven by the global wind system.

H

H$_s$ Significant Wave Height.

H5 500 millibar level height that is near 5,500 meters (18,000 ft) in a standard atmosphere.

H7 700 millibar level height that is near 3,000 meters (10,000 ft) in a standard atmosphere.

H8 750 millibar level height.

HABITAT Place where a population (e.g., human, animal, plant, microorganism) lives and its surroundings, both living and non-living; all non-living (physical) aspects of the aquatic ecosystem, although living components like aquatic macrophytes and riparian vegetation also are usually included.

HABITAT CAPABILITY Estimated carrying capacity of an area to support a wildlife, fish, or sensitive plant population.

HABITAT INDICATOR Physical, chemical, or biological attribute measured to characterize the conditions necessary to support an organism, population, community, or ecosystem in the absence of stressors (e.g. pollutants).

HABITAT QUALITY INDEX Qualitative description of instream habitat and riparian conditions surrounding the reach sampled.

HAGUE LINE North Atlantic boundary between the U.S. and Canada fishing waters as determined by the World Court in The Hague, Netherlands.

HAIL Showery precipitation in the form of irregular pellets or balls of ice more than 5 mm in diameter, falling from a cumulonimbus cloud.

HAIL CONTAMINATION Limitation in NEXRAD rainfall estimates whereby abnormally high reflectivitys associated with hail are converted to rainfall rates and rainfall accumulations.

HAIL INDEX Indication of whether the thunderstorm structure of each storm identified is conducive to the production of hail.

HAIL SIZE Diameter of hailstones.

HAIL SPIKE Area of reflectivity extending away from the radar immediately behind a thunderstorm with extremely large hail.

HAINES INDEX (LOWER ATMOSPHERE STABILITY INDEX)
Index composed of a stability term

and a moisture term that indicates potential for large fire growth.

HALF-LIFE Time required for a pollutant to lose one-half of its original coconcentrationor; time required for half of the atoms of a radioactive element to undergo self-transmutation or decay; time required for the elimination of half a total dose from the body.

HALO Bright circles or arcs centered on the Sun or Moon, caused by the refraction or reflection of light by ice crystals suspended in the Earth's atmosphere and exhibiting prismatic coloration ranging from red inside to blue outside.

HALOGEN Type of incandescent lamp with higher energy-efficiency that standard ones.

HALON Bromine-containing compounds with long atmospheric lifetimes whose breakdown in the stratosphere causes depletion of ozone.

HALOPHYTIC Salt-loving; plants that thrive in soils that contain salt and/or sodium; plant that grows in salty or alkaline soil.

H-ALPHA Absorption line of neutral hydrogen that falls in the red part of the visible spectrum and is convenient for solar observations.

HAMMER MILL High-speed machine that uses hammers and cutters to crush, grind, chip, or shred solid waste.

HANGER Support for pipe.

HANGING (ICE) DAM Mass of ice composed mainly of frazil or broken ice deposited underneath an ice cover in a region of low flow velocity.

HARD WATER Alkaline water containing dissolved salts that interfere with some industrial processes and prevent soap from sudsing; water considered hard if it has a hardness greater than the typical hardness of water from the region.

HARDNESS Water-quality indication of the concentration of alkaline salts in water, mainly calcium and magnesium; property of water that causes the formation of an insoluble residue when the water is used with soap and a scale in vessels in which water has been allowed to evaporate; physical-chemical characteristic that commonly is recognized by the increased quantity of soap required to produce lather.

HARDPAN Hard, impervious layer, composed chiefly of clay, that is cemented by relatively insoluble materials, that does not become plastic when mixed with water, and definitely limits the downward movement of water and roots.

HARMONIC MEAN FLOW Number of daily flow measurements divided by the sum of the reciprocals of the flows (i.e. reciprocal of the mean of reciprocals).

HARROW Agricultural tool that loosens and works the ground surface.

HARVEST Number of fish that are caught and kept.

HATCHERY Place for hatching fish eggs.

HAUL DISTANCE Average distance a grading material is moved from cut to fill; distance measured along the center line or most direct practical route between the center of the mass of excavation and the center of mass of the fill as finally placed; distance material is moved.

HAULER Garbage collection company that offers complete refuse removal service.

HAUNCHES (HAUNCH) Outside areas between the springline and the bottom of a pipe; sides of the lower third of the circumference of a pipe.

HAZARD Something that creates the potential for adverse consequences such as loss of life, property damage, and adverse social and environmental impacts; potential for radiation, a chemical or other pollutant to cause human illness or injury; inherent toxicity of a compound.

HAZARD (HZ) Situation that creates the potential for adverse consequences such as loss of life, property damage, or other adverse impacts.

HAZARD ASSESSMENT Evaluating the effects of a stressor or determining a margin of safety for an organism by comparing the concentration which causes toxic effects with an estimate of exposure to the organism.

HAZARD CLASS Group of materials that share a common major hazardous property such as radioactivity or flammability.

HAZARD CLASSIFICATION Rating on the potential consequences of failure.

HAZARD COMMUNICATION STANDARD OSHA regulation that requires chemical manufacturers, suppliers, and importers to assess the hazards of the chemicals that they make, supply, or import, and to inform employers, customers, and workers of these hazards through MSDS information.

HAZARD EVALUATION Component of risk evaluation that involves gathering and evaluating data on the types of health injuries or diseases that may be produced by a chemical and on the conditions of exposure under which such health effects are produced.

HAZARD IDENTIFICATION Determining if a chemical or a microbe can cause adverse health effects in humans and what those effects might be.

HAZARD POTENTIAL Possible adverse incremental consequences that result from failure or misoperation.

HAZARD POTENTIAL CLASSIFICATION System that categorizes according to the degree of adverse incremental consequences of a failure or misoperation.

HAZARD QUOTIENT Ratio of estimated site-specific exposure to a single chemical from a site over a specified period to the estimated daily exposure level, at which no adverse health effects are likely to occur.

HAZARD RATIO Term used to compare an animal's daily dietary intake of a pesticide to its (Lethal Dose) LD 50 value.

HAZARDOUS AIR POLLUTANTS Air pollutants (e.g., asbestos, beryllium, mercury, benzene, coke oven emissions, radionuclides, and vinyl chloride) that are not covered by ambient air quality standards but which, as defined in the Clean Air Act, may present a threat of adverse human health effects or adverse environmental effects.

HAZARDOUS CHEMICAL Any hazardous material requiring an MSDS under OSHA's Hazard Communication Standard.

HAZARDOUS MATERIALS Materials that pose the potential for grave, immediate, future, and genetic injury and illness when handled without proper equipment and precautions.

HAZARDOUS MATERIALS RESPONSE TEAM (HMRT) Team of specially trained personnel who respond to a hazardous materials incident.

HAZARDOUS RANKING SYSTEM Principal screening tool to evaluate risks to public health and the environment associated with abandoned or uncontrolled hazardous waste sites.

HAZARDOUS SEAS WARNING Warning for wave heights and/or wave steepness values meeting or exceeding locally defined warning criteria.

HAZARDOUS SEAS WATCH Watch for an increased risk of a hazardous seas warning event to meet Hazardous Seas Warning criteria but its occurrence, location, and/or timing is still uncertain.

HAZARDOUS SUBSTANCE Any material that poses a threat to human health and/or the environment (i.e., substances that are toxic, corrosive, ignitable, explosive, or chemically reactive); any substance designated to be reported if a designated quantity of the substance is spilled in the waters of the United States or is otherwise released into the environment.

HAZARDOUS WASTE By-products of society that can pose a substantial or potential hazard to human health or the environment when improperly managed.

HAZARDOUS WASTE INDICATORS Indicators that may signal the presence of hazardous materials include stressed vegetation or unusual lack of vegetation; dead or sick domestic stock, wildlife, or birds; fish kills or otherwise unexplained stream sterility or diminished species and numbers of flora and fauna; unusual coloration or discoloration of the land surface; and acrid or other chemical odors.

HAZARDOUS WASTE LANDFILL Excavated or engineered site where hazardous waste is deposited and covered.

HAZARDOUS WASTE MINIMIZATION Reducing the amount of toxicity or waste produced by a facility via source reduction or environmentally sound recycling.

HAZARDOUS WASTE SITES Any spill, authorized or unauthorized dumping, abandoned, or inactive waste disposal sites containing or suspected of containing hazardous materials.

HAZARDOUS WEATHER OUTLOOK Narrative statement produced by the National Weather Service to provide information regarding the potential of significant weather expected during the next 1 to 5 days.

HAZARDS ANALYSIS Procedures used to identify potential sources of release of hazardous materials from fixed facilities or transportation accidents; procedures used to determine the vulnerability of a geographical area to a release of hazardous materials; procedures used to compare hazards to determine which present greater or lesser risks to a community.

HAZARDS ASSESSMENT Provides emergency managers, planners, forecasters and the public advance notice of potential hazards related to climate, weather and hydrological events.

HAZARDS IDENTIFICATION Providing information on which facilities have extremely hazardous substances, what those chemicals are, how much there is at each facility, how the chemicals are stored, and whether they are used at high temperatures.

HAZE (HZ) Aggregation in the atmosphere of very fine, widely dispersed, solid or liquid particles, or both, giving the air an opalescent appearance that subdues colors.

H-COMPONENT OF THE GEOMAGNETIC FIELD Components of the geomagnetic field at the surface of the Earth.

HDRAIN Hourly Digital Rainfall Product of the WSR-88D.

HEAD Differential of pressure causing flow in a fluid system; height of water above a specified point; vertical distance between two points in a fluid; vertical distance that would statically result from the velocity of a moving fluid; difference between the pool height and tailwater height; height above a datum plane (such as sea level) of the column of water that can be supported by the hydraulic pressure at a given point in a groundwater system; back-pressure against a pump.

HEAD DEPENDENT BOUNDARY See MIXED BOUNDARY.

HEAD LOSS Energy lost (decrease in total head) from a flowing fluid due to friction, transitions, bends, etc.

HEAD RACE Channel which directs water to a water wheel; a forebay.

HEAD, STATIC Vertical distance between two points in a fluid.

HEAD, VELOCITY Vertical distance that would statically result from the velocity of a moving fluid.

HEADER See MANIFOLD.

HEADING In a tunnel, a digging face and its work area.

HEADSPACE Vapor mixture trapped above a solid or liquid in a sealed vessel.

HEADWALL Upstream wall.

HEADWARD EROSION Erosion which occurs in the upstream end of the valley of a stream, causing it to lengthen its course in such a direction.

HEADWATER Streams at the source of a river; source and upper part of a stream or reservoir; water upstream of a dam or powerhouse; water upstream from a point on a stream; small streams that come together to form a river.

HEADWATER ADVISORY PROGRAM (ADVIS) Program that uses the Antecedent Precipitation Index (API) method of estimating runoff, unit hydrograph theory and stage-discharge ratings to produce hydrologic forecasts for headwater basins.

HEADWATER ADVISORY TABLE Pre-computed matrix of values allows a forecaster to ascertain an anticipated crest or rise on a small river or stream for a variety of rainfall events and soil moisture conditions.

HEADWATER BASIN Basin at the headwaters of a river.

HEALTH ADVISORY Non-regulatory levels of contaminants in drinking water that may be used as guidance in the absence of regulatory limits.

HEALTH ADVISORY LEVEL Non-regulatory health-based reference level of chemical traces (usually in ppm) in drinking water at which there are no adverse health risks when ingested over various periods of time.

HEALTH ASSESSMENT Evaluation of available data on existing or potential risks to human health posed by a Superfund site.

HEAP Soil carried above the sides of a body or bucket.

HEAT ADVISORY Issued within 12 hours of the onset of the following conditions: heat index of at least 105 °F but less than 115 °F for less than 3 hours per day, or nighttime lows above 80 °F for 2 consecutive days.

HEAT EXHAUSTION Mild form of heat stroke, characterized by faintness, dizziness, and heavy sweating.

HEAT INDEX (HI) Accurate measure of how hot it really feels when the Relative Humidity (RH) is added to the actual air temperature.

HEAT ISLAND EFFECT Elevated temperatures over an urban area caused by structural and pavement heat fluxes, and pollutant emissions.

HEAT LIGHTNING Lightning that occurs at a distance such that thunder is no longer audible.

HEAT PIPE Device that can quickly transfer heat from one point to another.

HEAT PUMP Electric device with both heating and cooling capabilities that extracts heat from one medium at a lower (the heat source) temperature and transfers it to another at a higher temperature (the heat sink), thereby cooling the first and warming the second; takes thermal energy from the groundwater and transfers it to the space being heated; removes heat from a building and transfers it to the ground water.

HEAT PUMP/AIR CONDITIONING RETURN FLOW WELLS Class V wells that reinject groundwater that has been passed through a heat exchanger to heat or cool buildings.

HEAT RECOVERY SYSTEM Process by which the exhaust air preheats the supply air when the outdoor air is cooler than the inside air; when the outside air is warmer than the inside air, the exhaust air will cool the supply air.

HEAT STROKE Condition resulting from excessive exposure to intense heat, characterized by high fever, collapse, and sometimes convulsions or coma.

HEAT WAVE Period of abnormally and uncomfortably hot and unusually humid weather.

HEATING DEGREE DAYS (HDD) Form of degree day used to estimate energy requirements for heating.

HEATING, VENTILATION, AND AIR CONDITIONING (HVAC) Controls the ambient environment (temperature, humidity, air flow, and air filtering) of a building and must be planned for and operated along with other data center components such as computing hardware, cabling, data storage, fire protection, physical security systems, and power.

HEAVE Upward movement of land surfaces or structures due to subsurface expansion of soil or rock, or vertical faulting of rock; upward movement of soil caused by expansion or displacement resulting from phenomena such as moisture absorption, removal of overburden, driving of piles, frost action, and loading of an adjacent area.

HEAVY FREEZING SPRAY Accumulation of freezing water droplets on a vessel at a rate of 2 cm per hour or greater caused by some appropriate combination of cold water, wind, cold air temperature, and vessel movement.

HEAVY FREEZING SPRAY WARNING Warning for an accumulation of freezing water droplets on a vessel at a rate of 2 cm per hour or greater caused by some appropriate combination of cold water, wind, cold air temperature, and vessel movement.

HEAVY FREEZING SPRAY WATCH Watch for an increased risk of a heavy freezing spray event to meet Heavy Freezing Spray Warning criteria but its occurrence, location, and/or timing is still uncertain.

HEAVY METALS Metallic elements with high atomic weights (e.g., mercury, chromium, cadmium, arsenic, and lead) that can damage living things at low concentrations and tend to accumulate in the food chain.

HEAVY SNOW Snowfall accumulating to 4″ or more in depth in 12 hours or less; snowfall accumulating to 6″ or more in depth in 24 hours or less.

HEAVY SNOW WARNING Issued by the National Weather Service when snowfall of 6 inches (15 cm) or more in 12 hours or 8 inches (20 cm) or more in 24 hours is imminent or occurring.

HEAVY SURF ADVISORY Advisory issued by the National Weather Service for fast moving deep water waves which can result in big breaking waves in shallow water.

HECTARE Measure of area in the metric system similar to an acre (1 hectare = 10,000 m^2 = 2.4711 acres).

HECTOPASCAL (HPA) Unit of pressure equal to a millibar (1 hPa = 1 mb).

HEEL Junction of the upstream face of a gravity or arch dam with the ground surface.

HEIGHT Height above sea level of a pressure level.

HEIGHT, ABOVE GROUND Maximum height from natural ground surface to the top of a dam.

HEIGHT, HYDRAULIC Vertical difference between the maximum design water level and the lowest point in the original streambed.

HEIGHT, STRUCTURAL Vertical distance between the lowest point of the excavated foundation to the top of the dam.

HELICAL Spiral.

HELICITY Property of a moving fluid which represents the potential for helical flow (i.e. flow which follows the pattern of a corkscrew) to evolve.

HEPTACHLOR Insecticide that was banned on some food products.

HERBACEOUS With characteristics of an herb; a plant with no persistent woody stem above ground.

HERBICIDE Chemical pesticide designed to control or destroy plants, weeds, or grasses.

HERBIVORE Animal that feeds on plants.

HERPETOFAUNA Reptiles and amphibians.

HERTZ (HZ) Number of complete electromagnetic cycles or waves in one second of an electrical or electronic circuit; unit of frequency equal to one cycle per second.

HETEROGENEITY Characteristic of a medium in which material properties vary from point to point everywhere.

HETEROTROPHIC ORGANISMS Species that are dependent on organic matter for food.

HIC Hydrologist In Charge.

HIGH (HI) Region of high pressure; anticyclone.

HIGH CLOUDS Clouds that have bases between 16,500 and 45,000 feet in the mid-latitudes.

HIGH DAM Dam over 300 feet high.

HIGH END EXPOSURE (DOSE) ESTIMATE Estimate of exposure, or dose level received anyone in a defined population that is greater than the 90th percentile of all individuals in that population, but less than the exposure at the highest percentile in that population.

HIGH ENERGY EVENT Flares (class two or more) with outstanding Centimetric Bursts and SID.

HIGH FREQUENCY (HF) Portion of the radio frequency spectrum between 3 and 30 MHz.

HIGH HAZARD Hazard classification in which more than 6 lives would be in jeopardy and excessive economic loss (urban area including extensive community, industry, agriculture, or outstanding natural resources) would occur as a direct result of failure.

HIGH INTENSITY DISCHARGE Generic term for mercury vapor, metal halide, and high pressure sodium lamps and fixtures.

HIGH RISK (OF SEVERE THUNDERSTORMS) Severe weather is expected to affect more than 10 percent of the area.

HIGH SEAS FORECAST (HSF) Marine forecasts for the major oceans of the world.

HIGH SURF Large waves breaking on or near the shore resulting from swells spawned by a distant storm.

HIGH SURF ADVISORY Issued when breaking wave action poses a threat to life and property within the surf zone.

HIGH SURF WARNING Issued when breaking wave action results in an especially heightened threat to life and property within the surf zone.

HIGH TIDE Maximum height reached by each rising tide.

HIGH WIND Sustained wind speeds of 40 mph or greater lasting for 1 hour or longer; winds of 58 mph or greater for any duration.

HIGH WIND ADVISORY Issued by the National Weather Service when high wind speeds may pose a hazard.

HIGH WIND WARNING Issued by the National Weather Service when high wind speeds may pose a hazard or is life threatening.

HIGH WIND WATCH Issued by the National Weather Service when there is the potential of high wind speeds developing that may pose a hazard or is life threatening.

HIGH-DENSITY POLYETHYLENE Material used to make plastic bottles and other products that produces toxic fumes when burned.

HIGHEST DOSE TESTED Highest dose of a chemical or substance tested in a study.

HIGH-LEVEL NUCLEAR WASTE FACILITY Plant designed to handle disposal of used nuclear fuel, high-level radioactive waste, and plutonium waste.

HIGH-LEVEL RADIOACTIVE WASTE (HLRW) Waste generated in core fuel of a nuclear reactor, found at nuclear reactors or by nuclear fuel reprocessing.

HIGH-LINE JUMPERS Pipes or hoses connected to fire hydrants and laid on top of the ground to provide emergency water service for an isolated portion of a distribution system.

HIGH-PRECIPITATION STORM (HP STORM) Thunderstorm in which heavy precipitation (often including hail) falls on the trailing side of the mesocyclone.

HIGH-PRECIPITATION SUPERCELL (HP SUPERCELL) See HIGH-PRECIPITATION STORM.

HIGH-PRESSURE GATE Gate consisting of a rectangular leaf encased in a body and bonnet and equipped with a hydraulic hoist for moving the gate leaf.

HIGH-RISK COMMUNITY Community located within the vicinity of numerous sites of facilities or other potential sources of envienvironmental exposure/health hazards which may result in high levels of exposure to contaminants or pollutants.

HIGH-SPEED STREAM Feature of the solar wind having velocities that are about double average solar wind values.

HIGH-TO-LOW-DOSE EXTRAPOLATION Process of prediction of low exposure risk to humans and animals from the measured high-exposure-high-risk data involving laboratory animals.

HILSENHOFF?S BIOTIC INDEX (HBI) Indicator of organic pollution that uses tolerance values to weight taxa abundances that usually increases with pollution.

HINGE CRACK Crack caused by significant changes in water level.

HISTORIC FLOWS Collection of recorded flow data for a stream during the period of time in which steam gauges were in operation.

HLS Hurricane Local Statement.

HMD Hemispheric Map Discussion.

HOAR FROST Deposit of interlocking crystals formed by direct sublimation on objects, usually those of small diameter freely exposed to the air, such as tree branches, plants, wires, poles, etc.

HOD Hydrologist on Duty.

HODOGRAPH Polar coordinate graph which shows the vertical wind profile of the lowest 7000 meters of the atmosphere.

HOE Shovel that digs by pulling a broom-and-stick-mounted bucket toward itself.

HOGBACK Ridge formed by erosion of resistant, steeply inclined sedimentary layers.

HOIST Mechanism by which a bucket or blade is lifted, or the process of lifting it.

HOLDING POND Pond or reservoir, usually made of earth, built to store polluted runoff.

HOLDING TIME Maximum amount of time a sample may be stored before analysis.

HOLLOW GRAVITY DAM (CELLULAR GRAVITY DAM) Dam which has the outward appearance of a gravity dam but is of hollow construction; dam constructed of concrete and/or masonry on the outside but having a hollow interior and relying on its weight for stability.

HOLLOW JET VALVE Valve having a closing member that moves upstream to shut off the flow; device for regulating high-pressure outlets.

HOLLOW STEM AUGER DRILLING Conventional drilling method that uses augurs to penetrate the soil.

HOMEOWNER WATER SYSTEM Any water system which supplies piped water to a single residence.

HOMOGENEITY Characteristic of a medium in which material properties are identical everywhere.

HOMOGENEOUS EARTHFILL DAM Embankment dam construction throughout of more or less uniform earth materials, except for possible inclusion of internal drains or blanket drains.

HOMOLOGOUS FLARES Solar flares that occur repetitively in the same active region, with essentially the same position and with a common pattern of development.

HOOD CAPTURE EFFICIENCY Ratio of the emissions captured by a hood and directed into a control or disposal device, expressed as a percent of all emissions.

HOOK ECHO Radar reflectivity pattern characterized by a hook-shaped extension of a thunderstorm echo, usually in the right-rear part of the storm (relative to its direction of motion).

HOPPER Storage bin or a funnel that is loaded from the top, and discharges through a door or chute in the bottom.

HORIZON Distant line along which the Earth and sky appear to meet.

HORNBLENDE Black blade-like mineral common in igneous and metamorphic rocks.

HORNFELS Fine-grained, gray-green metamorphic rock produced by "baking" of sedimentary rocks by an igneous intrusion in which sedimentary features may still be preserved.

HORSEHEAD (FALSE SET) Temporary support for forepoles used in tunneling soft ground.

HORSEPOWER (HP) Measurement of power that includes the factors of force and speed; force required to lift 33,000 pounds one foot in one minute.

HORST See FAULT-BLOCK.

HOST Organism, typically a bacterium, into which a gene from another organism is transplanted; animal infected or parasitized by another organism.

HOURLY PRECIPITATION DATA (HPD) Contains data on nearly 3,000 hourly precipitation stations.

HOUSEHOLD HAZARDOUS WASTE Hazardous products used and disposed of by residential as opposed to industrial consumers.

HOUSEHOLD WASTE (DOMESTIC WASTE) Solid waste, composed of garbage and rubbish, which normally originates in a private home or apartment house.

HOWELL BUNGER VALVE See FIXED CONE VALVE.

HPC Hydrometeorological Prediction Center.

HR Hour.

HRL Hydrological Research Laboratory at the Office of Hydrology (OH).

HRS Hours.

HSA (HYDROLOGIC SERVICE AREA) Geographical area assigned to Weather Service Forecast Office's/Weather Forecast Office's that embraces one or more rivers.

HSB Hydrologic Systems Branch in the Office of Hydrology (OH).

HTC Hydrometeorological Training Council.

HUB Enlarged end of a bell and spigot cast-iron pipe.

HUMAN ENVIRONMENT Natural and physical environment and the relationship of people with that environment, including all combinations of physical, biological, cultural, social, and economic factors in a given area.

HUMAN EQUIVALENT DOSE Dose that when administered to humans produces an effect equal to that produced by a dose in animals.

HUMAN EXPOSURE EVALUATION Describing the nature and size of the population exposed to a substance and the magnitude and duration of their exposure.

HUMAN HEALTH ADVISORY Guidance provided, in the absence of regulatory limits, to describe acceptable contaminant levels in drinking water or edible fish.

HUMAN HEALTH RISK Likelihood that a given exposure or series of exposures may have damaged or will damage the health of individuals.

HUMIDITY Measure of the water vapor content of the air.

HUMIDITY RECOVERY Change in relative humidity over a given period of time.

HUMMOCK Hillock of broken ice which has been forced upward by pressure.

HUMMOCKED ICE Ice piled haphazardly one piece over another to form an uneven surface.

HUMUS Decayed organic matter; dark fluffy swamp soil composed chiefly of decayed vegetation; brown or black material formed by the partial decomposition of vegetable or animal matter; organic portion of the soil remaining after prolonged microbial decomposition.

HURRICANE Tropical cyclone in the Atlantic, Caribbean Sea, Gulf of Mexico, or eastern Pacific, which the maximum 1-minute sustained surface wind is 64 knots (74 mph) or greater.

HURRICANE FORCE WIND WARNING Warning for sustained winds, or frequent gusts, of 64 knots (74 mph) or greater, either predicted or occurring, and not directly associated with a tropical cyclone.

HURRICANE FORCE WIND WATCH Watch for an increased risk of a hurricane force wind event for sustained surface winds, or frequent gusts, of 34 knots 64 knots (74 mph) or greater, but its occurrence, location, and/or timing is still uncertain.

HURRICANE LOCAL STATEMENT Public release prepared by local National Weather Service offices in or near a threatened area giving specific details for its county/parish warning area on weather conditions, evacuation decisions made by local officials and other precautions necessary to protect life and property.

HURRICANE SEASON Part of the year having a relatively high incidence of tropical cyclones where in the Atlantic, Caribbean, and Gulf of Mexico, and central North Pacific, the hurricane season is the period from June through November and in the eastern Pacific, May 15 through November 30.

HURRICANE WARNING Announcement that hurricane conditions (sustained winds of 74 mph or higher) are expected somewhere within the specified coastal area.

HURRICANE WATCH Announcement that hurricane conditions (sustained winds of 74 mph or higher) are possible within the specified coastal area.

HV Have.

HVY Heavy.

HWRF Hurricane Weather Research and Forecasting Model.

HWVR However.

HYDER FLARE Filament-associated two-ribbon flare, often occurring in spotless regions.

HYDRAUCONE Draft tube in which the emerging water impinges on a plate.

HYDRAULIC Powered by water; having to do with water in motion.

HYDRAULIC CONDUCTIVITY Rate at which water can move through a permeable medium (i.e. coefficient of permeability); capacity of a rock to transmit water; constant of proportionality which relates the rate of groundwater flow to the hydraulic head gradient; volume of water at the existing kinematic viscosity that will move in unit time under unit hydraulic gradient through a unit area measured at right angles to the direction of low.

HYDRAULIC EFFICIENCY Ability of hydraulic structure or element to conduct water with minimum energy loss; efficiency of a pump or turbine to impart energy to or extract energy from water.

HYDRAULIC FILL Fill material that is transported and deposited using water.

HYDRAULIC FILL DAM Dam constructed of materials, often dredged, that are conveyed and placed by suspension in flowing water.

HYDRAULIC FILL STRUCTURE Dam or impoundment made of hydraulic fill.

HYDRAULIC FLOW Atmospheric flow that is similar in character to the flow of water over an obstacle.

HYDRAULIC FRACTURING Creation of fractures within a reservoir that contains oil or natural gas to increase flow and maximize production.

HYDRAULIC GRADE LINE (HGL) Line whose plotted ordinate position represents the sum of pressure head plus elevation head for the various positions along a given fluid flow path, such as along a pipeline or a groundwater streamline; equal to the water surface in open channels; lies below the energy grade line by an amount equal to the velocity head at the section.

HYDRAULIC GRADIENT Change of hydraulic head per unit of distance in a given direction; the direction of groundwater flow due to changes in the depth of the water table; change in total hydraulic head per unit distance of flow at a given point and in the direction of groundwater flow.

HYDRAULIC HEAD Height of the free surface of a body of water above a given point beneath the surface; height of the water level at the headwork, or an upstream point, of a waterway, and the water surface at a given point downstream; height of a hydraulic grade line above the center line of a pressure pipe, at a given point; height above a datum plane of the column of water than can be supported by the hydraulic pressure at a given point in a groundwater system.

HYDRAULIC HEIGHT Height to which the water rises behind the dam, and is the difference between the lowest point in the original streambed at the axis or the centerline crest of the dam, or the invert of the lowest outlet works, whichever is lower, and the maximum controllable water surface.

HYDRAULIC JUMP Constant discharge that experiences a change in flow depth from below critical depth to above critical depth where velocity changes from supercritical to subcritical; method for dissipating energy in an open channel, sewer, or spillway; steady disturbance in the lee of a mountain, where the airflow passing over the mountain suddenly changes from a region of low depth and high velocity to a region of high depth and low velocity.

HYDRAULIC MEAN DEPTH See HYDRAULIC RADIUS.

HYDRAULIC MODEL Physical scale model of a river used for engineering studies.

HYDRAULIC PERMEABILITY Flow of water through a unit cross-sectional area of soil normal to the direction of flow when the hydraulic gradient is unity.

HYDRAULIC PROPERTIES Properties of solid and rock that govern the entrance of water and the capacity to hold, transmit and deliver water (e.g., porosity, effective porosity, specific retention, permeability and direction of maximum and minimum permeability).

HYDRAULIC RADIUS Ratio of flow area to wetted perimeter; right cross-sectional area of a stream of water divided by the length of that part of its periphery in contact with its containing conduit.

HYDRAULICS Having to do with the mechanical properties of water in motion and the application of these properties in engineering.

HYDRIC Characterized by, or thriving in, an abundance of moisture.

HYDRIC SOIL Soil that is wet long enough to periodically produce anaerobic conditions, thereby influencing the growth of plants.

HYDROCARBONS (HC) Chemical compounds that consist entirely of carbon and hydrogen.

HYDROCHLOROFLUOROCARBONS (HCFCS) Compounds containing carbon, hydrogen, chlorine, and fluorine; have potential to destroy stratospheric ozone.

HYDROELECTRIC PLANT Electric power plant using falling water as its motive force; power plant that produces electricity from the power of rushing water turning turbine-generators.

HYDROELECTRIC POWER Electrical energy produced by flowing water.

HYDROELECTRIC POWER WATER USE Use of water in the generation of electricity at plants where the turbine generators are driven by falling water.

HYDROELECTRIC UNIT (HYDROUNIT) Electric generator and turbine combination which is driven by water flow, thereby converting mechanical energy to electric energy.

HYDROFLUOROCARBONS (HFCS) Man-made compounds containing hydrogen, fluorine, and carbon; do not have the potential to destroy stratospheric ozone.

HYDROGEN SULFIDE Gas emitted during organic decomposition; byproduct of oil refining and burning.

HYDROGEOCHEMISTRY Chemistry of groundwater and surface water.

HYDROGEOLOGIC CONDITIONS Conditions stemming from the interaction of groundwater and the surrounding soil and rock.

HYDROGEOLOGIC CYCLE Natural process of recycling water from the atmosphere down to (and through) the Earth and back to the atmosphere again.

HYDROGEOLOGIST Person who studies and works with ground water.

HYDROGEOLOGY Geology of ground water, with particular emphasis on the chemistry and movement of water.

HYDROGRAPH Graphical representation of the stage or discharge as a function of time at a particular point on a watercourse; time-discharge curve of the unsteady flow of water; graph showing, for a given point on a stream, river, or conduit, the discharge, stage, velocity, available power, rate of runoff, or other property of water with respect to time.

HYDROGRAPH SEPARATION Process where the storm hydrograph is separated into baseflow components and surface runoff components.

HYDROGRAPH, BREACH OR DAM FAILURE Flood hydrograph resulting from a dam breach.

HYDROGRAPH, FLOOD Graph showing, for a given point on a stream, the discharge, height, or other characteristic of a flood with respect to time.

HYDROGRAPH, UNIT See UNIT HYDROGRAPH.

HYDROGRAPHIC SURVEY Instrumental survey to measure and determine characteristics of streams and other bodies of water within an area, including such things as location, areal extent, and depth of water in lakes or the ocean; width, depth, and course of streams; position and elevation of high water marks; location and depth of wells.

HYDROGRAPHY Scientific study of physical aspects of all waters on the Earth's surface.

HYDROLOGIC BOUNDARIES Physical boundaries of a hydrologic system.

HYDROLOGIC BUDGET Accounting of the inflow to, outflow from, and storage in, a hydrologic unit, such as a drainage basin, aquifer, soil zone, lake, reservoir, or irrigation project.

HYDROLOGIC CYCLE Movement or exchange of water between the atmosphere and Earth; natural recycling process powered by the Sun that causes water to evaporate into the atmosphere, condense and return to Earth as precipitation.

HYDROLOGIC EQUATION Equation balancing the hydrologic budget, water inventory equation (Inflow = Outflow + Change in Storage) which expresses the basic principle that during a given time interval the total inflow to an area must equal the total outflow plus the net change in storage.

HYDROLOGIC INDEX STATIONS Continuous-record gaging stations that have been selected as representative of streamflow patterns for their respective regions.

HYDROLOGIC MODEL Conceptual or physically-based procedure for numerically simulating a process or processes which occur in a watershed.

HYDROLOGIC REGIME Characteristic behavior and total quantity of water involved in a drainage basin.

HYDROLOGIC SERVICE AREA (HSA)
Geographical area assigned to Weather Service Forecast Office's/Weather Forecast Office's that embraces one or more rivers.

HYDROLOGIC SERVICES General term referring to the operations, products, verbal communications, and related forms of support provided by the National Weather Service for the Nation's streams, reservoirs, and other areas affected by surface water.

HYDROLOGIC UNIT Geographical area representing part or all of a surface drainage basin or distinct hydrologic feature such as a reservoir, lake, etc.; geographic area representing part or all of a surface drainage basin or distinct hydrologic feature as delineated by the U.S. Geological Survey on State Hydrologic Unit Maps where each hydrologic unit is assigned a number.

HYDROLOGIC UNIT CODE Number used to identify a geographic area representing part or all of a surface drainage basin or distinct hydrologic feature.

HYDROLOGISTS Individuals who study the applied science of hydrology and solve hydrologic problems.

HYDROLOGY Study of water: its properties, distribution, and behavior; science that treats the properties, occurrence, circulation, and distribution of the waters of the Earth and interaction with the land surface and underlying soils and rocks.

HYDROLYSIS Decomposition of organic compounds by interaction with water.

HYDROMET System of data collection platforms that gathers hydrometeorological data, and transmits that data via GOES Satellite to a computer downlink.

HYDROMETEOR Particle of condensed water (liquid, snow, ice, graupel, hail) in the atmosphere.

HYDROMETEOROLOGICAL AUTOMATED DATA SYSTEM (HADS)
Software that replaced GDDS to process and distribute the GOES DCP data and CADAS data collected from DCP's and LARCS.

HYDROMETEOROLOGICAL REPORT (HMR)
Series of hydrometeorological reports published by the National Weather Service (NWS) addressing meteorological issues related mainly to developing estimates of probable maximum precipitation used in the determination of the probable maximum flood for design of water control structures.

HYDROMETEOROLOGICAL TECHNICIANS
Individuals whose duties include data collection, quality control, gauge network maintenance, as well as the gathering and disseminating of data and products at National Weather Service Forecast Offices.

HYDROMETEOROLOGISTS Individuals who have the combined knowledge in the fields of both meteorology and hydrology which enables them to study and solve hydrologic problems where meteorology is a factor.

HYDROMETEOROLOGY Interdisciplinary science involving the study and analysis of the interrelationships between the atmospheric and land phases of water as it moves through the hydrologic cycle.

HYDROMETER Device for measuring the specific gravity of fluids.

HYDRONIC Ventilation system using heated or cooled water pumped through a building.

HYDROPHILIC Having a strong affinity (liking) for water; opposite of hydrophobic; not capable of uniting with or absorbing water.

HYDROPHYTE Any plant growing in water or on a substrate that is at least periodically deficient in oxygen as a result of excessive water content.

HYDROPNEUMATIC Water system, usually small, in which a water pump is automatically controlled by the pressure in a compressed air tank.

HYDROSPHERE Region that includes all the Earth's liquid water, frozen water, floating ice, frozen upper layer of soil, and the small amounts of water vapor in the Earth's atmosphere.

HYDROSTATIC Relating to pressure or equilibrium of fluids; pressures and forces resulting from the weight of a fluid at rest.

HYDROSTATIC HEAD Measure of pressure at a given point in a liquid in terms of the vertical height of a column of the same liquid which would produce the same pressure.

HYDROSTATIC PRESSURE Pressure exerted by the water at any given point in a body of water at rest.

HYETOGRAPH Graphical representation of rainfall intensity versus time that is often represented by a bar graph.

HYGROMETER Instrument which measures the humidity of the air.

HYGROSCOPIC Absorbing or attracting moisture from the air.

HYPERCONCENTRATED FLOW Moving mixture of sediment and water between 40 and 80 percent water by volume.

HYPERSENSITIVITY DISEASES Diseases characterized by allergic responses to pollutants; diseases most clearly associated with indoor air quality are asthma, rhinitis, and pneumonic hypersensitivity.

HYPOCENTER Point or focus within the Earth which is the center of an earthquake and the origin of its elastic waves; location within the Earth where the sudden release of energy is initiated; location where the slip responsible for an earthquake originates; focus of an earthquake.

HYPOLIMNETIC Pertaining to the lower, colder portion of a lake or reservoir which is separated from the upper, warmer portion (epilimnion) by the thermocline.

HYPOLIMNION Lower, or bottom, layer of a lake or reservoir with essentially uniform colder temperatures; lowest layer in a thermally stratified lake or reservoir.

HYPORHEIC ZONE Ground water habitats created by the movement of river water from the active channel to areas to the side and beneath the active channel; uniquely adapted organisms that can provide food for fish live in the groundwater habitat.

HYPOTHERMIA Rapid, progressive mental and physical collapse that accompanies the lowering of body temperature.

HYPOXIA/HYPOXIC WATERS Waters with dissolved oxygen concentrations of less than 2 ppm, the level generally accepted as the minimum required for most marine life to survive and reproduce.

HYPSOGRAPHY Elevation measurement system based on a sea level datum.

I

ICE AGE Time of widespread glaciation.

ICE BOOM Floating structure designed to retain ice.

ICE BRIDGE Continuous ice cover of limited size extending from shore to shore like a bridge.

ICE CRYSTALS Barely visible crystalline form of ice that has the shape of needles, columns or plates.

ICE FOG Type of fog, composed of suspended particles of ice; partly ice crystals but chiefly (especially when dense) ice particles, formed by direct freezing of supercooled water droplets with little growth directly from the vapor.

ICE GORGE Gorge or opening left in a jam after it has broken.

ICE JAM Stationary accumulation that restricts or blocks streamflow.

ICE NUCLEUS Any particle that serves as a nucleus in the formation of ice crystals in the atmosphere.

ICE PELLETS (IP) Pellets of ice composed of frozen or mostly frozen raindrops or refrozen partially melted snowflakes.

ICE PUSH Compression of an ice cover particularly at the front of a moving section of ice cover.

ICE RUN Flow of ice in a river.

ICE SHOVE On-shore ice push caused by wind, and currents, changes in temperature, etcetera.

ICE STORM Occasions when damaging accumulations of ice are expected during freezing rain situations.

ICE STORM WARNING Issued by the National Weather Service when freezing rain produces a significant and possibly damaging accumulation of ice.

ICE TWITCH Downstream movement of a small section of an ice cover.

ICEBERG Piece of a glacier which has broken off and is floating in the sea.

ICELANDIC LOW Semi-permanent, subpolar area of low pressure in the North Atlantic Ocean.

ICHTHYOLOGY Scientific study of fish.

ICING Coating of ice on a solid object.

IDEAL GAS LAWS Thermodynamic laws applying to perfect gases.

IDENTIFICATION CODE (EPA I.D. NUMBER) Unique code assigned to each generator, transporter, and treatment, storage, or disposal facility by regulating agencies to facilitate identification and tracking of chemicals or hazardous waste.

IFLOWS Integrated Flood Observing and Warning System.

IFR Instrument Flight Rules.

IGNEOUS Rock formed by molten magma; rock that forms from the solidification of molten rock or magma.

IGNEOUS ROCKS Rocks that have solidified from molten or partly molten material.

IGNITABLE Capable of burning or causing a fire.

IMHOFF CONE Clear, cone-shaped container used to measure the volume of settleable solids in a specific volume of water.

IMMEDIATELY DANGEROUS TO LIFE AND HEALTH (IDLH) Maximum level to which a healthy individual can be exposed to a chemical for 30 minutes and escape without suffering irreversible health effects or impairing symptoms.

IMMINENT HAZARD One that would likely result in unreasonable adverse effects on humans or the environment or risk unreasonable hazard to an endangered species during the time required for a pesticide registration cancellation proceeding.

IMMINENT THREAT High probability that exposure is occurring.

IMMISCIBILITY Inability of two or more substances or liquids to readily dissolve into one another, such as soil and water.

IMMOBILIZE To hold by a strong chemical attraction.

IMPACT Estimated loss associated with the risk; change in the chemical, physical, or biological quality or condition of a water body caused by external sources.

IMPAIRED Condition of the quality of water that has been adversely affected for a specific use by contamination or pollution.

IMPAIRMENT Detrimental effect on the biological integrity of a water body caused by impact that prevents attainment of the designated use.

IMPELLER Rotary pump member using centrifugal force to discharge a fluid into outlet passages; rotating set of vanes in a pump designed to pump or lift water.

IMPERMEABILITY Incapacity of a rock to transmit a fluid.

IMPERMEABLE Material that does not permit water to move through quickly; not easily penetrated; property of a material or soil that does not allow, or allows only with great difficulty, the movement or passage of water.

IMPERMEABLE BOUNDARY Conceptual representation of a natural feature such as a fault or depositional contact that places a boundary of significantly less-permeable material laterally adjacent to an aquifer.

IMPERMEABLE LAYER Layer of solid material, such as rock or clay, which does not allow water to pass through.

IMPERVIOUS Not permeable; not allowing liquid to pass through; ability to repel water; resistant to movement of water.

IMPERVIOUS CORE See CORE.

IMPERVIOUS MATERIAL Relatively waterproof soils.

IMPINGE To collide or strike.

IMPLEMENTATION MONITORING Documents whether or not management practices were applied as designed.

IMPLEMENTATION PROCESS Procurement, installation, and implementation of all the components of an Early Warning System; includes inspecting and exercising the system.

IMPORT Water piped or channeled into an area.

IMPORTS Municipal solid waste and recyclables that have been transported to a state or locality for processing or final disposition (but that did not originate in that state or locality).

IMPOUNDMENT Body of water created by a dam, dike, floodgate, or other barrier.

IMPROVED SINKHOLE Naturally occurring karst depression or other natural crevice found in volcanic terrain and other geologic settings which have been modified for the purpose of directing and emplacing fluids into the subsurface.

IMPROVEMENT Structural measures for the betterment, modernization, or enhancement of an existing facility or system to improve the social, economic, and environmental benefits of the project.

IMPT Important.

IMPULSE (IMPL) General term for any large-scale or mesoscale disturbance capable of producing upward motion (lift) in the middle or upper parts of the atmosphere.

IN HG Inches of Mercury.

IN SITU Original location; unmoved unexcavated; remaining at the site or in the subsurface.

IN VITRO Testing or action outside an organism (e.g., inside a test tube or culture dish).

IN VIVO Testing or action inside an organism.

INACTIVE CAPACITY (INACTIVE STORAGE) Reservoir capacity exclusive of and above the dead capacity from which the stored water is normally not available because of operating agreements or physical restrictions.

INACTIVE LAYER Depth of material beneath the active layer.

INACTIVE STORAGE Storage volume of a reservoir between the crest of the invert of the lowest outlet and the minimum operating level.

INACTIVE STORAGE CAPACITY Portion of capacity below which the reservoir is not normally drawn, and which is provided for sedimentation, recreation, fish and wildlife, aesthetic reasons, or for the creation of a minimum controlled operational or power head in compliance with operating agreements or restrictions.

INCH-DEGREES Product of inches of rainfall multiplied by the temperature in degrees above freezing (Fahrenheit

Scale), used as a measure of the snow melting capacity of rainfall.

INCHES OF MERCURY (IN HG) Unit of atmospheric pressure used in the United States.

INCHES OF RUNOFF Volume of water from runoff of a given depth over the entire drainage.

INCIDENT COMMAND POST Facility located at a safe distance from an emergency site, where the incident commander, key staff, and technical representatives can make decisions and deploy emergency manpower and equipment.

INCIDENT COMMAND SYSTEM (ICS) Combination of facilities, equipment, personnel, procedures, and communications operating within a common organizational structure to effectively respond to an emergency or disaster.

INCIDENT COMMANDER Person in charge of on-scene coordination of a response to an incident, usually a senior officer in a fire department.

INCINERATION Treatment technology involving destruction of waste by controlled burning at high temperatures.

INCINERATION AT SEA Disposal of waste by burning at sea on specially-designed incinerator ships.

INCINERATOR Furnace for burning waste under controlled conditions.

INCIPIENT MOTION Flow condition at which a given size bed particle just begins to move.

INCISED STREAM Stream that has cut its channel into the bed of the valley through degradation.

INCLINED STAFF GAUGE Gage that is placed on the slope of a stream bank and graduated so that the scale reads directly in vertical depth.

INCLINOMETER An instrument that measures offset from the vertical.

IN-CLOUD LIGHTNING (IC) Lightning that takes place within the cloud.

INCOMPATIBLE WASTE Waste unsuitable for mixing with another waste or material because it may react to form a hazard.

INCR Increase.

INCREMENTAL LOSS OF LIFE Measurement for assessing damage after a dam fails; difference between the projected loss of life had the dam not been built and the projected loss of life with the dam in place.

INDC Indicate.

INDEMNIFICATION Legal requirement that EPA pay certain end-users, dealers, and distributors for the cost of stock on hand at the time a pesticide registration is suspended.

INDEPENDENT POWER PRODUCER (IPP) Non-utility owned electric resources; non-utility power generator that is not a regulated utility, government agency, or qualifying facility (QF) under the Public Utility Regulatory Policies Act of 1978 (PURPA).

INDEX OF BIOTIC INTEGRITY (IBI) Aggregated number, or index, based on several attributes or metrics of a fish community that provides an assessment of biological conditions.

INDEX OF WETNESS Ratio of precipitation for a given year over the mean annual precipitation.

INDEX PERIOD Sampling period during which selection is based on the temporal behavior of the indicator and the practical considerations for sampling.

INDEX PRECIPITATION Index that can be used to adjust for bias in regional precipitation, often quantified as the expected annual precipitation.

INDIAN COUNTRY All land within the limits of any Indian reservation under the jurisdiction of the United States government, notwithstanding the issuance of any patent, and, including rights-of-way running through the reservation; all dependent Indian communities with the borders of the United States whether within the originally or subsequently acquired territory thereof, and whether within or without the limits of a state; all Indian allotments, the Indian titles to which have not been extinguished, including rights-of-ways running through the same.

INDIAN SUMMER Unseasonably warm period near the middle of autumn, usually following a substantial period of cool weather.

INDICATOR Organism, species, or community which indicates certain environmental conditions; substance that shows a visible change, usually of color, at a desired point in a chemical reaction; device that indicates the result of a measurement; pressure gauge or a moveable scale.

INDICATOR SITES Stream sampling sites located at outlets of drainage basins with relatively homogeneous land use and physiographic conditions where most indicator-site basins have drainage areas ranging from 20 to 200 square miles.

INDIGENOUS Native to a given area.

INDIGENOUS SPECIES Species that originally inhabited a particular geographic area.

INDIRECT DISCHARGE Introduction of pollutants from a non-domestic source into a publicly owned waste-treatment system.

INDIRECT FLOOD DAMAGE Expenditures made as a result of the flood (other than repair) such as relief and rescue work, removing silt and debris, etc.

INDIRECT HIT Locations that do not experience a direct impact from a tropical cyclone, but do experience hurricane force winds (either sustained or gusts) or tides of at least 4 feet above normal.

INDIRECT SOURCE Any facility or building, property, road or parking area that attracts motor vehicle traffic and, indirectly, causes pollution.

INDOOR AIR Breathable air inside a habitable structure or conveyance.

INDOOR AIR POLLUTION Chemical, physical, or biological contaminants in indoor air.

INDOOR CLIMATE Temperature, humidity, lighting, air flow and noise levels in a habitable structure or conveyance.

INDURATED Cemented, hardened, or a rocklike condition.

INDUSTRIAL CONSUMPTION Quantity of water consumed in a municipality or district for mechanical, trade, and manufacturing purposes, in a given period, generally one day.

INDUSTRIAL POLLUTION PREVENTION Combination of industrial source reduction and toxic chemical use substitution.

INDUSTRIAL PROCESS WASTE Residues produced during manufacturing operations.

INDUSTRIAL SLUDGE Semi-liquid residue or slurry remaining from treatment of industrial water and wastewater.

INDUSTRIAL SOURCE REDUCTION Practices that reduce the amount of any hazardous substance, pollutant, or contaminant entering any waste stream or otherwise released into the environment that also reduces the threat to public health and the environment associated with such releases.

INDUSTRIAL SOURCES Non-municipal, or industrial sources, often generate wastewater that is discharged to surface waters.

INDUSTRIAL WASTE Unwanted materials from an industrial operation.

INDUSTRIAL WASTE DAM Embankment dam, usually built in stages, to create storage for the disposal of waste products from an industrial process.

INDUSTRIAL WATER USE (WITHDRAWALS) Water used for industrial purposes such as fabrication, processing, washing, and cooling, and includes such industries as steel, chemical and allied products, paper and allied products, mining, and petroleum refining.

INEFFECTIVE FLOW When high ground or some other obstruction such as a levee prevents water from flowing into a subsection, the area up to that point is ineffective for conveying flow and is not used for hydraulic computations until the water surface exceeds the top elevation of the obstruction.

INERT INGREDIENT Pesticide components such as solvents, carriers, dispersants, and surfactants that are not active against target pests.

INERTIAL SEPARATOR Device that uses centrifugal force to separate waste particles.

INFECTIOUS AGENT Any organism, such as a pathogenic virus, parasite, or bacterium, that is capable of invading body tissues, multiplying, and causing disease.

INFECTIOUS WASTE Hazardous waste capable of causing infections in humans (e.g., contaminated animal waste, human blood and blood products, isolation waste, pathological waste, needles, scalpels or broken medical instruments).

INFILTRATION Movement of water through the soil surface into the soil; downward movement of water from the atmosphere into soil or porous rock; flow of a liquid into a substance through pores of small openings; gradual flow or movement of water into and through the pores of a soil; penetration of water through the ground surface into sub-surface soil; penetration of water from the soil into sewer or other pipes through defective joints, connections, or manhole walls;

technique of applying large volumes of waste water to land to penetrate the surface and percolate through the underlying soil.

INFILTRATION CAPACITY Maximum rate at which water can enter the soil at a particular point under a given set of conditions.

INFILTRATION CAPACITY CURVE
Graph showing the time-variation of infiltration capacity; shows the time-variation of the infiltration rate which would occur if the supply were continually in excess of infiltration capacity.

INFILTRATION GALLERY Horizontal conduit for intercepting and collecting groundwater by gravity flow; subsurface groundwater collection system, typically shallow in depth, constructed with open-jointed or perforated pipes that discharge collected water into a water-tight chamber from which the water is pumped to treatment facilities and into the distribution system.

INFILTRATION INDEX Average constant value of infiltration.

INFILTRATION RATE Quantity of water that can enter the soil in a specified time interval; rate of water entry into the soil expressed as a depth of water per unit of time in inches per hour or feet per day; rate at which groundwater enters an infiltration ditch or gallery, drain, sewer, or other underground conduit.

INFLECTION POINT Point on a hydrograph separating the falling limb from the recession curve; any point on the hydrograph where the curve changes concavity.

INFLOW Water that flows into a body of water; amount of water entering a reservoir; entry of extraneous rain water into a sewer system from sources other than infiltration, such as basement drains, manholes, storm drains, and street washing.

INFLOW BANDS Bands of low clouds, arranged parallel to the low-level winds and moving into or toward a thunderstorm.

INFLOW DESIGN FLOOD (IDF)
Flood flow above which the incremental increase in downstream water surface elevation due to failure of a dam or other water impounding structure is no longer considered to present an unacceptable threat to downstream life or property; flood hydrograph used in the design of a dam and its appurtenant works particularly for sizing the spillway and outlet works and for determining maximum storage, height of dam, and freeboard requirements.

INFLOW JETS Local jets of air near the ground flowing inward toward the base of a tornado.

INFLOW NOTCH Radar signature characterized by an indentation in the reflectivity pattern on the inflow side of the storm.

INFLOW STINGER Beaver tail cloud with a stinger-like shape.

INFLOWING LOAD CURVE See RATING CURVE.

INFLUENT Water, wastewater, or other untreated liquid flowing into a reservoir, basin, or treatment plant.

INFLUENT SEEPAGE Movement of gravity water in the zone of aeration from the ground surface toward the water table.

INFLUENT STREAM Any watercourse in which all, or a portion of the surface water flows back into the ground namely the, vadose zone, or zone of aeration.

INFORMATION COLLECTION REQUEST (ICR)
Description of information to be gathered in connection with rules, proposed rules, surveys, and guidance documents that contain information-gathering requirements.

INFRARED SATELLITE IMAGERY
Satellite imagery senses surface and cloud top temperatures by measuring the wavelength of electromagnetic radiation emitted from these objects.

INHALABLE PARTICLES All dust capable of entering the human respiratory tract.

INHIBITION CONCENTRATION (IC)
Point estimate of the toxicant concentration that would cause a given percent reduction in a non-lethal biological measurement of the test organisms, such as reproduction or growth.

INITIAL COMPLIANCE PERIOD (WATER)
First full three-year compliance period which begins at least 18 months after promulgation.

INITIAL CONDITIONS Conditions prevailing prior to an event; specified values for the dependent variable (e.g., water levels, velocities, etc.) at the beginning of the model simulation; first estimate of the variables the model is trying to solve.

INITIAL DETENTION Volume of water on the ground, either in depressions or in transit, at the time active runoff begins.

INITIAL LOSS Rainfall preceding the beginning of surface runoff (includes interception, surface wetting, evaporation and infiltration unless otherwise specified).

INITIAL MOISTURE DEFICIENCY
Quantity, usually expressed in depth of water in inches upon a unit area, by which the actual water content of a given soil zone (usually the root zone) in such area is less than the field capacity of such zone at the beginning of the rainy season.

INITIAL PHASE Of a geomagnetic storm, that period when there may be an increase of the middle-latitude horizontal intensity.

INITIAL RESPONDERS (FIRST RESPONDERS)
Individuals who are likely to witness or discover a hazardous substance release and who have been trained to initiate an emergency response sequence by notifying the proper authorities of the release.

INITIAL SEED EXAMINATION
First formal onsite examination.

INITIAL WATER DEFICIENCY
Quantity, usually expressed in depth of water in inches upon a unit area, by which the actual water content of a given soil zone (usually the root zone) in such area is less than the field capacity of such zone at the beginning of the rainy season.

INJECTATE Fluids injected into an underground injection well.

INJECTION Subsurface discharge of fluids through a well.

INJECTION WELL Well into which fluids are injected for purposes such as waste disposal, improving the recovery of crude oil, or solution mining.

INJECTION ZONE Geological formation, group of formations, or part of a formation that receives fluids through a well.

INLAND FRESHWATER WETLANDS Swamps, marshes, and bogs found inland beyond the coastal saltwater wetlands.

INLD Inland.

INLET CHANNEL (INLET STRUCTURE) Concrete lined portion of spillway between approach channel and gate or crest structure.

IN-LINE FILTRATION Pretreatment method in which chemicals are mixed by the flowing water.

INNOVATIVE TECHNOLOGIES New or inventive methods to treat effectively hazardous waste and reduce risks to human health and the environment.

INNOVATIVE TREATMENT TECHNOLOGIES Technologies whose routine use is inhibited by lack of data on performance and cost.

INOCULUM Bacteria or fungi injected into compost to start biological action; medium containing organisms, usually bacteria or a virus, that is introduced into cultures or living organisms.

INORGANIC Containing no carbon; matter other than plant or animal; substances that are of mineral origin.

INORGANIC CHEMICALS Chemical substances of mineral origin, not of basically carbon structure.

INORGANIC CONTAMINANTS Mineral-based compounds such as metals, nitrates, and asbestos.

INORGANIC SOIL Soil with less than 20 percent organic matter in the upper 16 inches.

INSECTICIDE Pesticide compound specifically used to kill or prevent the growth of insects.

IN-SITU FLUSHING Introduction of large volumes of water, at times supplemented with cleaning compounds, into soil, waste, or groundwater to flush hazardous contaminants from a site.

IN-SITU FOSSIL FUEL RECOVERY WELLS Class V wells that are used to recover lignite, coal, tar sands, and oil shale by injecting water, air, oxygen, solvents, combustibles, or explosives into underground or oil shale beds to free fossil fuels so they can be extracted.

IN-SITU OXIDATION Technology that oxidizes contaminants dissolved in ground water, converting them into insoluble compounds.

IN-SITU STRIPPING Treatment system that removes or "strips" volatile organic compounds from contaminated ground or surface water by forcing an airstream through the water and causing the compounds to evaporate.

IN-SITU VITRIFICATION Technology that treats contaminated soil in place at extremely high temperatures, at or more than 3000 °F.

INSOLATION Incoming solar radiation; solar heating; sunshine.

INSPECTION AND MAINTENANCE (I/M) Activities to ensure that vehicles' emission controls work properly; applies to wastewater treatment plants and other anti-pollution facilities and processes.

INSTABILITY Tendency for air parcels to accelerate when they are displaced from their original position.

INSTALLED CAPACITY Total of the capacities shown on the nameplates of the generating units in a power plant.

INSTANTANEOUS DISCHARGE Volume of water that passes a point at a particular instant of time.

INSTANTANEOUS MAXIMUM LIMIT Maximum allowable concentration of a pollutant determined from the analysis of any discrete or composite sample collected, independent of the flow rate and the duration of the sampling event.

INSTANTANEOUS UNIT HYDROGRAPH Theoretical, ideal, unit hydrograph that has an infinitesimal duration.

INSTITUTIONAL WASTE Waste generated at institutions such as schools, libraries, hospitals, prisons, etc.

INSTITUTIONALIZED POPULATIONS People in schools, hospitals, nursing homes, prisons, federal buildings, or other facilities that require special care or consideration during emergencies by virtue of their dependency on others for appropriate protection.

INSTREAM FLOW REQUIREMENTS Amount of water flowing through a defined stream channel needed to sustain instream values(e.g., flows designated for fish and wildlife).

INSTREAM USE Use of water that does not require withdrawal or diversion from its natural watercourse (e.g., navigation, recreation, hydroelectric power generation, water quality improvement, fish propagation, recreation).

INSTRUMENT FLIGHT RULES General weather conditions pilots can expect at the surface and applies to the weather situations at an airport during which a pilot must use instruments to assist takeoff and landing.

INSTRUMENT SHELTER Boxlike structure designed to protect temperature measuring instruments from exposure to direct sunshine, precipitation, and condensation, while at the same time providing adequate ventilation.

INSTRUMENTATION Any device used to monitor the performance of the structure during its construction and throughout its useful life; arrangement of devices installed into or near dams that provide for measurements that can be used to evaluate the structural behavior and performance parameters of the structure.

INSULATED Stream or reach of a stream that neither contributes water to the zone of saturation nor receives water from it (i.e. separated from the zones of saturation by an impermeable bed).

INTAKE Any structure through which water can be drawn into a waterway; any structure in a reservoir, dam, or river through which water can be discharged.

INTAKE STRUCTURE Concrete portion of an outlet works, including trash racks and/or fish screens, upstream from the tunnel or conduit portions; entrance to an outlet works.

INTANGIBLE FACTORS Factors that affect a decision, but that cannot be expressed in monetary terms.

INTANGIBLE FLOOD DAMAGE Estimates of the damage done by disruption of business, danger to health, shock, and loss of life and in general all costs not directly measurable which require a large element of judgment for estimating.

INTEGRATED DRAINAGE Drainage developed during geomorphic maturity in an arid region, characterized by coalescence of drainage basins as a result of headward erosion in the lower basins or spilling over from the upper basins.

INTEGRATED EARLY WARNING SYSTEM Flood warning system tied into adjacent systems so that multiple users can access the basin data and monitor the event.

INTEGRATED EXPOSURE ASSESSMENT Cumulative summation (over time) of the magnitude of exposure to a toxic chemical in all media.

INTEGRATED PEST MANAGEMENT (IPM) Mixture of chemical and other, non-pesticide, methods to control pests.

INTEGRATED RESOURCE PLANNING (IRP) Public planning process and framework within which the costs and benefits of both demand and supply side resources are evaluated to develop the least total cost mix of utility resource options.

INTEGRATED RISK INFORMATION SYSTEM (IRIS) EPA's electronic data base containing the latest descriptive and quantitative regulatory information on chemical constituents.

INTEGRATED WASTE MANAGEMENT Using a variety of practices to handle municipal solid waste (e.g., source reduction, recycling, incineration, and landfilling).

INTEGRATOR OR MIXED-USE SITE Stream sampling site located at an outlet of a drainage basin that contains multiple environmental settings.

INTENSITY SCALE Arbitrary scale to describe the degree of shaking at a particular place.

INTENSITY, SEISMIC Numerical index describing the effects of an earthquake on man, manmade structures, or other features of the Earth's surface.

INTENSIVE FIXED SITES Basic Fixed Sites with increased sampling frequency during selected seasonal periods and analysis of dissolved pesticides for 1 year.

INTERBASIN TRANSFER Physical transfer of water from one watershed to another.

INTERCEPTING (CURTAIN) DRAIN
Drain constructed at the upper end of the area to be drained, to intercept surface or groundwater flowing toward the protected area from higher ground, and carry it away from the area.

INTERCEPTION Capture of precipitation above the ground surface (e.g., by vegetation or buildings) and returned to the atmosphere through evaporation.

INTERCEPTION STORAGE REQUIREMENTS
Water caught by plants, before rainfall reaches the ground, at the onset of a rainstorm.

INTERCEPTOR SEWERS Large sewer lines that, in a combined system, control the flow of sewage to the treatment plant.

INTERCONNECTED SYSTEM
System consisting of two or more individual power systems normally operating with connecting tie lines.

INTERFACE Boundary between two substances (i.e., liquid and solid, liquid and gas, or two liquids); contact zone between two fluids of different chemical or physical makeup.

INTERFACIAL TENSION Strength of the film separating two immiscible fluids (e.g., oil and water) usually measured in dynes per centimeter or millidynes per centimeter.

INTERFLOW Lateral movement of water in the upper layer of soil; lateral motion of water through the upper layers until it enters a stream channel.

INTERIM (PERMIT) STATUS Period during which treatment, storage and disposal facilities are temporarily permitted to operate while awaiting a permanent permit.

INTERLOCK Electrical switch, usually magnetically operated.

INTERMEDIATE SEED EXAMINATION
Onsite examination performed approximately every 3 years, or more frequently if conditions dictate, to evaluate the operation, performance, and existing condition of all features.

INTERMEDIATE SYNOPTIC TIMES
Times of 0300, 0900, 1500, and 2100 UTC.

INTERMEDIATE ZONE Subsurface water zone below the root zone and above the capillary fringe.

INTERMITTENT CAPACITY
Load carrying capability of a generator having less than full availability for meeting loads over specific periods of time.

INTERMITTENT STREAM (SEASONAL STREAM)
Stream that flows periodically; stream that flows part of the time; stream that flows only when it receives water from source (e.g., rainfall runoff, springs, melting snow).

INTERMONTANE Situated between or surrounded by mountains, mountain ranges, or mountainous regions.

INTERNAL COMBUSTION PLANT
Plant in which the prime mover is an internal combustion engine.

INTERNAL DOSE Amount of a substance penetrating the absorption barriers (e.g., skin, lung tissue, gastrointestinal tract) of an organism through either physical or biological processes.

INTERNAL DRAINAGE Movement of water down through soil to porous aquifers or to surface outlets at lower elevations; drainage within a basin that has no outlet (e.g. ocean).

INTERNAL EROSION Formation of voids within soil or soft rock caused by the mechanical or chemical removal of material by seepage.

INTERNAL VIBRATION Method of consolidating soil in which vibrators are used within a thoroughly wetted soil mass to consolidate the soil to the desired density.

INTERNATIONAL BOUNDARY COMMISSION SURVEY DATUM Geodetic datum established at numerous monuments along the United States-Canada boundary by the International Boundary Commission.

INTERNATIONAL DATE LINE Line of longitude located at 180 degrees East or West (with a few local deviations) where the date changes by a day; west of the line is one day later than east of the line.

INTERPLANETARY MAGNETIC FIELD (IMF) Magnetic field carried with the solar wind.

INTERPOLATE To estimate a value within an interval between two known values.

INTERRUPTIBLE DEMANDS Demands that, by contract, can be interrupted if the supplying system lacks capacity.

INTERRUPTIBLE LOAD Program activities that can interrupt consumer load at times of peak load.

INTERSTATE CARRIER WATER SUPPLY Source of water for drinking and sanitary use on planes, buses, trains, and ships operating in more than one state.

INTERSTATE WATERS Waters that flow across or form part of state or international boundaries (e.g., the Great Lakes, the Mississippi River, or coastal waters).

INTERSTICE Opening or space in a rock or soil.

INTERSTITIAL FLOW Portion of surface water that infiltrates the streambed and moves through pores in subsurface.

INTERSTITIAL MONITORING Continuous surveillance of the space between the walls of an underground storage tank.

INTERTIDAL Alternately flooded and exposed by tides.

INTERTROPICAL CONVERGENCE ZONE (ITCZ) Region where the northeasterly and southeasterly trade winds converge, forming an often continuous band of clouds or thunderstorms near the equator.

INTOLERANT ORGANISMS Organisms that are not adaptable to human alterations to the environment and thus decline in numbers where human alterations occur.

INTRADOS Curved downstream surface of horizontal arch elements of an arch dam.

INTRASEASONAL OSCILLATION Oscillation with variability on a timescale less than a season.

INTRASTATE PRODUCT Pesticide products once registered by states for sale and use only in the state.

INTRINSIC PERMEABILITY Relative ease with which a porous medium can transmit a liquid under a hydraulic gradient or potential gradient.

INTRUSION Feature (landform, vegetation, or structure) that is generally considered out of context because of excessive contrast and disharmony with characteristic landscape.

INTS Intense.

INTST (INTSFY) Intensify.

INUNDATE To cover with impounded waters or floodwaters.

INUNDATION Process of covering normally dry areas with flood waters.

INUNDATION MAP Map delineating the area that would be flooded by a particular flood event; map delineating the area that would be inundated in the event of a dam failure.

INVERSE METHOD Method of calibrating model using a computer code to systematically vary inputs or input parameters to minimize residuals or residual statistics.

INVERSION Layer of warm air that prevents the rise of cooling air and traps pollutants beneath it.

INVERT Lowest portion of the inside of any horizontal pipe; lowest point of the channel inside a pipe, conduit, or canal; lowest point of an underground excavation or the lowest point of the interior of a circular conduit, pipe, or tunnel.

INVERTEBRATE Animal having no backbone or spinal column.

INVERTEBRATES Animals without a vertebral column (e.g., spiders, crabs, or worms).

INVESTOR OWNED UTILITY (IOU) Company owned by stockholders for profit that provides utility services.

ION An electrically charged atom, group of atoms or molecule that has lost or gained one or more electrons.

ION EXCHANGE TREATMENT Common water-softening method often found on a large scale at water purification plants that remove some organics and radium by adding calcium oxide or calcium hydroxide to increase the pH to a level where the metals will precipitate out.

IONIZATION CHAMBER Device that measures the intensity of ionizing radiation.

IONIZING RADIATION Radiation that can strip electrons from atoms (e.g., alpha, beta, and gamma radiation).

IONOSPHERE Complex atmospheric zone of ionized gases that extends between 50 and 400 miles (80 to 640 kilometers) above the Earth's surface; located between the mesosphere and the exosphere and is included as part of the thermosphere.

IONOSPHERIC STORM Disturbance in the F region of the ionosphere, which occurs in connection with geomagnetic activity.

IOWA VANE Flow deflector that will divert water in a way calculated to attract fish.

IPV Improve.

IR Infrared Satellite Imagery.

IRIDESCENCE Brilliant spots or borders of colors in clouds, usually red and green, caused by diffraction of light by small cloud particles.

IRIDESCENT CLOUDS Clouds that exhibit brilliant bright spots, bands, or borders of colors, usually red and green, observed up to about 30 degrees from the Sun.

IRRADIATED FOOD Food subject to brief radioactivity, usually gamma rays, to kill insects, bacteria, and mold, and to permit storage without refrigeration.

IRRADIATION Exposure to radiation of wavelengths shorter than those of visible light (gamma, x-ray, or ultraviolet), for medical purposes, to sterilize milk or other foodstuffs, or to induce polymerization of monomers or vulcanization of rubber.

IRRETRIEVABLE Commitments that are lost for a period of time.

IRREVERSIBLE Commitments that cannot be reversed.

IRREVERSIBLE EFFECT Effect characterized by the inability of the body to partially or fully repair injury caused by a toxic agent.

IRRIGABLE ACREAGE (AREA) FOR SERVICE Acreage classified as irrigable for which project works have been constructed and project water is available.

IRRIGABLE AREA NOT FOR SERVICE Part of the total irrigable area for which service eventually will be available but for which project facilities have not yet been constructed.

IRRIGABLE LAND Arable land under a specific plan for which a water supply is or can be made available and which is provided with or planned to be provided with irrigation, drainage, flood protection, and other facilities as necessary for sustained irrigation.

IRRIGATED ACREAGE Irrigable acreage actually irrigated in any one year (i.e., irrigated cropland harvested, irrigated pasture, cropland planted but not harvested, and acreage in irrigation rotation used for soil-building crops).

IRRIGATION Supplying water (not supplied by rainfall) to land in order to grow crops, grass, trees, and other plants.

IRRIGATION CHECK Small dike or dam used in the furrow alongside an irrigation border to make the water spread evenly across the border.

IRRIGATION DISTRICT Cooperative, self-governing public corporation having taxing power to obtain and distribute water for irrigation of lands within the district.

IRRIGATION EFFICIENCY Ratio (expressed as a percent) of the average depth of irrigation water that is beneficially used to the average depth of irrigation water applied; amount of water stored in the crop root zone compared to the amount of irrigation water applied.

IRRIGATION REQUIREMENT Quantity of water (not supplied by rainfall) required to grow crops, grass, trees, and other plants.

IRRIGATION RETURN FLOW Surface and subsurface water which leaves the field following application of irrigation water.

IRRIGATION WATER USE Water application on lands to assist in the growing of crops and pastures or to maintain vegetative growth.

IRRIGATION WITHDRAWALS Withdrawals of water for application on land to assist in the growing of crops and pastures or to maintain recreational lands.

IRRITANT Substance that can cause irritation of the skin, eyes, or respiratory system.; chemicals which inflame living tissue by chemical action at the site of contact, causing pain or swelling.

ISALLOBAR Line of equal change in atmospheric pressure during a specified time period.

ISENTROPIC ANALYSIS Looking at the atmosphere in 3-dimensions instead of looking at constant pressure surfaces in 2-dimensions.

ISENTROPIC LIFT Lifting of air that is traveling along an upward-sloping isentropic surface.

ISENTROPIC SURFACE Two-dimensional surface containing points of equal potential temperature.

ISLAND Mid-channel bar that has permanent woody vegetation, is flooded once a year, on average, and remains stable except during large flood events.

ISOBAR Line connecting points of equal pressure.

ISOBARIC CHART Weather map representing conditions on a surface of equal atmospheric pressure.

ISOBARIC PROCESS Any thermodynamic change of state of a system that takes a place at constant pressure.

ISOBATH Imaginary line on the Earth's surface or a line on a map connecting all points which are the same vertical distance above the upper or lower surface of a water-bearing formation or aquifer.

ISOCHROME Line on a chart connecting equal times of occurrence of an event; curve showing the distribution of the excess hydrostatic pressure at a given time during a process of consolidation.

ISOCONCENTRATION More than one sample point exhibiting the same isolate concentration.

ISODOP Contour of constant Doppler velocity values.

ISODROSOTHERM Line connecting points of equal dew point temperature.

ISOHEIGHT Contour depicting vertical height of some surface above a datum plane.

ISOHEL Line on a weather map connecting points receiving equal sunlight.

ISOHYET Line that connects points of equal precipitation or rainfall intensity.

ISOHYETAL LINE Line drawn connecting points that receive the same amount of precipitation.

ISOLATED National Weather Service convective precipitation descriptor for a 10 percent chance of measurable precipitation.

ISOLD Isolated.

ISOPLETH Line connecting points with equal values of a particular atmospheric variable (temperature, dew point, etc.); line or area represented by an isoconcentration.

ISOTACH Line connecting points of equal wind speed.

ISOTHERM Line connecting points of equal temperature.

ISOTHERMAL Having a constant temperature.

ISOTOPE Variation of an element that has the same atomic number of protons but a different weight because of the number of neutrons.

ISOTROPIC Having the same characteristics in all directions.

ISOTROPIC MASS Mass having the same property (or properties) in all directions.

ISOTROPY Condition in which the hydraulic or other properties of an aquifer are the same in all directions.

ISSUANCE TIME Time the product is transmitted.

J

J/KG Joules per kilogram.

JANUARY THAW Period of mild weather popularly supposed to recur each year in late January.

JAR TEST Laboratory procedure that simulates a water treatment plant's coagulation/flocculation units with differing chemical doses, mix speeds, and settling times to estimate the minimum or ideal coagulant dose required to achieve certain water quality goals.

JEOPARDY OPINION U.S. Fish and Wildlife Service (FWS) or National Marine Fisheries Service (NMFS) opinion that an action is likely to jeopardize the continued existence of a listed species or result in the destruction or adverse modification of critical habitat.

JET Fast-moving wind current surrounded by slower moving air.

JET MAX (JET STREAK) Point or area of relative maximum wind speeds within a jet stream.

JET STREAM (JSTR) Relatively strong winds concentrated in a narrow stream in the atmosphere, normally referring to horizontal, high-altitude winds.

JET STREAM CIRRUS Filamentous cirrus that appears to radiate from a point in the sky, and exhibits characteristics associated with strong vertical wind shear, such as twisted or curved filaments.

JET WIND SPEED PROFILE Vertical wind speed profile characterized by a relatively narrow current of high winds with slower moving air above and below.

JET-FLOW GATE Gate consisting of a wheel-mounted leaf moved vertically by a motor-driven screw hoist; high pressure gate resembling a ring follower gate in general configuration, but designed for regulating flow with minimal cavitation damage.

JETTING Method of compacting soil using a hose or other device, with a high velocity stream of water, worked down through the depth of soil placed; drilling with high pressure water or air jets.

JETTING PUMP Water pump that develops very high discharge pressure.

JETTY Pier or other structure built out into a body of water to influence the current or tide, or to protect a harbor or shoreline.

JOB HAZARD ANALYSIS (JHA) Study of a job or activity to identify hazards or potential accidents associated with each step or task, and develop solutions that will eliminate, nullify, or prevent such hazards or accidents.

JOGGING Frequent starting and stopping of an electric motor.

JOINT Point where two things are joined.

JOINT AND SEVERAL LIABILITY Liability for Superfund site cleanup and other costs on the part of more than one potentially responsible party.

JOINT INFORMATION CENTER Single location where public information officials gather to collaborate on, and coordinate the release of, emergency public information.

JOINT METERS Embedded instrument that uses electrical principles to measure movement across a joint or crack.

JOINT USE CAPACITY Reservoir capacity assigned to flood control purposes during certain periods of the year and to conservation purposes during other periods of the year.

JOKULHLAUP Icelandic term meaning glacier dammed lake outburst flood.

JOULE Measure of energy, work or quantity of heat.

JTWC Joint Typhoon Warning Center.

JURISDICTION Boundary of authorization; level of management responsibility an entity has for a specific area using its rules and regulations.

JUVENILE Young fish older than 1 year but not capable of reproduction.

JUVENILE WATER Water formed chemically within the Earth and brought to the surface in intrusive rock.

K

K AMS Cold Air Mass.

K CORONA White-light corona that portion which is caused by sunlight scattered by electrons in the hot outer atmosphere of the Sun.

KARST Geologic formation that results from dissolution and collapse of carbonate rocks such as limestone, dolomite, and gypsum, and that is characterized by closed depressions or sinkholes, caves, and underground drainage.

KATABATIC WIND Wind that is created by air flowing downhill.

KATAFRONT Front where the warm air descends the frontal surface (except in the low layers of the atmosphere).

KEETCH-BYRUM DROUGHT INDEX
Index used to gauge the severity of drought in deep duff and organic soils.

KELVIN TEMPERATURE SCALE
Absolute temperature scale in which a change of 1 Kelvin equals a change of 1 °C.

KELVIN WAVES Fluctuations in wind speed at the ocean surface at the Equator result in eastward propagating waves.

KELVIN-HELMHOLTZ WAVES
Vertical waves in the air associated with wind shear across statically-stable regions.

KETTLE Steep-sided hole or depression, commonly without surface drainage, formed by the melting of a large detached block of stagnant ice that had been buried in the glacial drift.

KETTLE LAKE Body of water occupying a kettle.

KILL Stream or creek.

KILO Prefix meaning "thousand".

KILOGRAM (KG) One thousand grams.

KILOPASCAL Unit (kPa) for measuring atmospheric pressure.

KILOVOLT (KV) One thousand volts.

KILOVOLT-AMPERE (KVA) 1000 volt-amperes and approximately $^{89}/_{100}$ of a kilowatt.

KILOWATT (KW) Electrical unit of work or power; One thousand watts.

KILOWATTHOUR (KWH) Power demand of 1,000 watts for one hour; unit of energy equivalent to one thousand watt-hours.

K-INDEX Measure of the thunderstorm potential based on vertical temperature lapse rate, moisture content of the lower atmosphere, and the vertical extent of the moist layer.

KINETIC ENERGY Energy possessed by a moving object or water body; energy of a body with respect to the motion of the body; one-half the product of a body's mass and the square of its speed.

KINETIC RATE COEFFICIENT Number that describes the rate at which a water constituent such as a biochemical oxygen demand or dissolved oxygen rises or falls, or at which an air pollutant reacts.

KLYSTRON Electron tube used as a low-power oscillator or a high-power amplifier at ultrahigh frequencies.

KNOT (KT) Unit of speed used in navigation equal to 1 nautical mile (the length of 1 minute latitude) per hour.

KNUCKLES Lumpy protrusions on the edges, and sometimes the underside, of a thunderstorm anvil.

KRIGING Geostatistical interpolation procedure for estimating spatial distributions of model inputs from scattered observations.

KT See KNOT.

KTS Knots.

L

LA NIÑA Periodic cooling of surface ocean waters in the eastern tropical Pacific along with a shift in convection in the western Pacific further west than the climatological average.

LAB PROCESS WATER Water used for laboratory experiments and procedures.

LABORATORY ANIMAL STUDIES Investigations using animals as surrogates for humans.

LABORATORY REPORTING LEVEL (LRL) Twice the yearly determined long-term method detection level (LT-MDL).

LACCOLITH Igneous intrusion that squeezes between sedimentary layers and domes the overlying layers.

LACUSTRINE Pertaining to, produced by, or formed in a lake.

LACUSTRINE HABITAT Lake or reservoir wetland habitat.

LAG Time from beginning to peak; center of mass of rainfall to the hydrograph peak; time it takes a flood wave to move downstream.

LAG TIME See LAG.

LAGGING In tunneling, planking placed against the dirt or rock walls and ceiling, outside the ribs.

LAGOMORPHS Order of mammals including rabbits, hares, and pikas.

LAGOON Shallow pond where sunlight, bacterial action, and oxygen work to purify wastewater; storage of wastewater or spent nuclear fuel rods; shallow body of water, often separated from the sea by coral reefs or sandbars.

LAKE BREEZE Thermally produced wind blowing during the day from the surface of a large lake to the shore, caused by the difference in the rates of heating of the surfaces of the lake and of the land.

LAKE EFFECT SNOW Snow showers that are created when cold, dry air passes over a large warmer lake, such as one of the Great Lakes, and picks up moisture and heat.

LAKE EFFECT SNOW ADVISORY Issued by the National Weather Service when pure lake effect snow may pose a hazard or it is life threatening.

LAKE EFFECT SNOW SQUALL Local, intense, narrow band of moderate to heavy snow squall that can extend long distances inland.

LAKE EFFECT SNOW WARNING Issued by the National Weather Service when pure lake effect snow may pose a hazard or it is life threatening.

LAKE EFFECT STORM Fall or winter storm that produces heavy but localized precipitation as a result of temperature differences between the air over snow-covered ground and the air over the open waters of a lake.

LAKESHORE FLOOD ADVISORY See COASTAL/LAKESHORE FLOOD ADVISORY.

LAKESHORE FLOOD WATCH See COASTAL/LAKESHORE FLOOD WATCH.

LAMINAR Smooth; non-turbulent.

LAMINAR FLOW Streamline flow in which successive flow particles follow similar path lines and head loss is proportional to the first power of the velocity.

LAND APPLICATION Discharge of wastewater onto the ground for treatment or reuse.

LAND BAN Phasing out of land disposal of most untreated hazardous wastes.

LAND BREEZE Coastal breeze at night blowing from land to sea, caused by the difference in the rates of cooling of their respective surfaces.

LAND CLASSIFICATION Bureau of Reclamation systematic placing of lands into classes based on their suitability for sustained irrigated farming with Class 1 being the most productive.

LAND DISPOSAL RESTRICTIONS Rules that require hazardous wastes to be treated before disposal on land to destroy or immobilize hazardous constituents that might migrate into soil and ground water.

LAND FARMING (OF WASTE) Disposal process in which hazardous waste deposited on or in the soil is degraded naturally by microbes.

LAND OWNERSHIP Land held in title.

LAND RETIREMENT Permanent removal of land from agricultural production.

LANDFALL Intersection of the surface center of a tropical cyclone with a coastline.

LANDFILL Open area where trash is buried; facility in which solid waste from municipal and/or industrial sources is disposed.

LANDHOLDER Individual or legal entity owning or leasing land subject to the acreage limitation provisions.

LANDSCAPE Traits, patterns, and structure of a specific geographic area, including its biological composition, its physical environment, and its anthropogenic or social patterns; area where interacting ecosystems are grouped and repeated in similar form.

LANDSCAPE CHARACTERIZATION Documentation of the traits and patterns of the essential elements of the landscape.

LANDSCAPE ECOLOGY Study of the distribution patterns of communities and ecosystems, the ecological processes that affect those patterns, and changes in pattern and process over time.

LANDSCAPE INDICATOR Measurement of the landscape, calculated from mapped or remotely sensed data, used to describe spatial patterns of land use and land cover across a geographic area.

LANDSLIDE Failure of a sloped bank of soil or rock in which the movement of the mass takes place along a surface of sliding; unplanned descent (movement) of a mass of earth or rock down a slope.

LANDSPOUT Tornado that does not arise from organized storm-scale rotation and therefore is not associated with a wall cloud (visually) or a mesocyclone (on radar).

LAND-SURFACE DATUM (LSD) Datum plane that is approximately at land surface at each ground-water observation well.

LAND-USE STUDY Network of existing shallow wells in an area having a relatively uniform land use.

LANGELIER INDEX (LI) Index reflecting the equilibrium pH of a water with respect to calcium and alkalinity; used in stabilizing water to control both corrosion and scale deposition.

LAPSE RATE Rate of change of an atmospheric variable (e.g. temperature) with height.

LARGE (SYNOPTIC) SCALE Size scale referring generally to weather systems with horizontal dimensions of several hundred miles or more.

LARGE CONSTRUCTION ACTIVITY Includes clearing, grading, and excavating resulting in a land disturbance that will disturb equal to or more than five acres of land or will disturb less than five acres of total land area but is part of a larger common plan of development or sale that will ultimately disturb equal to or more than five acres.

LARGE QUANTITY GENERATOR Person or facility generating more than 2200 pounds of hazardous waste per month.

LARGE WATER SYSTEM Water system that services more than 50,000 customers.

LARGE-CAPACITY SEPTIC SYSTEMS Class V wells that dispose of sanitary waste through a septic tank and are used by multiple dwellings, business establishments, communities, and regional business establishments for the injection of wastes.

LARVAL FISH Immature stage that develops from the fertilized egg before assuming the characteristics of the adult.

LASER INDUCED FLUORESCENCE Method for measuring the relative amount of soil and/or groundwater with an in-situ sensor.

LAST UPDATE Time and date in which the forecast was issued or updated.

LATENCY Time from the first exposure of a chemical until the appearance of a toxic effect.

LATENT HEAT Amount of heat given up or absorbed when a substance changes from one state to another; heat absorbed or released during a change of phase at constant temperature and pressure.

LATENT HEAT FLUX Amount of heat energy that converts water from liquid to vapor (evaporation) or from vapor to liquid (condensation) across a specified cross-sectional area per unit time.

LATERAL Channel that conveys water from a canal to a farm, municipality, etc.

LATERAL MORAINE Ridge-like pile of sediment along the side of a glacier.

LATERAL SEWERS Pipes that run under city streets and receive the sewage from homes and businesses.

LATITUDE (LAT) Location north or south in reference to the equator, which is designated at zero (0) degrees.

LAUNDERING WEIR Sedimentation basin overflow weir.

LAVA Fluid, molten igneous rock erupted on the Earth's surface.

LAYERED HAZE Haze produced when air pollution from multiple line, area or point sources is transported long distances to form distinguishable layers of discoloration in a stable atmosphere.

LD50 See LETHAL DOSE 50.

LDS Lightning Detection System.

LDT Local Daylight Time.

LEACH To remove components from the soil by the action of water trickling through.

LEACHATE Water that collects contaminants as it trickles through wastes, pesticides or fertilizers; material that pollutes water as it seeps through solid waste.

LEACHATE COLLECTION SYSTEM
System that gathers leachate and pumps it to the surface for treatment.

LEACHING Process by which soluble constituents are dissolved and filtered through the soil by a percolating fluid; removal of materials in solution from soil or rock; movement of pesticides or nutrients from land surface to ground water.

LEACHING FIELD (LEACHING CESSPOOL)
Lined or partially lined underground pit into which raw household water (sewage) is discharged and from which the liquid seeps into the surrounding soil.

LEACHING REQUIREMENT
Quantity of irrigation water required for transporting salts through the soil profile to maintain a favorable salt balance in the root zone for plant development.

LEAD (PB) Heavy metal that is hazardous to health if breathed or swallowed.

LEAD SERVICE LINE Service line made of lead which connects the water to the building inlet and any lead fitting connected to it.

LEAD TIME Time available to determine if conditions at a dam and reservoir, or in the basin, warrant declaration of a specific emergency classification level.

LEADER Streamer that initiates the first phase of each stroke of a lightning discharge.

LEADER SPOT In a magnetically bipolar or multipolar sunspot group, the western part precedes and the main spot in that part is called the leader.

LEAKAGE Free flow loss of water through a hole or crack; flow of water from one hydrogeologic unit to another; natural loss of water from artificial structures as a result of hydrostatic pressure.

LEAKY AQUIFER Aquifers that lose or gain water through adjacent less permeable layers.

LEASE OF POWER PRIVILEGE Contractual right given to a non-federal entity to utilize, consistent with project purposes, water power head and storage from Reclamation projects for electric power generation.

LEE Side or part that is sheltered or turned away from the wind, such as with a mountain.

LEE WAVE Wavelike effect, characterized by severe updrafts and downdrafts, that occurs in the lee of a mountain range when rapidly flowing air is lifted up the steep front of a mountain range.

LEESIDE LOW Extratropical cyclones that form on the downwind (lee) side of a mountain chain.

LEEWARD Side away from the wind.

LEFT ABUTMENT See ABUTMENT.

LEFT EXIT REGION (LEFT FRONT QUADRANT) Area downstream from and to the left of an upper-level jet max (as would be viewed looking along the direction of flow).

LEFT FRONT QUADRANT (LEFT EXIT REGION) Area downstream from and to the left of an upper-level jet max (as would be viewed looking along the direction of flow).

LEFT MOVER Thunderstorm which moves to the left relative to the steering winds, and to other nearby thunderstorms.

LEFT OR RIGHT DESIGNATION Made with the observer looking in the downstream direction.

LEFT OVERBANK See OVERBANK.

LEG Side post in tunnel timbering.

LEGIONELLA Genus of bacteria, some species of which have caused a type of pneumonia called Legionnaires Disease.

LENGTH Distance in the direction of flow between two specific points along a river, stream, or channel.

LENGTH OF DAM See CREST LENGTH.

LENTIC Standing waters, such as lakes, ponds, and marshes.

LENTIC SYSTEM Non-flowing or standing body of fresh water (i.e., lake or pond).

LENTICULAR CLOUD Very smooth, round or oval, lens-shaped cloud that is often seen, singly or stacked in groups, near or in the lee of a mountain ridge.

LEPC Local Emergency Planning Committee.

LETHAL CONCENTRATION Point estimate of the toxicant concentration that would be lethal to a given percentage of the test organisms during a specified period.

LETHAL CONCENTRATION 50 (LC50) Concentration of a pollutant or effluent at which 50 percent of the test organisms die; common measure of acute toxicity.

LETHAL DOSE 50 (LD50) Dose of a toxicant that will kill 50 percent of test organisms within a designated period of time.

LETHAL DOSE LOW (LDLO) Lowest dose in an animal study at which lethality occurs.

LETTER OF AGREEMENT (LOA) See MEMORANDUM OF UNDERSTANDING.

LEVEE (DIKE) Natural or manmade earthen barrier along the edge of a stream, lake, or river; natural or manmade barrier that helps keep rivers from overflowing their banks; long, narrow embankment usually built to protect land from flooding.

LEVEL To cause to conform to a specified grade.

LEVEL OF CONCERN (LOC) Concentration in air of an extremely hazardous substance above which there may be serious immediate health effects to anyone exposed to it for short periods.

LEVEL OF FREE CONVECTION (LFC) Level at which a parcel of saturated air becomes warmer than the surrounding air and begins to rise freely.

LEVEL OF PROTECTION (LOP) Level of loading selected to which corrective actions will be designed to prevent dam failure.

LFT Lift.

LFWS Local Flood Warning System.

LGT Light.

LGWV Long Wave.

LID (CAP) Layer of warm air several thousand feet above the Earth's surface which suppresses or delays the development of thunderstorms.

LIFE CYCLE Various stages an animal passes through from egg fertilization to death.

LIFE CYCLE OF A PRODUCT All stages of a product's development, from extraction of fuel for power to production, marketing, use, and disposal.

LIFE ZONE Major area of plant and animal life; region characterized by particular plants and animals and distinguished by temperature differences.

LIFETIME AVERAGE DAILY DOSE Figure for estimating excess lifetime cancer risk.

LIFETIME EXPOSURE Total amount of exposure to a substance that a human would receive in a lifetime (usually assumed to be 70 years).

LIFR Low Instrument Flight Rules.

LIFT Step or bench in a multiple layer excavation or fill; in a sanitary landfill, a compacted layer of solid waste and the top layer of cover material.

LIFT LINE Horizontal construction joint created when new concrete is placed on previously placed concrete.

LIFTED INDEX (LI) Measure of atmospheric instability obtained by computing the temperature that air near the ground would have if it were lifted to some higher level and comparing that temperature to the actual temperature at that level.

LIFTING CONDENSATION LEVEL (LCL) Level at which a parcel of moist air becomes saturated when it is lifted dry adiabatically.

LIFTING STATION See PUMPING STATION.

LIGHT BRIDGE Observed in white light, a bright tongue or streaks penetrating or crossing sunspot umbrae.

LIGHT NON-AQUEOUS PHASE LIQUID (LNAPL) Non-aqueous phase liquid with a specific gravity less than 1.0.

LIGHT-ATTENUATION COEFFICIENT (EXTINCTION COEFFICIENT) Measure of water clarity.

LIGHT-EMITTING DIODE Long-lasting illumination technology which requires very little power.

LIGHTNING Visible electrical discharge produced by a thunderstorm.

LIGHTNING ACTIVITY LEVELS (LALS) LAL 1 - No thunderstorms; LAL 2 - Few building cumulus with isolated thunderstorms; LAL 3 - Much building cumulus with scattered thunderstorms with light to moderate rain; LAL 4 - Thunderstorms common with moderate to heavy rain reaching the ground; LAL 5 - Numerous thunderstorms with moderate to heavy rain reaching the ground; LAL 6 - Dry lightning (same as LAL 3 but without the rain).

LIGHTNING CHANNEL Irregular path through the air along which a lightning discharge occurs.

LIGHTNING DISCHARGE Series of electrical processes by which charge is transferred along a channel of high ion density between electrical charge centers of opposite sign.

LIGHTNING STROKE Any of a series of repeated electrical discharges comprising a single lightning discharge (strike).

LIKELY In probability of precipitation statements, the equivalent of a 60 or 70 percent chance.

LIKELY (LKLY) In probability of precipitation statements, the equivalent of a 60 or 70 percent chance.

LIMB Edge of the solar disk.

LIMB FLARE Solar flare seen at the edge (Limb) of the Sun.

LIMESTONE Sedimentary rock composed mostly of the mineral calcite and often containing marine fossils.

LIMESTONE SCRUBBING Use of a limestone and water solution to remove gaseous stack-pipe sulfur before it reaches the atmosphere.

LIMIT OF DETECTION (LOD) Minimum concentration of a substance being analyzed test that has a 99 percent probability of being identified.

LIMITED AUTOMATIC REMOTE COLLECTOR (LARC) Electronic device that interfaces a river or precipitation gauge with a telephone line making it possible for remote computers to call a gaging site and retrieve data.

LIMITED DEGRADATION Environmental policy permitting some degradation of natural systems but terminating at a level well beneath an established health standard.

LIMITING FACTOR Condition whose absence or excessive concentration, is incompatible with the needs or tolerance of a species or population and which may have a negative influence on their ability to thrive.

LIMNETIC Deepwater zone (greater than 2 meters deep); subsystem of the Lacustrine System of the U.S. Fish and Wildlife Service wetland classification system.

LIMNOLOGICAL CONDITIONS Conditions on freshwater lakes.

LIMNOLOGY Study of the physical, chemical, hydrological, and biological aspects of fresh water bodies.

LINDANE Pesticide that causes adverse health effects in domestic water supplies and is toxic to freshwater fish and aquatic life.

LINE ECHO WAVE PATTERN (LEWP) Radar echo pattern formed when a segment of a line of thunderstorms surges forward at an accelerated rate.

LINE ITEM Specified amount in a budget to spend on a particular activity.

LINE SOURCE Array of pollutant sources along a defined path that can be treated in dispersion models as an aggregate uniform release of pollutants along a line.

LINEAMENT Rectilinear topographic feature.

LINER Relatively impermeable barrier designed to keep leachate inside a landfill; insert or sleeve for sewer pipes to prevent leakage or infiltration.

LINING Any protective material used to line the interior surface of a conduit, pipe, or tunnel to provide water-tightness, prevent erosion, reduce friction, or support the periphery of the structure.

LIPID One of a family of compounds that are insoluble in water and that make up one of the principal components of living cells.

LIPID SOLUBILITY Maximum concentration of a chemical that will dissolve in fatty substances.

LIPOPHILIC High affinity for lipids (fats).

LIQUEFACTION Changing a solid into a liquid.

LIQUID INJECTION INCINERATOR System that relies on high pressure to prepare liquid wastes for incineration by breaking them up into tiny droplets to allow easier combustion.

LIQUID LIMIT (LL) Moisture content corresponding to the arbitrary limit between the liquid and plastic states of consistency of a soil.

LIQUID WATER EQUIVALENT Liquid content of solid precipitation that has accumulated on the ground (snow depth).

LISTED SPECIES Any species of fish, wildlife, or plant officially designated by an agency as being endangered or threatened.

LISTED WASTE Wastes listed as hazardous under RCRA but which have not been subjected to the Toxic Characteristics Listing Process because the dangers they present are considered self-evident.

LITHOLOGY Description of rocks, based on color, mineral composition and grain size.

LITHOMETEOR Dry particles that hang suspended in the atmosphere, such as dust, smoke, sand, and haze.

LITHOSPHERE That part of the Earth which is composed predominantly of rocks (either coherent or incoherent, and including the disintegrated rock materials known as soils and subsoils), together with everything in this rocky crust.

LITTER Highly visible portion of solid waste carelessly discarded outside the regular garbage and trash collection and disposal system; leaves and twigs fallen from forest trees.

LITTORAL Pertaining to the shore; shallow-water zone (less than 2 meters deep); subsystem of the Lacustrine System of the U.S. Fish and Wildlife Service wetland classification system.

LITTORAL ZONE Area on, or near the shore of a body water; zone or strip of land along the shoreline between the high and low water marks; portion of a body of fresh water extending from the shoreline lakeward to the limit of occupancy of rooted plants.

LIVE CAPACITY (LIVE STORAGE)
That part of the total reservoir capacity which can be withdrawn by gravity; equal to the total capacity less the dead capacity; sum of the active-and the inactive storage.

LIVE STORAGE See LIVE CAPACITY.

LIVESTOCK WATER USE Water for livestock watering, feed lots, dairy operations, fish farming, and other on-farm needs.

LIVV Lifted Index Vertical Velocity.

LIXIVIANT Solution of water or steam, possibly mixed with other chemicals, that is injected through an injection well into a formation to extract minerals in solution mining operations.

LLWS Low Level Wind Shear.

LMTD limited.

LN Line.

LOAD Amount of electrical capacity or energy delivered or required at a given point; power output of an engine or power plant under given circumstances; material that is moved or carried by streams; general term that refers to a material or constituent in solution, in suspension, or in transport.

LOAD ALLOCATIONS (LA) Portion of a receiving water TMDL that is attributed either lo one of its existing or future non-point sources of pollution or to natural background sources.

LOAD FACTOR Ratio of an average load to the maximum load; average load carried by an engine, machine, or plant, expressed as a percentage of its maximum capacity.

LOAD FOLLOWING Generating capacity having fast response time to varying loads over a wide range of unit loading by automatic generation control (AGC) from load frequency control equipment.

LOADED GUN (SOUNDING) Sounding characterized by extreme instability but containing a cap.

LOAM Soft, easily worked soil containing sand, silt, and clay.

LOCAL CAPABILITIES ASSESSMENT Evaluation and report performed by members of the Early Warning System design team and/or community planners to assess the warning and evacuation capabilities of local jurisdictions located downstream from Reclamation dams.

LOCAL CLIMATOLOGICAL DATA (LCD) National Climatic Data Center (NCDC) publication is produced monthly and annually for some 270 United States cities and it's territories that summarizes temperature, relative humidity, precipitation, cloudiness, wind speed and direction observation.

LOCAL CONVECTIVE WIND Local thermally driven winds arising over a comparatively small area and influenced by local terrain.

LOCAL EDUCATION AGENCY (LEA) Exists primarily to operate schools or to contract for educational services, including primary and secondary public and private schools.

LOCAL EMERGENCY OPERATIONS PLAN (LEOP) General planning document, required by law, that describes the responsibilities and actions to be performed in the event of an emergency and/or disaster.

LOCAL EMERGENCY PLANNING COMMITTEE (LEPC) Committee made up of local officials, citizens, and industry representatives charged with development and maintenance of emergency response plans for the local emergency planning district.

LOCAL FLOOD WARNING SYSTEM (LWFS) General designator for a network of stream and rain gauges implemented by a community or local government to monitor hydrologic events as they occur.

LOCAL FLOODING Flooding conditions over a relatively limited (localized) area.

LOCAL INFLOW/OUTFLOW POINT Points along any river segment at which water and sediment enter or exit that segment as a local flow.

LOCAL LIMITS Conditional discharge limits imposed by municipalities upon industrial or commercial facilities that discharge to the municipal sewage treatment system.

LOCAL OFFICIALS/AUTHORITIES Personnel authorized by election or job title to carry out the operation, planning, mitigation, preparedness, response, and recovery functions of the emergency management programs at local levels.

LOCAL ORGANIZATION Local government agency or office having the principal or lead role in emergency planning and preparedness.

LOCAL RUNOFF Water running off a local area, such as rainfall draining into a nearby creek.

LOCAL SCOUR Erosion caused by an abrupt change in flow direction or velocity.

LOCAL STORM REPORT (LSR) Product issued by local National Weather Service offices to inform users of reports of severe and/or significant weather-related events.

LOCKOUT Placement of a lockout device on an energy isolating device, in accordance with an established procedure, ensuring that the energy isolating device and the equipment being controlled cannot be operated until the lockout device is removed; clearance procedure in which physical locks replace Safety Tags and prevent operating switches, controls, etc.

LOCKOUT DEVICE Device that utilizes a positive means such as a lock, either key or combination type, to hold an energy isolating device in the safe position and prevent the energizing of a machine or equipment.

LOESS Wind-deposited silt; uniform Aeolian deposit of silty material having an open structure and relatively high cohesion due to cementation of clay or calcareous material at grain contacts; homogeneous, fine-grained sediment made up primarily of silt and clay, and deposited over a wide area.

LOESSIAL Medium-textured materials (usually silt or very fine sand) transported and deposited by wind action.

LOFTING Pattern of plume dispersion in a stable boundary layer topped by a neutral layer, in which the upper part of the plume disperses upward while the lower part of the plume undergoes little dispersion.

LOG (SAFETY) BOOM Chain of logs, drums, or pontoons secured end to end and floating on the surface of a reservoir so as to divert floating debris, trash, and logs; net-like device installed in a reservoir, upstream of the principal spillway, to prevent logs, debris and boaters from entering a water discharge facility or spillway; floating structure used to protect the face of a dam by deflecting floating material and waves away from the dam; device used to prevent floating debris from obstructing spillways and intakes; .

LOGBOOK (OPERATING LOG) Dated, written record of performed operation and maintenance items or observations pertinent to a structure.

LONG RANGE NAVIGATION (LORAN) System of long range navigation whereby latitude and longitude are determined from the time displacement of radio signals from two or more fixed transmitters.

LONGITUDE Location east or west in reference to the Prime Meridian, which is designated as zero (0) degrees longitude.

LONGITUDINAL Pertaining to or extending along the long axis, or length, of a structure; lengthwise.

LONG-TERM METHOD DETECTION LEVEL (LT-MDL) Detection level derived by determining the standard deviation of a minimum of 24 method detection limit

(MDL) spike sample measurements over an extended period of time.

LONG-TERM MONITORING Data collection over a period of years or decades to assess changes in selected hydrologic conditions.

LONG-TERM RETENTION Retention of data for 5 years to satisfy requirements for local studies and to support litigation.

LONG-TERM STORAGE DAMS Reservoirs used for recreational use or storage of irrigation, municipal or industrial water.

LONG-THROATED FLUMES Control discharge rate in a throat that is long enough to cause nearly parallel flow lines in the region of flow control.

LONGWAVE RADIATION Term used to describe the infrared energy emitted by the Earth and atmosphere at wavelengths between about 5 and 25 micrometers.

LONGWAVE TROUGH Trough in the prevailing westerly flow aloft which is characterized by large length and (usually) long duration.

LOOP PROMINENCE SYSTEM (LPS) System of loop prominences associated with major flares.

LOOPING Pattern of plume dispersion in an unstable atmosphere, in which the plume undergoes marked vertical oscillations as it is alternately affected by rising convective plumes and the subsiding motions between the plumes.

LOOSE YARDS Measurement of soil or rock after it has been loosened by digging or blasting.

LOPRES Low Pressure.

LOSING STREAM Stream or reach that contributes water to a zone of saturation.

LOSS Difference between the volume of rainfall and the volume of runoff.

LOTIC Flowing body of fresh water (i.e., rivers and streams).

LOTIC SYSTEM See LOTIC.

LOW Region of low pressure, marked as “L” on a weather map.

LOW DAM Dam up to 100 feet high.

LOW DENSITY MATERIAL Material having a low weight per unit volume either as it occurs in its natural state or after compacting it.

LOW DENSITY POLYETHYLENE (LOPE) Plastic material used for both rigid containers and plastic film applications.

LOW EMISSIVITY (LOW-E) WINDOWS Window technology that lowers the amount of energy loss through windows by inhibiting the transmission of radiant heat while still allowing sufficient light to pass through.

LOW FREQUENCY Portion of the radio frequency spectrum from 30 to 300 kHz.

LOW HAZARD Downstream hazard classification for dams in which no lives are in jeopardy and minimal economic loss (undeveloped agriculture, occasional uninhabited structures, or minimal outstanding natural resources) would occur as a result of failure of the dam.

LOW IMPACT DEVELOPMENT (LID) Sustainable landscaping approach that can be used to replicate or restore natural watershed functions and/or address targeted watershed goals and objectives.

LOW LEVEL JET (LLJ) Region of relatively strong winds in the lower part of the atmosphere.

LOW LEVEL OUTLET (BOTTOM OUTLET) Opening at a low level from a reservoir generally used for emptying or for scouring sediment and sometimes for irrigation releases.

LOW NOX BURNERS One of several combustion technologies used to reduce emissions of Nitrogen Oxides.

LOW PRESSURE SYSTEM Area of a relative pressure minimum that has converging winds and rotates in the same direction as the Earth.

LOW TIDE Minimum height reached by each falling tide.

LOW WATER ADVISORY Advisory to describe water levels which are significantly below average levels over the Great Lakes, coastal marine zones, and any tidal marine area, waterway, or river inlet within or adjacent to a marine zone that would potentially be impacted by low water conditions creating a hazard to navigation.

LOWER DETECTION LIMIT Smallest signal above background noise an instrument can reliably detect.

LOWER EXPLOSIVE LIMIT (LEL) Concentration of a compound in air below which the mixture will not catch on fire.

LOWEST ACCEPTABLE DAILY DOSE Largest quantity of a chemical that will not cause a toxic effect, as determined by animal studies.

LOWEST ACHIEVABLE EMISSION RATE Rate of emissions that reflects the most stringent emission limitation in the implementation plan of any state for such source unless the owner or operator demonstrates such limitations are not achievable or the most stringent emissions limitation achieved in practice, whichever is more stringent.

LOWEST OBSERVED ADVERSE EFFECT LEVEL (LOAEL) Lowest concentration of an effluent or toxicant that results in statistically significant adverse health effects as observed in chronic or subchronic human epidemiology studies or animal exposure.

LOW-FLOW FREQUENCY CURVE Graph showing the magnitude and frequency of minimum flows for a period of given length.

LOW-HEAD HYDROELECTRIC Hydroelectric power that operates with a head of 20 meters (approximately 66 feet) or less.

LOWLAND FLOODING Inundation of low areas near the river, often rural, but may also occur in urban areas.

LOW-LEVEL RADIOACTIVE WASTE (LLRW) Wastes less hazardous than most of those associated with a nuclear reactor (i.e., generated by hospitals, research laboratories, and certain industries).

LOW-PRECIPITATION STORM (LP STORM) Supercell thunderstorm characterized by a relative lack of visible precipitation.

LRG Large.

LST Local Standard Time.

LTD Limited.

LYSIMETER Device to measure the quantity or rate of downward water movement through a block of soil usually undisturbed, or to collect such percolated water for analysis as to quality.

M

M 3000 Optimum high frequency radio wave with a 3000 km range, which reflects only once from the ionosphere (single hop transmission)

M1 AND M2 CURVES Curves representing mild sloping water surface profiles.

MACKERAL SKY Cirrocumulus clouds with small vertical extent and composed of ice crystals.

MACROBURST Convective downdraft with an affected outflow area of at least 2 miles wide and peak winds lasting between 5 and 20 minutes.

MACROCLIMATE Climate representative of relatively large area.

MACROHABITAT Extensive habitat presenting considerable variation of the environment, containing a variety of ecological niches and supporting a large number and variety of complex flora and fauna.

MACROPHYTES Plant large enough to be seen by the naked eye.

MACROPORES Secondary soil features such as root holes or desiccation cracks that can create significant conduits for movement of NAPL and dissolved contaminants, or vapor-phase contaminants.

MACROSCALE Large scale, characteristic of weather systems several hundred to several thousand kilometers in diameter.

MADDEN-JULIAN OSCILLATION (MJO) Tropical rainfall exhibits strong variability on time scales shorter than the seasonal El Niño-Southern Oscillation (ENSO).

MAGFLARE Geomagnetic and/or cosmic storm has been associated with this flare.

MAGMA Molten or fluid rock material from which igneous rock is derived.

MAGNETIC BAY Relatively smooth excursion of the H (horizontal) component of the geomagnetic field away from and returning to quiet levels.

MAGNETIC SEPARATION Use of magnets to separate ferrous materials from mixed municipal waste stream.

MAGNETOGRAM Solar magnetograms are a graphic representation of solar magnetic field strengths and polarity.

MAGNETOPAUSE Boundary layer between the solar wind and the magnetosphere.

MAGNETOSPHERE Magnetic cavity surrounding the Earth, carved out of the passing solar wind by virtue of the geomagnetic field, which prevents, or at least impedes, the direct entry of the solar wind plasma into the cavity.

MAGNITUDE Rating of a given earthquake, independent of the place of observation; measure of the strength of an earthquake, or the strain energy released by it, as determined by seismographic observations; how much of a pollutant (or pollutant parameter) expressed as a concentration or toxic unit is allowable.

MAGNITUDE, BODY WAVE (MB)
Magnitude of an earthquake measured as the common logarithm of the maximum displacement amplitude (microns) and period (seconds) of the body waves.

MAGNITUDE, RICHTER OR LOCAL (ML)
Magnitude of an earthquake measured as a common logarithm of the displacement amplitude, in microns, of a standard Wood-Anderson seismograph located on firm ground 100 km from the epicenter and having a magnification of 2800, a natural period 0.8 second, and a damping coefficient of 80 percent.

MAGNITUDE, SURFACE WAVE (MS)
Magnitude of an earthquake measured as the common logarithm of the resultant of the maximum mutually perpendicular horizontal displacement amplitudes, in microns, of the 20 second period surface waves.

MAIN Principal pipe artery to which branches may be connected.

MAIN CHANNEL Deepest or central part of the bed of a stream, containing the main current.

MAIN CHANNEL POOL Reach of a stream or river with a low bed elevation, relative to rapids or riffles.

MAIN PHASE That period when the horizontal magnetic field at middle latitudes is generally decreasing.

MAIN STEM Reach of a river/stream formed by the tributaries that flow into it; principal trunk of a river or a stream.

MAIN SYNOPTIC TIMES Times of 0000, 0600, 1200, and 1800 UTC; standard synoptic times.

MAINSTREAM (MAINSTEM)
Main course of a stream where the current is the strongest.

MAINTENANCE All routine and extraordinary work necessary to keep the facilities in good repair and reliable working order to fulfill the intended designed project purposes.

MAINTENANCE MANAGEMENT SYSTEM
Any organized system used to ensure that all operations and maintenance activities (e.g., maintenance, inspection, operational testing) at a facility is accomplished and documented.

MAJOR FACILITY See MAJORS.

MAJOR FLARE This flare is the basis for the forecast of geomagstorm, cosmic storm and/or protons in the Earth's vicinity.

MAJOR FLOODING General term including extensive inundation and property damage.

MAJOR GEOMAGNETIC STORM Storm for which the Ap index was greater than 49 and less than 100.

MAJOR HURRICANE Hurricane which reaches Category 3 (sustained winds greater than 110 mph) on the Saffir/Simpson Hurricane Scale.

MAJOR IONS Constituents commonly present in water in concentrations exceeding 1.0 milligram per liter.

MAJOR MODIFICATION Term used to define modifications of major stationary sources of emissions.

MAJORS Larger publicly owned treatment works (POTWs) with flows equal to at least one million gallons per day (mgd) or servicing a population equivalent to 10,000 persons.

MAMMATUS CLOUDS Rounded, smooth, sack-like protrusions hanging from the underside of a cloud (usually a thunderstorm anvil).

MANAGEMENT INDICATOR SPECIES
Any species, group of species, or species habitat element selected to focus management attention for the purpose of resource production, population recovery, maintenance of population viability, or ecosystem diversity .

MANAGEMENT INDICATORS
Plant and animal species, communities, or special habitats that are selected for emphasis in planning and that are monitored during forest-plan implementation to assess the effects of management activities on their populations and the populations of other species with similar habitat needs that they may represent.

MANAGERIAL CONTROLS Methods of non-point source pollution control based on decisions about managing agricultural wastes or application times or rates for agrochemicals.

MANDATORY RECYCLING Programs which by law require consumers to separate trash so that some or all recyclable materials are recovered for recycling rather than going to landfills.

MANIFEST One-page form used by haulers transporting waste that lists EPA identification numbers, type and quantity of waste, the generator it originated from, the transporter that shipped it, and the storage or disposal facility to which it is being shipped.

MANIFEST SYSTEM Tracking of hazardous waste from "cradle-to-grave" (generation through disposal) with accompanying documents known as manifests.

MANIFOLD (HEADER) Large pipe to which a series of smaller pipes are connected.

MAN-MADE (ANTHROPOGENIC) BETA PARTICLE AND PHOTON EMITTERS
All radionuclides emitting beta particles and/or photons listed in Maximum Permissible Body Burdens and Maximum Permissible Concentrations of Radionuclides in Air and Water for Occupational Exposure.

MANNING'S EQUATION Empirical equation commonly applied in water surface profile calculations that defines the relationship between surface roughness, discharge, flow geometry, and rate of friction loss for a given stream location.

MANNING'S ROUGHNESS (N) COEFFICIENT Coefficient of roughness that accounts for energy loss due to the friction between the bed and the water; coefficient used to describe the relative roughness of a channel and overbank areas.

MANOMETER Instrument for measuring pressure.

MANTLE Thick layer of rock deep within the Earth that separates the Earth's crust above from the Earth's core below.

MANUAL SEPARATION Hand sorting of recyclable or compostable materials in waste.

MANUFACTURER'S FORMULATION List of substances or component parts as described by the maker of a coating, pesticide, or other product containing chemicals or other substances.

MANUFACTURING USE PRODUCT Any product intended (labeled) for formulation or repackaging into other pesticide products.

MAP Two-dimensional representation of all or part of the Earth's surface showing selected natural or manmade features or data.

MARBLE Metamorphic rock formed by the "baking" and recrystallization of limestone.

MARC VELOCITY SIGNATURE Doppler radar-velocity based precursor towards forecasting the initial onset of damaging straight-line winds in a linear Quasi-Linear Convective System (QLCS) or bowing convective system.

MARE'S TRAIL Name given to thin, wispy cirrus clouds composed of ice crystals that appear as veil patches or strands, often resembling a horse's tail.

MARGIN OF EXPOSURE (MOE) Ratio of the no-observed adverse-effect-level to the estimated exposure dose.

MARGIN OF SAFETY Maximum amount of exposure producing no measurable effect in animals (or studied humans) divided by the actual amount of human exposure in a population.

MARGINAL LAND Land which, in its natural state, is not well suited for a particular purpose, such as raising crops.

MARGINAL VISUAL FLIGHT RULES (MVFR) General weather conditions pilots can expect at the surface; criteria means a ceiling between 1,000 and 3,000 feet and/or 3 to 5 miles visibility.

MARINE INVERSION Temperature inversion produced when cold marine air underlies warmer air.

MARINE OPTICAL BUOY (MOBY) Measures solar radiation to calibrate satellite ocean color instruments.

MARINE PUSH Replacement of the current air mass with air from off the ocean.

MARINE REPORT (MAREP) Voluntary marine observation program of the National Weather Service whose goal is to solicit meteorological and oceanographic observations in plain language from recreational and small commercial mariners who are not part of Voluntary Observing Ship program.

MARINE SANITATION DEVICE Any equipment or process installed on board a vessel to receive, retain, treat, or discharge sewage.

MARINE WEATHER STATEMENT National Weather Service product to provide mariners with details on significant or potentially hazardous conditions not otherwise covered in existing marine warnings and forecasts.

MARINE WETLAND Wetlands that are exposed to waves and currents of the open ocean and to water having a salinity greater than 30 parts per thousand.

MARINE ZONE Specific, defined over-water areas contained in the various National Weather Service marine forecasts.

MARITIME AIR MASS Air mass influenced by the sea.

MARITIME POLAR AIR MASS (mP) Air mass characterized by cold, moist air.

MARITIME TROPICAL AIR MASS (mT) An air mass characterized by warm, moist air.

MARKET-BASED RATES Rates for power or electric service that are established in an unregulated, competitive market.

MARKETER Agent for generation projects who markets power on behalf of the generator.

MAROB Voluntary marine observation program of the National Weather Service in the early stages of development whose goal is to solicit meteorological and oceanographic observations in coded format from recreational and small commercial mariners who are not part of the more in-depth Voluntary Observing Ship program.

MARS Voluntary marine observation program of the National Weather Service whereby U.S. Coast Guard Sector Stations report marine weather conditions from several shore locations within their operating area.

MARSH Type of wetland that does not accumulate appreciable peat deposits and is dominated by herbaceous vegetation; water-saturated, poorly drained area, intermittently or permanently water covered, having aquatic and grasslike vegetation.

MASONRY DAM Dam constructed mainly of stone, brick, or concrete blocks pointed with mortar.

MASS CARE CENTER Facility for providing emergency lodging and care for people made temporarily homeless by an emergency or disaster.

MASS CONCRETE Large volume of concrete cast-in-place, generally as a monolithic structure.

MASS CURVE Graph of the cumulative values of a hydrologic quantity (i.e., precipitation and runoff), generally as ordinate, plotted against time or date as abscissa.

MASS-BASED STANDARD Discharge limit that is measured in a mass unit such as pounds per day.

MASSIF Compact portion of a mountain range, containing one or more summits.

MASSIVE HEAD BUTTRESS DAM
Buttress dam in which the buttress is greatly enlarged on the upstream side to span the gap between buttresses.

MASTIC Soft sealing material.

MATERIAL CATEGORY In the asbestos program, broad classification of materials into thermal surfacing insulation, surfacing material, and miscellaneous material.

MATERIAL SAFETY DATA SHEET (MSDS)
Worksheet required by the U.S. Occupational Safety and Health Administration (OSHA) containing information about hazardous chemicals in the workplace.

MATERIAL TYPE Classification of suspect material by its specific use or application (e.g., pipe insulation, fireproofing, and floor tile).

MATERIALS RECOVERY FACILITY (MRF)
Facility that processes residentially collected mixed recyclables into new products available for market.

MATHEMATICAL MODEL Model that uses mathematical expressions (i.e. a set of equations) to represent a physical process.

MATURITY (STREAM) Stage in the evolutionary erosion of land areas in which the flat uplands have been widely dissected by deep river valleys; stage in the development of a stream at which it has reached its maximum efficiency, when velocity is just sufficient to carry the sediment delivered to it by tributaries.

MAX Maximum.

MAX PARCEL LEVEL (MPL)
Highest attainable level that a convective updraft (e.g. thunderstorm) can reach.

MAXIMALLY (MOST) EXPOSED INDIVIDUAL
Person with the highest exposure in a given population.

MAXIMUM ACCEPTABLE TOXIC CONCENTRATION
Range (or geometric mean) between the No Observable Adverse Effect Level and the Lowest Observable Adverse Effects Level.

MAXIMUM AVAILABLE CONTROL TECHNOLOGY (MACT)
Emission standard for sources of air pollution requiring the maximum reduction of hazardous emissions, taking cost and feasibility into account.

MAXIMUM CONTAMINANT LEVEL (MCL)
Maximum permissible level of a contaminant in water that is delivered to any user of a public water system.

MAXIMUM CONTAMINANT LEVEL GOAL (MCLG)
Level of a contaminant in drinking water below which there is no known or expected risk to health.

MAXIMUM CONTROLLABLE WATER SURFACE
Highest reservoir water surface elevation at which gravity flows from the reservoir can be completely shut off.

MAXIMUM CREDIBLE EARTHQUAKE (MCE)
Largest hypothetical earthquake that may be reasonably expected to occur along a given fault or other seismic source could produce under the current tectonic setting.

MAXIMUM DEMAND Greatest of all demands of the load that has occurred within a specified period of time.

MAXIMUM DESIGN EARTHQUAKE (MDE) Earthquake selected for design or evaluation of the structure.

MAXIMUM EXPOSURE RANGE Estimate of exposure or dose level received by an individual in a defined population that is greater than the 98th percentile dose for all individuals in that population, but less than the exposure level received by the person receiving the highest exposure level.

MAXIMUM FLOOD CONTROL LEVEL Highest elevation of the flood control storage.

MAXIMUM PROBABLE FLOOD Largest flood for which there is any reasonable expectancy in this climatic era; flood that may be expected from the most severe combination of critical meteorological and hydrologic conditions that are reasonably possible in the drainage basin under study.

MAXIMUM RESIDUE LEVEL Enforceable limit on food pesticide levels in some countries.

MAXIMUM SPILLWAY DISCHARGE Spillway discharge (cfs) when reservoir is at maximum designed water surface elevation.

MAXIMUM SUSTAINED SURFACE WIND Highest one-minute average wind (at an elevation of 10 meters with an unobstructed exposure) associated with that weather system at a particular point in time.

MAXIMUM TEMPERATURE Highest temperature recorded during a specified period of time.

MAXIMUM TOLERATED DOSE Maximum dose that an animal species can tolerate for a major portion of its lifetime without significant impairment or toxic effect other than carcinogenicity.

MAXIMUM UNAMBIGUOUS RANGE Range from the radar at which an echo can be known unquestionably as being at that range.

MAXIMUM UNAMBIGUOUS VELOCITY Highest radial velocity that can be measured unambiguously by a pulsed Doppler radar.

MAXIMUM UNIT WEIGHT Dry unit weight defined by the peak of a compaction curve.

MAXIMUM WATER SURFACE (MAXIMUM POOL) Highest acceptable water surface elevation with all factors affecting the safety of the structure considered; highest water surface elevation resulting from a computed routing of the inflow design flood through the reservoir under established operating criteria.

MAXIMUM WAVE Highest wave in a wave group.

MAXIMUM WIND Most severe wind for generating waves that is reasonably possible at a particular reservoir.

MAXT High temperature for the day.

MB Millibar.

MCC Mesoscale Convective Complex.

MCV Mesoscale Cyclonic Vortices.

MDT Moderate; Mountain Daylight Time.

MEAN Arithmetic average of a set of observations.

MEAN ANNUAL TEMPERATURE
Average temperature for the entire year at any given location.

MEAN AREAL PRECIPITATION (MAP)
Average rainfall over a given area expressed as an average depth over the area.

MEAN CONCENTRATION OF SUSPENDED SEDIMENT
Time-weighted concentration of suspended sediment passing a stream cross section during a given time period.

MEAN DAILY TEMPERATURE
Average of the highest and lowest temperatures during a 24-hour period.

MEAN DEPTH Average depth of water in a stream channel or conduit; equal to the cross-sectional area divided by the surface width.

MEAN DISCHARGE (MEAN)
Arithmetic mean of individual daily mean discharges of a stream during a specific period.

MEAN DOPPLER VELOCITY
Reflectivity-weighted average velocity of targets in a given pulse resolution volume.

MEAN HIGH TIDE Average altitude of all high tides recorded at a given place over a 19-year period.

MEAN LAYER CAPE (MLCAPE)
Calculated using a parcel consisting of Mean Layer values of temperature and moisture from the lowest 100 mb above ground level.

MEAN LAYER LIFTED INDEX (MLLI)
Lifted Index (LI) calculated using a parcel consisting of Mean Layer values of temperature and moisture from the lowest 100 mb above ground level.

MEAN LOW TIDE Average altitude of all low tides recorded at a given place over a 19-year period.

MEAN LOW WATER (MLW) Average of all the low water heights observed over the National Tidal Datum Epoch.

MEAN LOWER LOW WATER (MLLW)
Average of the lower low water height of each tidal day observed over the National Tidal Datum Epoch.

MEAN NUMBER Average number.

MEAN SEA LEVEL (MSL) Arithmetic mean of hourly heights observed over the National Tidal Datum Epoch; arithmetic mean of hourly water elevations observed over a specific 19-year tidal epoch.

MEAN WAVE DIRECTION (MWD)
Direction corresponding to energy of the dominant period.

MEANDER Winding of a stream channel; big bend and loops in a river channel as the river snakes through a flat land area.

MEANDER AMPLITUDE Distance between points of maximum curvature of successive meanders of opposite phase in a direction normal to the general course of the meander belt, measured between centerlines of channels.

MEANDER BELT Area between lines drawn tangential to the extreme limits of fully developed meanders.

MEANDER BREADTH Distance between the lines used to define the meander belt.

MEANDER LENGTH Distance of the meanders between corresponding points of successive meanders.

MEANDERING STREAM Alluvial stream characterized in planform by a series of pronounced alternating bends.

MEASURE OF EXPOSURE Measurable characteristic of a stressor (e.g., specific amount of mercury in a body of water) used to help quantify the exposure of an ecological entity or individual organism.

MEASURED CEILING Ceiling classification applied when the ceiling value has been determined by an instrument, such as a ceilometer or ceiling light, or by the known heights of unobscured portions of objects, other than natural landmarks, near the runway.

MEASUREMENT ENDPOINT Measurable characteristic of ecological entity that can be related to an assessment endpoint.

MEASURING POINT (MP) Arbitrary permanent reference point from which the distance to water surface in a well is measured to obtain water level.

MECHANICAL AERATION Use of mechanical energy to inject air into water to cause a waste stream to absorb oxygen.

MECHANICAL INTEGRITY Absence of significant leakage within the injection tubing, casing, or packer (known as internal mechanical integrity), or outside of the casing (known as external mechanical integrity).

MECHANICAL SEPARATION Using mechanical means to separate waste into various components.

MECHANICAL TURBULENCE Random irregularities of fluid motion in air caused by buildings or other non-thermal processes.

MEDIA Specific environments (e.g., air, water, soil) that are the subject of regulatory concern and activities.

MEDIAN Middle value in a distribution, above and below which lie an equal number of values; 50th percentile.

MEDICAL SURVEILLANCE Periodic comprehensive review of a worker's health status.

MEDIUM Environmental vehicle by which a pollutant is carried to the receptor (e.g., air, surface water, soil, or groundwater).

MEDIUM FREQUENCY (MF) Portion of the radio frequency spectrum from 0.3 to 3 MHz.

MEDIUM RANGE In forecasting, three to seven days in advance.

MEDIUM-HEIGHT DAM Dam between 100 and 300 feet high.

MEDIUM-SIZE WATER SYSTEM Water system that serves 3,300 to 50,000 customers.

MEDIUM-THICK ARCH DAM Arch dam with a base thickness to structural height ratio between 0.2 and 0.3.

MEGA Million.

MEGA (MILLION) ELECTRONVOLT (MEV) Unit of energy used to describe the total energy carried by a particle or photon.

MEGAHERTZ Unit of frequency.

MEGAWATT (MW) One million watts of electrical power (capacity).

MEGAWATT-HOUR (MWH) One million watt-hours of electrical energy.

MEGG Merging.

MELTING LEVEL Altitude which ice crystals and snowflakes melt as they descend through the atmosphere.

MELTING POINT Temperature at which a solid material changes to a liquid.

MEMBRANE (DIAPHRAGM) Sheet or thin zone or facing, made of a flexible impervious material (e.g., asphaltic concrete, plastic concrete, steel, wood, copper, plastic, etc.).

MEMBRANE FILTER Thin microporous material of specific pore size used to filter bacteria, algae, and other very small particles from water.

MEMORANDUM OF UNDERSTANDING (MOU) Formal document that states the intentions and/or responsibilities of the signatory parties.

MENISCUS Curved top of a column of liquid in a small tube caused by surface tension; curved surface of the liquid at the open end of a capillary column.

MERCURY (HG) Naturally occurring element that is found in air, water, and soil; heavy metal that can accumulate in the environment and is highly toxic if breathed or swallowed.

MERCURY BAROMETER Instrument for measuring atmospheric pressure.

MERIDIAN Imaginary line on the Earth's surface passing through both geographic poles and through any given point on the planet; line of longitude.

MERIDIONAL FLOW Large-scale atmospheric flow in which the north-south component (i.e., longitudinal, or along a meridian) is pronounced.

MEROMICTIC LAKE Lake where partial content remains unmixed with main body.

MESOCLIMATE Climate of a small area of the Earth's surface which may differ from the general climate of the district.

MESOCYCLONE (MESO) Storm-scale region of rotation, typically around 2-6 miles in diameter and often found in the right rear flank of a supercell (or often on the eastern, or front, flank of an HP storm).

MESOHIGH Relatively small area of high atmospheric pressure that forms beneath a thunderstorm.

MESOLOW (SUB-SYNOPTIC LOW) Mesoscale low-pressure center.

MESONET Regional network of observing stations (usually surface stations) designed to diagnose mesoscale weather features and their associated processes.

MESOPAUSE Top of the mesosphere, corresponding to the level of minimum temperature in the atmosphere found at 70 to 80 km.

MESOPHYTE Any plant growing where moisture and aeration conditions lie between the extremes of "wet" and "dry."

MESOSCALE Size scale referring to weather systems smaller than synoptic-scale systems but larger than storm-scale systems.

MESOSCALE CONVECTIVE COMPLEX (MCC) Large Mesoscale Convective System (MCS), generally round or oval-shaped, which normally reaches peak intensity at night.

MESOSCALE CONVECTIVE SYSTEM (MCS) Complex of thunderstorms which becomes organized on a scale larger than the individual thunderstorms, and normally persists for several hours or more.

MESOSCALE DISCUSSION (MD) When conditions actually begin to shape up for severe weather, SPC (Storm Prediction Center) often issues a Mesoscale Discussion (MCD) statement anywhere from roughly half an hour to several hours before issuing a weather watch.

MESOSCALE HIGH WINDS High winds usually follow the passage of organized convective systems and are associated with wake depressions or strong mesohighs.

MESOSPHERE Atmospheric shell between about 20 km and about 70 to 80 km, extending from the top of the stratosphere (the stratopause) to the upper temperature minimum that defines the mesopause (the base of the thermosphere).

MESOTROPHIC Reservoirs and lakes which contain moderate quantities of nutrients and are moderately productive in terms of aquatic animal and plant life.

METABOLITE Substance produced in or by biological processes.

METADATA Information that describes the content, quality, condition, and other characteristics of data.

METALIMNION Middle layer of a thermally stratified lake or reservoir with a rapid temperature decrease with depth.

METAMORPHIC Rock compressed or changed by pressure, heat, or water; rock formed from a preexisting rock that is altered ("baked") by high temperatures and pressures, causing minerals to recrystallize but not melt.

METAMORPHIC ROCK Rock that has formed in the solid state in response to pronounced changes of temperature, pressure, and chemical environment.

METAMORPHIC STAGE Stage of development that an organism exhibits during its transformation from an immature form to an adult form.

METAR International code (Aviation Routine Weather Report) used for reporting, recording and transmitting weather observations.

METEOGRAM Graphical depiction of trends in meteorological variables such as temperature, dew point, wind speed and direction, pressure, etc.

METEORIC WATER Water derived from precipitation.

METEOROLOGICAL HOMOGENEITY Climates and orographic influences that are alike or similar.

METEOROLOGIST Person who studies meteorology.

METEOROLOGY Science dealing with the atmosphere and its phenomena.

METHANE Colorless, non-poisonous, flammable gas created by anaerobic decomposition of organic compounds; greenhouse gas that is emitted during the production and transport of coal, natural gas, and oil or from the decomposition of organic wastes in municipal solid waste landfills and the raising of livestock.

METHANOL Alcohol that can be used as an alternative fuel or as a gasoline additive.

METHOD 18 EPA test method which uses gas chromatographic techniques to measure the concentration of volatile organic compounds in a gas stream.

METHOD 24 EPA reference method to determine density, water content and total volatile content (water and VOC) of coatings.

METHOD 25 EPA reference method to determine the VOC concentration in a gas stream.

METHOD CODE One-character code that identifies the analytical or field method used to determine a value stored in the National Water Information System (NWIS).

METHOD COMPARABILITY Characteristics that allow data produced by multiple methods to meet or exceed the data-quality objectives of primary or secondary data users.

METHOD DETECTION LIMIT (MDL) Minimum concentration of a substance that can be measured and reported with 99-percent confidence that the analyte concentration is greater than zero.

METHOD OF CUBATURES Method of computing discharge in tidal estuaries based on the conservation of mass equation.

METHOD VALIDATION Process of substantiating a method to meet certain performance criteria for sampling and (or) analytical and (or) data handling operations (ITFM)

METHOXYCHLOR Pesticide that causes adverse health effects in domestic water supplies and is toxic to freshwater and marine aquatic life.

METHYL ORANGE ALKALINITY Measure of the total alkalinity in a water sample in which the color of methyl orange reflects the change in level.

METHYLENE BLUE ACTIVE SUBSTANCES (MBAS) Indicate the presence of detergents (anionic surfactants).

METRIC Biological attribute, some feature or characteristic of the biotic assemblage, that reflects ambient conditions, especially the influence of human actions on these conditions.

METRO Metropolitan.

MG Milligram; metric unit of mass; one thousandth of a gram.

MI Mile.

MIC Meteorologist In Charge.

MICA Group of minerals that form thin, platy flakes, typically with shiny surfaces, especially common in metamorphic rocks.

MICRO Small; prefix meaning "one millionth".

MICROBAROGRAPH Instrument designed to continuously record a barometer's reading of very small changes in atmospheric pressure.

MICROBIAL GROWTH Amplification or multiplication of microorganisms such as bacteria, algae, diatoms, plankton, and fungi.

MICROBIAL PESTICIDE Microorganism that is used to kill a pest, but is of minimum toxicity to humans.

MICROBURST Convective downdraft with an affected outflow area of less than 2.5 miles wide and peak winds lasting less than 5 minutes.

MICROCLIMATE Climate of a small area, particularly that of the living space of a certain species, group or community; climate around a tree or shrub or a stand of trees.

MICROENVIRONMENTAL METHOD Method for sequentially assessing exposure for a series of microenvironments that can be approximated by constant concentrations of a stressor.

MICROENVIRONMENTS Well-defined surroundings such as the home, office, or kitchen that can be treated as uniform in terms of stressor concentration.

MICROGRAMS PER GRAM (MG/G) Unit expressing the concentration of a chemical constituent as the mass (micrograms) of the element per unit mass (gram) of material analyzed.

MICROGRAMS PER KILOGRAM (MG/KG) Unit expressing the concentration of a chemical constituent as the mass (micrograms) of the constituent per unit mass (kilogram) of the material analyzed; equivalent to 1 part per billion.

MICROGRAMS PER LITER (MG/L) Unit expressing the concentration of constituents in solution as weight (micrograms) of solute per unit volume (liter) of water; equivalent to one part per billion.

MICROHABITAT Small, specialized, and effectively isolated location.

MICRON Unit of length equal to one millionth of a meter; equal to one thousandth of a millimeter.

MICROORGANISMS Tiny living organisms that can be seen only with the aid of a microscope.

MICROSCALE Pertaining to meteorological phenomena, such as wind circulations or cloud patterns, that are less than 2 km in horizontal extent.

MICROSIEMENS PER CENTIMETER (MS/CM) Unit expressing the amount of electrical conductivity of a solution as measured between opposite faces of a centimeter cube of solution at a specified temperature.

MICROSYSTEM IRRIGATION Method of precisely applying irrigation water to the immediate root zone of the target plant at very low rates.

MICROWAVE Type of electromagnetic radiation with wavelengths between those of infrared radiation and radio waves.

MICROWAVE BURST Radiowave signal associated with optical and/or X-ray flares.

MID Middle.

MIDDLE (MID) LATITUDE AREAS Latitude belt roughly between 30 and 60 degrees North and South of the Equator.

MIDDLE CLOUDS (MID-LEVEL CLOUDS) Term used to signify clouds with bases between 6,500 and 23,000 feet.

MID-FLAME WIND Wind measured at the midpoint of the flames, considered to be most representative of the speed of the wind that is affecting fire behavior.

MIDGE Small fly in the family Chironomidae where larval (juvenile) life stages are aquatic.

MID-LEVEL COOLING Local cooling of the air in middle levels of the atmosphere (roughly 8 to 25 thousand feet), which can lead to destabilization of the entire atmosphere if all other factors are equal.

MIE SCATTERING Any scattering produced by spherical particles whose diameters are greater than 1/10 the wavelength of the scattered radiation.

MIGMATITE Rock composed of a complex mixture of metamorphic rock and igneous granitic rock.

MIGRATORY Moving from one area to another on a seasonal basis.

MIL Unit of length equal to 0.001 of an inch.

MILESTONE Measurable action, state, or goal which marks a point of achievement on the way to solving the problem.

MILITARY CREST Ridge that interrupts the view between a valley and a hilltop.

MILL Monetary cost and billing unit used by utilities; equal to 1/1000 of U.S. dollar (or 1/10 of one cent).

MILLI Prefix meaning "one thousandth".

MILLIBAR Unit of atmospheric pressure equal to 1/1000 bar; equal to 1000 dynes per square centimeter.

MILLIGRAM (MG) Mass equal one-thousandth of a gram.

MILLIGRAM PER LITER (MG/L) Equivalent to 1 part per million.

MILLIGRAMS PER CUBIC METER Mass in micrograms of a substance contained within a cubic meter of another substance or vacuum.

MILLIGRAMS PER LITER (MG/L) Unit expressing the concentration of chemical constituents in solution as weight (milligrams) of solute per unit volume (liter) of water; equivalent to one part per million.

MILLION ACRE-FEET (MAF) Volume of water that would cover 1 million acres to a depth of 1 foot.

MILLION GALLONS PER DAY (MGD, MGAL/D) Rate of flow of water commonly used for wastewater discharges; equivalent to 1.547 cubic feet per second, 133,680.56 cubic feet per day or 3.0689 acre-feet per day.

MILLISECOND DELAY (SHORT PERIOD DELAY) Type of delay cap with a definite but extremely short interval between initiation, or passing of current, and explosion.

MIN Minimum; minute.

MINE BACKFILL WELLS Class V wells that inject water, sand, mill tailings, or other mining byproducts in order to control subsidence caused by mining, to dispose of mining byproducts, or to fill sections of a mine.

MINE TAILINGS DAM Industrial waste dam in which the waste materials come from mining operations or mineral processing.

MINERAL SOIL Soil composed predominantly of mineral rather than organic materials; less than 20 percent organic material.

MINERS' INCH Rate of discharge through an orifice one inch square under a specific head.

MINIMIZATION Comprehensive program to minimize or eliminate wastes, usually applied to wastes at their point of origin.

MINIMUM DISCERNIBLE SIGNAL
Smallest input signal that will a produce a detectable signal at the output; minimal amount of back scattered energy that is required to produce a target on the radar screen.

MINIMUM FLOW Negotiated lowest flow in a regulated stream that will sustain an aquatic population of agreed-upon levels.

MINIMUM LEVEL (ML) Level at which the entire analytical system gives recognizable mass spectra and acceptable calibration points when analyzing for pollutants of concern; lowest point at which the calibration curve is determined.

MINIMUM OPERATING LEVEL
Lowest level to which the reservoir is drawn down under normal operating conditions.

MINIMUM REPORTING LEVEL (MRL)
Smallest measured concentration of a constituent that may be reliably reported using a given analytical method.

MINIMUM TEMPERATURE Lowest temperature recorded during a specified period of time.

MINING Removal of soil or rock having value because of its chemical composition.

MINING OF AN AQUIFER Withdrawal over a period of time of groundwater that exceeds the rate of recharge of the aquifer.

MINING WASTE Residues resulting from the extraction of raw materials from the Earth.

MINING WATER USE Water use during quarrying rocks and extracting minerals from the land.

MINOR FLOODING Indicating minimal or no property damage but possibly some public inconvenience.

MINOR GEOMAGNETIC STORM
Storm for which the Ap index was greater than 29 and less than 50.

MINOR SOURCE New emissions sources or modifications to existing emissions sources that do not exceed NAAQS emission levels.

MINOR TIDAL OVERFLOW Minor flooding caused by high tides that results in little if any damage.

MINORS Publicly owned treatment works with flows less than 1 million gallons per day.

MINT Minimum temperature.

MISC Miscellaneous.

MISCELLANEOUS ACM Interior asbestos-containing building material or structural components, members or fixtures (i.e., floor and ceiling tiles).

MISCELLANEOUS MATERIALS Interior building materials on structural components (i.e., floor and ceiling tiles).

MISCELLANEOUS SITE Site where streamflow, sediment, and/or water-quality data or water-quality or sediment samples are collected once, or more often on a random or discontinuous basis to provide better areal coverage for defining hydrologic and water-quality conditions over a broad area in a river basin.

MISCIBLE LIQUIDS Two or more liquids that can be mixed and will remain mixed under normal conditions.

MISG Missing.

MISOSCALE Scale of meteorological phenomena that ranges in size from 40 meters to about 4 kilometers.

MISSED DETECTION Situation that occurs when a test indicates that a tank is "tight" when in fact it is leaking.

MISSION OF THE HYDROLOGIC SERVICES PROGRAM To provide river and flood forecasts and warnings for the protection of life and property and to provide basic hydrologic forecast information for the Nation's economic and environmental well-being.

MISSION OF THE NATIONAL WEATHER SERVICE To provide weather and flood warnings, public forecasts and advisories for all of the United States, its territories, adjacent waters and ocean areas, primarily for the protection of life and property.

MIST Liquid particles measuring 40 to 500 micrometers (pm), are formed by condensation of vapor; visible aggregate of minute water particles suspended in the atmosphere that reduces visibility to less than 7 statute miles, but greater than or equal to $^5/_8$ statute miles.

MITIGATION (MEASURES) Methods or plans to reduce, offset, or eliminate adverse project impacts; actions taken to avoid, reduce, or compensate for the effects of human-induced environmental damage.

MIXED BOUNDARY Linear combination of head and flux at a boundary.

MIXED FACE Digging in dirt and rock in the same heading at the same time.

MIXED FUNDING Settlements in which potentially responsible parties and EPA share the cost of a response action.

MIXED GLASS Recovered container glass not sorted into categories (e.g., color and grade).

MIXED LAYER Atmospheric layer, usually the layer immediately above the ground, in which pollutants are well mixed by convective or shear-produced turbulence.

MIXED LIQUOR Mixture of activated sludge and water containing organic matter undergoing activated sludge treatment in an aeration tank.

MIXED METALS Recovered metals not sorted into categories.

MIXED MUNICIPAL WASTE Solid waste that has not been sorted into specific categories.

MIXED PAPER Recovered paper not sorted into categories.

MIXED PLASTIC Recovered plastic not sorted into categories.

MIXED PRECIPITATION Any of the following combinations of freezing and frozen precipitation: snow and sleet, snow and freezing rain, or sleet alone.

MIXING DEPTH Vertical distance between the ground and the altitude to which pollutants are mixed by turbulence caused by convective currents or vertical shear in the horizontal wind.

MIXING HEIGHTS Height to which a parcel of air, or a column of smoke, will rise, mix or disperse.

MIXING RATIO Ratio of the weight of water vapor in a specified volume (e.g. air parcel) to the weight of dry air in that same volume.

MIXING ZONE Area where an effluent discharge undergoes initial dilution and is extended to cover the secondary mixing in the ambient water body.

MOBILE INCINERATOR SYSTEMS Hazardous waste incinerators that can be transported from one site to another.

MOBILE SOURCE Any non-stationary source of air pollution (i.e., cars, trucks, motorcycles, buses, airplanes, and locomotives).

MODEL Assembly of concepts in the form of mathematical equations that portray an understanding of a natural phenomenon; physical or mathematical representation of a process that can be used to predict some aspect of the process.

MODEL CONSTRUCTION Process of transforming the conceptual model into a parameterized mathematical form.

MODEL GRID System of connected nodal points superimposed over the problem domain to spatially discretize the problem domain into cells (finite difference method) or elements (finite element method) for the purpose of numerical modeling.

MODEL INPUT Constitutive coefficients, system parameters, forcing terms, auxiliary conditions and program control parameters required to apply a computer code to a particular problem.

MODEL OUTPUT STATISTICS (MOS) Short range (6 to 60 hours) guidance package generated from the NGM, GFS, and ETA models for over 300 individual stations in the continental United States that use model output to forecast the probability of precipitation, high and low temperature, cloud cover, and precipitation amount for many cities across the USA.

MODEL PLANT Hypothetical plant design used for developing economic, environmental, and energy impact analyses as support for regulations or regulatory guidelines.

MODEL VERIFICATION Procedure of determining if a (site-specific) model's accuracy and predictive capability lie within acceptable limits of error by tests independent of the calibration data; using the set of parameter values and boundary conditions from a calibrated model to acceptably approximate a second set of field data measured under similar hydrologic conditions.

MODELING Use of mathematical equations to simulate and predict real events and processes; process of formulating a model of a system of process.

MODELING OBJECTIVES Purpose(s) of a model application.

MODERATE FLOODING Inundation of secondary roads; transfer to higher elevation necessary to save property.

MODERATE FREQUENCY FLOOD
Flood of lesser magnitude than the IDF, used for the service spillway design when supplemented by a separate auxiliary spillway.

MODERATE RISK (OF SEVERE THUNDERSTORMS)
Severe thunderstorms are expected to affect between 5 and 10 percent of the area.

MODIFICATION DECISION ANALYSIS (MDA)
Process of determining with confidence whether dam safety deficiencies exist for maximum loading conditions.

MODIFIED BIN METHOD Way of calculating the required heating or cooling for a building based on determining how much energy the system would use if outdoor temperatures were within a certain temperature interval and then multiplying the energy use by the time the temperature interval typically occurs.

MODIFIED HOMOGENEOUS EARTHFILL DAM
Homogeneous earthfill dam that uses pervious material specially placed in the embankment to control seepage.

MODIFIED MERCALLI SCALE
Earthquake intensity scale which has twelve divisions ranging from I (not felt by people) to XII (nearly total damage).

MODIFIED SOURCE Enlargement of a major stationary pollutant sources is often referred to as modification, implying that more emissions will occur.

MOHR CIRCLE Graphical representation of the stresses acting on the various planes at a given point.

MOIST ADIABAT Line on a Skew T-Log P chart that depicts the change in temperature of saturated air as it rises and undergoes cooling due to adiabatic expansion.

MOIST ADIABATIC LAPSE RATE (MALR)
Rate at which the temperature of a parcel of saturated air decreases as the parcel is lifted in the atmosphere.

MOIST-ADIABATIC (SATURATION-ADIABATIC PROCESS)
Adiabatic process for which the air is saturated and may contain liquid water.

MOISTURE Water diffused in the atmosphere or the ground; water vapor content in the atmosphere, or the total water, liquid, solid or vapor, in a given volume of air.

MOISTURE ADVECTION Transport of moisture by horizontal winds.

MOISTURE CONTENT (WATER CONTENT) Ratio of the weight of water in a soil sample to the weight of the dry soil, expressed as a percentage; water equivalent of snow on the ground.

MOISTURE CONVERGENCE Measure of the degree to which moist air is converging into a given area, taking into account the effect of converging winds and moisture advection.

MOISTURE RIDGE Axis of relatively high dew point values.

MOLECULE Smallest particle of a substance that retains the properties of the substance and is composed of one or more atoms.

MOLTEN SALT REACTOR Thermal treatment unit that rapidly heats waste in a heat-conducting fluid bath of carbonate salt.

MONITOR STAGE Stage which, when reached by a rising stream, represents the level where appropriate officials are notified of the threat of possible flooding.

MONITORING Testing that water systems must perform to detect and measure contaminants; periodic or continuous surveillance or testing to determine the level of compliance with statutory requirements and/or pollutant levels in various media or in humans, plants, and animals; repeated observation, measurement, or sampling at a site, on a scheduled or event basis, for a particular purpose.

MONITORING WELL Well designed for measuring water levels and testing ground-water quality; well drilled at a hazardous waste management facility or Superfund site to collect ground-water samples for the purpose of physical, chemical, or biological analysis to determine the amounts, types, and distribution of contaminants in the groundwater beneath the site.

MONOCLINE Bend or steplike fold in rock layers where all strata are inclined in the same direction.

MONOCLONAL ANTIBODIES Man-made (anthropogenic) clones of a molecule, produced in quantity for medical or research purposes; molecules of living organisms that selectively find and attach to other molecules to which their structure conforms exactly.

MONOCOQUE GATE Thin-shell radial gate in which the usual skin plate and cross-beam framework are replaced by a hollow shell having an approximate elliptical cross section.

MONOCYCLIC AROMATIC HYDROCARBONS Single-ring aromatic compounds.

MONOMICTIC Lakes and reservoirs which are relatively deep, do not freeze over, and undergo a single stratification and mixing cycle during the year.

MONOSTATIC RADAR Radar that uses a common antenna for both transmitting and receiving.

MONSOON Thermally driven wind arising from differential heating between a land mass and the adjacent ocean that reverses its direction seasonally.

MONTANE Of, pertaining to, or inhabiting cool upland slopes below the timber line.

MORAINE Mound, ridge, or other distinct accumulation of unsorted, unstratified glacial drift, predominantly till, deposited chiefly by direct action of glacier ice.

MORATORIUM During the negotiation process, a period during which agency and potentially responsible parties may reach settlement but no site response activities can be conducted.

MORBIDITY Rate of disease incidence.

MORNING GLORY Elongated cloud band, visually similar to a roll cloud, usually appearing in the morning hours, when the atmosphere is relatively stable.

MORNING GLORY SPILLWAY Circular or glory hole form of a drop inlet spillway.

MORTALITY Death rate.

MOST PROBABLE NUMBER (MPN) Estimate of microbial density per unit volume of water sample.

MOST UNSTABLE LIFTED INDEX (MULI) Lifted Index (LI) calculated using a parcel from the pressure level that results in the Most Unstable value (lowest value) of LI possible.

MOSTLY CLEAR See MOSTLY SUNNY.

MOSTLY CLOUDY When the 6/8th to 7/8ths of the sky is covered by with opaque (not transparent) clouds.

MOSTLY SUNNY When the 1/8th to 2/8ths of the sky is covered by with opaque (not transparent) clouds.

MOTOR EFFICIENCY Ratio of energy delivered by a motor to the energy supplied to it during a fixed period or cycle.

MOUNT WILSON MAGNETIC CLASSIFICATIONS Classification system for sunspots: Alpha - denotes a unipolar sunspot group; Beta - sunspot group having both positive and negative magnetic polarities, with a simple and distinct division between the polarities; Beta-Gamma - sunspot group that is bipolar but in which no continuous line can be drawn separating spots of opposite polarities; Delta - complex magnetic configuration of a solar sunspot group consisting of opposite polarity umbrae within the same penumbra; Gamma - complex active region in which the positive and negative polarities are so irregularly distributed as to prevent classification as a bipolar group.

MOUNTAIN WAVE Wavelike effect, characterized by updrafts and downdrafts, that occurs above and behind a mountain range when rapidly flowing air encounters the mountain range's steep front.

MOUNTAIN WIND SYSTEM System of diurnal winds that forms in a complex terrain area, consisting of mountain-plain, along-valley, cross-valley and slope wind systems.

MOUNTAINADO Vertical-axis eddy produced in a downslope windstorm by the vertical stretching of horizontal roll vortices produced near the ground by vertical wind shear.

MOUNTAIN-PLAIN WIND SYSTEM Closed, large-scale, thermally driven circulation between the mountains and the surrounding plain.

MOUTH Place where a stream discharges to a larger stream, a lake, or the sea.

MOV Move.

MOVABLE BED Stream bed made up of materials readily transportable by the stream flow; portion of a river channel cross section that is considered to be subject to erosion or deposition.

MOVABLE BED LIMITS Lateral limits of the movable bed that define where scour or deposition occur.

MOVABLE BED MODEL Model in which the bed and/or side material is erodible and transported in a manner similar to the prototype.

MOVEABLE BED STREAMS Streams where steep slopes and lack of vegetation result in a lot of erosion.

MOVG Moving.

MR More.

MRGL Marginal.

MRNG Morning.

MS CONNECTORS Surface delays for use when shooting with detonating cord.

MSG Message.

MSL Mean Sea Level.

MSLP Mean Sea Level Pressure.

MST Mountain Standard Time.

MSTR Moisture.

MTN (MT) Mountain.

MTNS (MTS) Mountains.

MUCK Mud rich in humus; stone, dirt, debris, or useless material; organic soil of very soft consistency; dark, finely divided, well-decomposed, organic matter forming a surface deposit in some poorly drained areas; finely blasted rock, particularly from underground.

MUCK SOILS Earth made from decaying plant materials.

MUD Any soil containing enough water to make it soft; mixture of soil and water in a fluid or weakly solid state.

MUD SLIDE Fast moving soil, rocks and water that flow down mountain slopes and canyons during a heavy downpour of rain.

MUDBALLS Round material that forms in filters and gradually increases in size when not removed by backwashing.

MUDFLAT Mud-covered, gently sloping tract of land alternately covered and left bare by water; muddy, nearly level bed of a dry lake.

MUDSTONE Fine-grained sedimentary rock formed from hardened clay and silt that lacks the thin layers typical of shale.

MUGGY Term for warm and excessively humid conditions.

MULCH Layer of material (e.g., wood chips, straw, leaves, etc.) placed around plants to hold moisture, prevent weed growth, and enrich or sterilize the soil.

MULTICELL THUNDERSTORM Thunderstorms organized in clusters of at least 2-4 short-lived cells.

MULTI-MEDIA APPROACH Joint approach to several environmental media, such as air, water, and land.

MULTIPLE ARCH DAM Buttress dam comprised of a series of arches for the upstream face.

MULTIPLE CHEMICAL SENSITIVITY Diagnostic label for people who suffer multi-system illnesses as a result of contact with, or proximity to, a variety of airborne agents and other substances.

MULTIPLE DOPPLER ANALYSIS Use of more than one radar to reconstruct spatial distributions of the 2D or 3D wind field, which cannot be measured from a single radar alone.

MULTIPLE USE Use of water or land for more than one purpose.

MULTIPLE VORTEX TORNADO Tornado in which two or more condensation funnels or debris clouds are present at the same time, often rotating about a common center or about each other.

MULTIPLE-PLATE SAMPLERS Artificial substrates of known surface area used for obtaining benthic invertebrate samples.

MULTIPLE-PURPOSE RESERVOIR (MULTIPURPOSE RESERVOIR) Reservoir constructed and equipped to provide storage and release of water for two or more purposes such as flood control, power development, navigation, irrigation, recreation, pollution abatement, domestic water supply, etc.; reservoir planned to operate for more than one purpose.

MULTIPURPOSE DAM Dam constructed for two or more purposes (e.g., storage, flood control, navigation, power generation, recreation, or fish and wildlife enhancement).

MULTIPURPOSE PROJECT Project designed for irrigation, power, flood control, municipal and industrial, recreation, and fish and wildlife benefits, in any combinations of two or more (contrasted to single-purpose projects serving only one need).

MULTI-SECTOR GENERAL PERMIT (MSGP) Authorizes the discharge of stormwater from industrial facilities, consistent with the terms of the permit, in areas of the United States where EPA manages the NPDES permit program.

MULTI-STAGE PUMP Pump that has more than one impeller.

MULTISTAGE REMOTE SENSING Strategy for landscape characterization that involves gathering and analyzing information at several geographic scales, ranging from generalized levels of detail at the national level through high levels of detail at the local scale.

MUNICIPAL (PUBLIC) WATER SYSTEM Water system that has at least five service connections or which regularly serves 25 individuals for 60 days.

MUNICIPAL DISCHARGE Discharge of effluent from waste water treatment plants which receive waste water from households, commercial establishments, and industries in the coastal drainage basin.

MUNICIPAL SEWAGE Wastes (mostly liquid) originating from a community that may be composed of domestic wastewaters and/or industrial discharges.

MUNICIPAL SLUDGE Semi-liquid residue remaining from the treatment of municipal water and wastewater.

MUNICIPAL SOLID WASTE Common garbage or trash generated by industries, businesses, institutions, and homes.

MUNICIPAL USE OF WATER Various uses of water in developed urban areas (i.e., domestic use, industrial use, street sprinkling, fire protection, etc.).

MUNICIPAL UTILITY (MUNICIPALLY-OWNED ELECTRIC SYSTEM) Utility that is owned and operated by a city.

MUNICIPALIZATION Process by which a municipal entity assumes responsibility for supplying utility service to its constituents.

MUSHROOM Thunderstorm with a well-defined anvil rollover.

MUSKEG Large expanses of peatlands or bogs in subarctic zones.

MUTAGEN Chemical or physical agent that induces a permanent change in the genetic material.

MUTAGENICITY Capacity of a chemical or physical agent to cause such permanent changes.

MVS Moves.

MWS Marine Weather Statement.

N

N North.

NACREOUS CLOUDS Clouds of unknown composition that have a soft, pearly luster and that form at altitudes about 25 to 30 km above the Earth's surface.

NADIR Point on any given observer's celestial sphere diametrically opposite of one's zenith.

NAME PLATE CAPACITY Equipment rating as required by the purchaser.

NAMEPLATE Power generation capacity of a generator that can be guaranteed under continuous operation.

NANOGRAMS PER LITER (NG/L, NG/L)
Unit expressing the concentration of chemical constituents in solution as mass (nanograms) of solute per unit volume (liter) of water.

NANOMETER (MILLIMICRON)
One millionth of a millimeter.

NANOTESLA (NT) Unit of magnetism equal to 10-9 tesla; equivalent to a gamma (10-5 gauss).

NAO INDEX Index measures the anomalies in sea level pressure between the Icelandic low pressure system and the Azores high pressure system in the North Atlantic Ocean.

NATIONAL AMBIENT AIR QUALITY STANDARDS (NAAQS)
National standards for the ambient concentrations in air of different air pollutants designed to protect human health and welfare.

NATIONAL CENTERS FOR ENVIRONMENTAL PREDICTION (NCEP)
Part of the National Weather Service which provides nationwide computerized and manual guidance to Warning and Forecast Offices concerning the forecast of basic weather elements.

NATIONAL CLIMATIC DATA CENTER
Agency that archives climatic data from the National Oceanic and Atmospheric Administration, as well as other climatological organizations.

NATIONAL DIGITAL FORECAST DATABASE (NDFD)
National Weather Service's access to gridded forecasts of sensible weather elements (e.g., wind, wave height) through the National Digital Forecast Database (NDFD).

NATIONAL DISASTER MEDICAL SYSTEM

System designed to deal with extensive medical care needs in very large disasters or emergencies.

NATIONAL EMISSIONS STANDARDS FOR HAZARDOUS AIR POLLUTANTS (NESHAPS)

Emissions standards set by EPA for an air pollutant not covered by NAAQS that may cause an increase in fatalities or in serious, irreversible, or incapacitating illness.

NATIONAL ENVIRONMENTAL PERFORMANCE PARTNERSHIP AGREEMENTS

System that allows states to assume greater responsibility for environmental programs based on their relative ability to execute them.

NATIONAL ENVIRONMENTAL POLICY ACT (NEPA)

Act requiring analysis, public comment, and reporting for environmental impacts of federal actions.

NATIONAL ENVIRONMENTAL SATELLITE, DATA, AND INFORMATION SERVICE (NESDIS)

Collects, processes, stores, analyzes, and disseminates various types of hydrologic, meteorological, and oceanic data.

NATIONAL ESTUARY PROGRAM

Program established to develop and implement conservation and management plans for protecting estuaries and restoring and maintaining their chemical, physical, and biological integrity, as well as controlling point and non-point pollution sources.

NATIONAL FIRE DANGER RATING SYSTEM

Uniform fire danger rating system used in the United States that focuses on the environmental factors that impact the moisture content of fuels.

NATIONAL FLOOD SUMMARY

National Weather Service daily product that contains nationwide information on current flood conditions.

NATIONAL GEODETIC VERTICAL DATUM (NGVD)

As corrected in 1929, a vertical control measure used as a reference for establishing varying elevations.

NATIONAL GEODETIC VERTICAL DATUM OF 1929 (NGVD 29)

Fixed reference adopted as a standard geodetic datum for elevations determined by leveling; formerly called "Sea Level Datum of 1929."

NATIONAL HURRICANE CENTER (NHC)

One of three branches of the Tropical Prediction Center (TPC) that maintains a continuous watch on tropical cyclones over the Atlantic, Caribbean, Gulf of Mexico, and the Eastern Pacific from 15 May through November 30.

NATIONAL HURRICANE OPERATIONS PLAN (NHOP)

Issued annually by the federal coordinator for Meteorological Services and Supporting Research that documents interdepartmental agreements relating to tropical cyclone observing, warning, and forecasting services.

NATIONAL INSTITUTE FOR OCCUPATIONAL SAFETY AND HEALTH (NIOSH)
Federal agency responsible for conducting research and making recommendations for the prevention of work-related disease and injury.

NATIONAL INTERAGENCY FIRE CENTER (NIFC)
Nation's support center for wildland firefighting.

NATIONAL MUNICIPAL PLAN
Policy created to bring all publicly owned treatment works (POTWs) into compliance with Clean Water Act requirements.

NATIONAL OCEANIC AND ATMOSPHERIC ADMINISTRATION (NOAA)
Mission is to describe and predict changes in the Earth's environment, and conserve and wisely manage the Nation's coastal and marine resources.

NATIONAL OIL AND HAZARDOUS SUBSTANCES CONTINGENCY PLAN (NOHSCP/NCP)
Federal regulation that guides determination of the sites to be corrected under both the Superfund program and the program to prevent or control spills into surface waters or elsewhere.

NATIONAL OPERATIONAL HYDROLOGIC REMOTE SENSING CENTER (NOHRSC)
Organization under the National Weather Service Office of Hydrology (OH) that mainly deals with snow mapping.

NATIONAL PARK SERVICE (NPS)
Preserves unimpaired the natural and cultural resources and values of the national park system for the enjoyment, education, and inspiration of this and future generations; agency within the U.S. Department of the Interior.

NATIONAL POLLUTANT DISCHARGE ELIMINATION SYSTEM (NPDES)
Permitting program of the Clean Water Act required for all point sources discharging pollutants into waters of the United States to protect human health and the environment.

NATIONAL PRETREATMENT STANDARD OR PRETREATMENT STANDARD
Any regulation promulgated that applies to a specific category of industrial users and provides limitations on the introduction of pollutants into publicly owned treatment works.

NATIONAL PRIORITIES LIST (NPL)
List of the most serious uncontrolled or abandoned hazardous waste sites identified for possible long-term remedial action under Superfund.

NATIONAL REGISTER OF HISTORIC PLACES
Federally maintained register of districts, sites, buildings, structures, architecture, archeology, and culture.

NATIONAL RESPONSE CENTER (NRC)
Operated by the U.S. Coast Guard, federal operations center that receives notifications of all releases of oil and hazardous substances into the environment.

NATIONAL RESPONSE TEAM (NRT)
Consists of representatives from sev-

eral federal agencies that provides program direction, planning and preparedness guidance, and review of regional response activities to nationally significant incidents of pollution.

NATIONAL SEVERE STORMS LABORATORY (NSSL)
Internationally known as Environmental Research Laboratories that lead the way in investigations of all aspects of severe weather.

NATIONAL WARNING SYSTEM (NAWAS)
Dedicated, commercially leased, nationwide voice telephone warning system operated on a 24-hour basis, with a National Warning Center and an Alternate National Warning Center staffed by attack warning officers.

NATIONAL WATER QUALITY ASSESSMENT (NAWQA) PROGRAM
Long term USGS program to assess the occurrence and distribution of water-quality conditions Nationwide.

NATIONAL WEATHER AND CROP SUMMARY
Product of the National Agricultural Statistics Service, Agricultural Statistics Board, and U.S. Department of Agriculture that contains weekly national agricultural weather summaries, including the weather's effect on crops; summaries and farm progress for 44 states and New England area.

NATIONAL WEATHER SERVICE (NWS)
Agency of the Federal Government within the Department of Commerce, National Oceanic and Atmospheric Administration, that provides weather, hydrologic, and climate forecasts and warnings for the United States, its territories, adjacent waters and ocean areas, for the protection of life and property and the enhancement of the national economy.

NATIONAL WEATHER SERVICE RIVER FORECAST MODEL VERSION 5 (NWSRFS V5.0)
System of data entry, data preprocessing, and forecast programs which are used by River Forecast Centers (RFC).

NATIONWIDE RIVERS INVENTORY (NRI)
Listing of more than 3,400 free-flowing river segments in the United States that are believed to possess one or more "outstandingly remarkable" natural or cultural values judged to be of more than local or regional significance.

NATIVE SPECIES Any animal and plant species originally in the United States.

NATURAL CONTROL Stream gaging control which is natural to the stream channel, in contrast to an artificial control structure by man.

NATURAL FLOODWAY Channel of a water course and those portions of the adjoining flood plain which are reasonably required to carry a selected probability flood.

NATURAL FREQUENCY (F) Frequencies of free vibration.

NATURAL LEVEE Long, broad, low ridge built by a stream on its flood plain along one or both banks of its channel in time of flood.

NATURAL PERIOD OF VIBRATION (T)
Time required for one cycle of the

simple harmonic motion in one of these characteristic patterns (shapes).

NATURAL SUBSTRATE Any naturally occurring immersed or submersed solid surface, such as a rock or tree, upon which an organism lives.

NATURALNESS Area which generally appears to have been affected primarily by the forces of nature, with the imprint of man's work substantially unnoticeable.

NAUTICAL DAWN Time at which the Sun is 12 degrees below the horizon in the morning; time at which there is just enough sunlight for objects to be distinguishable.

NAUTICAL DUSK Time at which the Sun is 12 degrees below the horizon in the evening; objects are no longer distinguishable.

NAUTICAL MILE Unit of distance used in marine navigation and marine forecasts, equal to 1.15 statute miles, 1852 meters or 1 minute of latitude.

NAUTICAL TWILIGHT Time after civil twilight, when the brighter stars used for celestial navigation have appeared and the horizon may still be seen; ends when the center of the Sun is 12 degrees below the horizon, and it is too difficult to perceive the horizon, preventing accurate sighting of stars.

NAVIGABLE WATER All surface water of the United States, including the territorial seas; waters sufficiently deep and wide for navigation by all, or specified vessels.

NAVTEX International automated medium frequency (518 kHz) direct-printing service for delivery of navigational and meteorological warnings and forecasts, as well as urgent marine safety information to ships.

NAVTEX FORECAST (NAV) National Weather Service marine forecast combining various Coastal Waters and Offshore forecasts, optimized to accommodate transmission via NAVTEX.

NAVY OPERATIONAL GLOBAL ATMOSPHERIC PREDICTION SYSTEM (NOGAPS) 144-hour numerical model of the atmosphere run by the U.S. Navy twice daily.

NC No Change.

NCAR National Center for Atmospheric Research.

NCCF NOAA Central Computer Facility.

NCDC National Climatic Data Center.

NDBC National Data Buoy Center.

NE Northeast.

NEAP TIDE Minimum tide occurring at the first and third quarters of the Moon.

NEARSHORE FORECAST (NSH) National Weather Service seasonal marine forecasts for an areas of the Great Lakes extending from a line approximating mean low water datum along the coast or an island, including bays, harbors, and sounds, out to 5 nm.

NEATLINES (OF STRUCTURE) Line which defines the limits of work, such as an excavation, cut stone, etc.; true face line of a building regardless of the projections of the stones; a line back of, or inside of, incidental projections.

NEC Necessary.

NECROSIS Death of plant or animal cells or tissues.

NEEDLE VALVE Any of a family of valves which regulate flow through the use of a needle moving into and out of an orifice.

NEG Negative.

NEGATIVE PRESSURE Pressure within a pipe that is less than atmospheric pressure.

NEGATIVE VORTICITY ADVECTION (NVA) Advection of lower values of vorticity into an area.

NEGATIVE-TILT TROUGH Upper level system which is tilted to the west with increasing latitude (i.e. with an axis from southeast to northwest).

NEKTON Consumers in the aquatic environment and consist of large, free-swimming organisms that are capable of sustained, directed mobility.

NELY Northeasterly.

NEMATOCIDE Chemical agent which is destructive to nematodes.

NEOPRENE Synthetic rubber with superior resistance to oils.

NEOTROPIC MIGRANT (NEOTROPICAL) Bird that migrates to tropical regions during the winter.

NEPHELOMETRIC Method of measuring turbidity in a water sample by passing light through the sample and measuring the amount of the light that is deflected.

NEPHELOMETRIC TURBIDITY UNIT (NTU) Unit of measure for the turbidity of water.

NERN Northeastern.

NET ALL-WAVE RADIATION Net or resultant value of the upward and downward longwave and shortwave radiative fluxes through a plane at the Earth-atmosphere interface; component of the surface energy budget.

NET CAPABILITY Maximum load-carrying ability of the equipment, exclusive of station use, under specified conditions for a given time interval, independent of the characteristics of the load.

NET CUT Cut required, less the fill required, at a particular station or part of a road.

NET ECONOMIC BENEFITS Economic benefits less economic costs.

NET FILL Fill required, less the cut required, at a particular station or part of a road.

NET METERING Method of crediting customers for electricity that they generate on site in excess of their own electricity consumption.

NET RAINFALL Portion of rainfall which reaches a stream channel or the concentration point as direct surface flow.

NETTING Concept in which all emissions sources in the same area that owned or controlled by a single company are treated as one large source, thereby allowing flexibility in controlling individual sources in order to meet a single emissions standard.

NEUTRAL LINE Line that separates longitudinal magnetic fields of opposite polarity.

NEUTRAL STABILITY Atmospheric condition that exists in unsaturated air when the environmental lapse rate equals the dry adiabatic rate, or in saturated air when the environmental lapse rate equals the moist adiabatic rate.

NEUTRALIZATION Decreasing the acidity or alkalinity of a substance by adding alkaline or acidic materials, respectively.

NEUTRON PROBE Instrument used to estimate soil moisture. Relates the rate of attenuation in pulsed neutron emissions to soil water content.

NEVE See FIRN.

NEW SOURCE Any stationary source built or modified after publication of final or proposed regulations that prescribe a given standard of performance.

NEW SOURCE PERFORMANCE STANDARDS (NSPS) Uniform national EPA air emission and water effluent standards which limit the amount of pollution allowed from new sources or from modified existing sources.

NEW SOURCE REVIEW (NSR) Clean Air Act requirement that State Implementation Plans must include a permit review that applies to the construction and operation of new and modified stationary sources in non-attainment areas to ensure attainment of national ambient air quality standards.

NEW YORK ROD Leveling rod marked with narrow lines, ruler-fashion.

NEWD Northeastward.

NEWTON Force which, when applied to a body having a mass of one kilogram, gives it an acceleration of one meter per second squared.

NEXT GENERATION RADAR (NEXRAD) National Weather Service network of about 140 Doppler radars operating nationwide.

NGT Night.

NIEVE PENITENTE Spike or pillar of compacted snow, firn or glacier ice, caused by differential melting and evaporation.

NIGHT Period of the day between dusk and dawn.

NIL None.

NIMBOSTRATUS (NS) Cloud of the class characterized by a formless layer that is almost uniformly dark gray; a rain cloud of the layer type, of low altitude, usually below 8000 ft (2400 m).

NIPPLE Short piece of pipe with male threads on each end.

NITRATE Compound containing nitrogen that can exist in the atmosphere or as a dissolved gas in water and which can have harmful effects on humans and animals; ion consisting of nitrogen and oxygen; plant nutrient that is very mobile in soils.

NITRIC OXIDE (NO) Gas formed by combustion under high temperature and high pressure in an internal combustion engine.

NITRIFICATION Process whereby ammonia in wastewater is oxidized to nitrite and then to nitrate by bacterial or chemical reactions.

NITRILOTRIACETIC ACID (NTA) Compound now replacing phosphates in detergents.

NITRITE Intermediate in the process of nitrification; nitrous oxide salts used in food preservation.

NITROGEN DIOXIDE Result of nitric oxide combining with oxygen in the atmosphere; major component of photochemical smog.

NITROGEN OXIDE Result of photochemical reactions of nitric oxide in ambient air; major component of photochemical smog.

NITROGENOUS WASTES Animal or vegetable residues that contain significant amounts of nitrogen.

NITROPHENOLS Synthetic organopesticides containing carbon, hydrogen, nitrogen, and oxygen.

NITROUS OXIDE Clear, colorless gas, with slightly sweet odor.

NLY Northerly.

NM Nautical Miles.

NMBR Number.

NMC National Meteorological Center.

NML Normal.

NMRS Numerous.

NO ACTION ALTERNATIVE Projected baseline condition, or future without; expected future condition if no action is taken.

NO FURTHER REMEDIAL ACTION PLANNED Determination following a preliminary assessment that a site does not pose a significant risk and so requires no further activity.

NO JEOPARDY OPINION Opinion that an action is not likely to jeopardize the continued existence of listed species or result in the destruction or adverse modification of critical habitat.

NO OBSERVABLE ADVERSE EFFECT LEVEL (NOAEL) Exposure level at which there are no statistically or biologically significant increases in the frequency or severity of adverse effects between the exposed population and its appropriate control.

NO TILL Planting crops without prior seedbed preparation, into an existing cover crop, sod, or crop residues, and eliminating subsequent tillage operations.

NOAA National Oceanic and Atmospheric Administration.

NOAA WEATHER RADIO Broadcasts National Weather Service warnings, watches, forecasts and other hazard information 24 hours a day.

NOAA WEATHER WIRE Mass dissemination via satellite of National Weather Service products to the media and public.

NOAEL No Observable Adverse Effect Level.

NOBLE METAL Chemically inactive metal (e.g. gold); does not corrode easily.

NOCTILUCENT CLOUDS Wavy, thin, bluish-white clouds that are best seen at twilight in polar latitudes.

NOCTURNAL Related to nighttime; occurring at night.

NOCTURNAL INVERSION Temperature inversion that develops during the night as a result of radiational cooling of the surface.

NOCTURNAL JET Maximum wind speed that occurs just above the nocturnal inversion at night.

NOCTURNAL THUNDERSTORMS Thunderstorms that develop after sunset.

NODE (NODAL POINT) In a numerical model, location in the discretized model domain where a dependent variable (i.e. hydraulic head) is computed.

NO-FLOW BOUNDARY Model boundary where the assigned flux is equal to zero.

NOISE Product-level or product-volume changes occurring during a test that are not related.

NO-MIGRATION PETITIONS Demonstrations that fluids injected into a Class I hazardous waste well will remain in the injection zone for 10,000 years or as long as the fluid remains hazardous or until the waste decomposes or otherwise is attenuated to non-hazardous levels before migrating from the injection zone.

NOMINAL DIAMETER Approximate measurement of the diameter of a pipe; usually not the exact inside diameter of the pipe.

NOMINAL DOLLARS Dollar value in the indicated year, not adjusted for inflation.

NOMINAL SIZE Approximate dimension(s) of standard materials.

NON-AQUEOUS PHASE LIQUID (NAPL) Contaminants that remain undiluted as the original bulk liquid in the subsurface(e.g. spilled oil).

NON-ATTAINMENT AREA Area that does not meet ambient air quality standards.

NON-BINDING ALLOCATIONS OF RESPONSIBILITY (NBAR) Process to propose a way for potentially responsible parties to allocate costs among themselves.

NON-COMMUNITY WATER SYSTEM Public water system that is not a community water system (e.g. water supply at national park).

NON-COMPLIANCE COAL Any coal that emits greater than 3.0 pounds of sulfur dioxide per million BTU when burned.

NON-CONSUMPTIVE WATER USES Water uses that do not substantially deplete water supplies (i.e., swimming, boating, waterskiing, fishing, etc.).

NON-CONTACT COOLING WATER Water used for cooling which does not come into direct contact with any raw material, product, byproduct, or waste.

NON-CONTACT COOLING WATER WELLS Class V wells that are used to inject non-contact cooling water that contains no additives and has not been chemically altered.

NON-CONTACT WATER RECREATION Recreational activities (i.e., fishing or boating) that do not include direct contact with the water.

NON-CONVENTIONAL POLLUTANT Any pollutant not statutorily listed or which is poorly understood by the scientific community (e.g., nitrogen and phosphorus).

NON-DEGRADATION Environmental policy which disallows any lowering of naturally occurring quality regardless of pre-established health standards.

NON-DISCHARGING TREATMENT PLANT Treatment plant that does not discharge treated wastewater into any stream or river.

NON-ENCAPSULATING CHEMICAL PROTECTIVE SUIT (NECP SUIT) Not gas or vapor tight.

NON-FERROUS METALS Non-magnetic metals (e.g., aluminum, lead, and copper).

NON-FILTERABLE Portion of the total residue retained by a filter.

NON-FIRM COMMERCIAL ENERGY Energy periodically available for sale at lower than firm power rates.

NON-FIRM POWER Power that is not available continuously and may be interruptible.

NON-FRIABLE ASBESTOS-CONTAINING MATERIALS Any material containing more than one percent asbestos that, when dry, cannot be crumbled, pulverized, or reduced to powder by hand pressure.

NON-HAZARDOUS INDUSTRIAL WASTE Industrial process waste in wastewater not considered municipal solid waste or hazardous waste.

NON-IONIZING ELECTROMAGNETIC RADIATION Radiation that does not change the structure of atoms but does heat tissue and may cause harmful biological effects; microwaves, radio waves, and low-frequency electromagnetic fields from high-voltage transmission lines.

NON-METHANE HYDROCARBON (NMHC) Sum of all hydrocarbon air pollutants except methane.

NON-METHANE ORGANIC GASES (NMOG) Sum of all organic air pollutants except methane.

NON-OVERFLOW DAM (SECTION) Dam or section of dam that is not designed to be overtopped.

NON-PERSISTENT EMERGENT PLANTS Emergent plants whose leaves and stems break down at the end of the growing season from decay or by the physical forces of waves and ice.

NON-POINT SOURCE Pollution source (of any water-carried material) that cannot be defined as originating from discrete points (e.g. pipe discharge).

NON-POINT SOURCE (NPS) CONTAMINANT Substance that pollutes or degrades water that comes from lawn or cropland runoff, the atmosphere, roadways, and other diffuse sources.

NON-POINT SOURCE (NPS) POLLUTION Pollution discharged over a wide land area, not from one specific location.

NON-POINT-SOURCE WATER POLLUTION Water contamination that originates from a broad area (such as leaching of agricultural chemicals from crop land) and enters the water resource diffusely over a large area.

NON-POTABLE Water that is unsafe or unpalatable to drink because it may contain objectionable pollution, contamination, minerals, or infective agents.

NON-PREFERENCE CUSTOMER Customer excluded by law from preference in the purchase of federally produced electrical energy.

NON-REIMBURSABLE Cost of constructing, operating, or maintaining a project that is borne by the taxpayer and is not reimbursed by any other individual, entity, or organization.

NON-ROAD EMISSIONS Pollutants emitted by combustion engines on farm and construction equipment, gasoline-powered lawn and garden equipment, and power boats and outboard motors.

NON-SELECTIVE HERBICIDE Kills or significantly retards growth of most higher plant species.

NON-TERRITORIAL COMMUNITIES Networks of associations around shared goals, values, and norms (i.e., agricultural, environmental, or recreational community).

NON-THRESHOLD EFFECTS Associated with exposure to chemicals that have no safe exposure levels (i.e. cancer).

NON-TRANSIENT, NON-COMMUNITY WATER SYSTEM Public water system which supplies water to 25 or more of the same people at least six months per year in places other than their residences (e.g., schools, office buildings, etc.).

NON-UNIFORM FLOW Velocity varies with position.

NON-UNIFORM SKY CONDITION Localized sky condition which varies from that reported in the body of the report.

NON-UNIFORM VISIBILITY Localized visibility which varies from that reported in the body of the report.

NON-UTILITY GENERATORS (NUG'S) Facilities for generating electricity that are not owned exclusively by an electric utility and which operate connected to an electric utility system.

NON-UTILITY POWER PRODUCER Corporation, person, agency, authority, or other legal entity or instrumentality that owns electric generating capacity and is not an electric utility.

NO-OBSERVED-ADVERSE-EFFECT-LEVEL (NOAEL) Tested dose of an effluent or a toxicant below which no adverse biological effects are observed, as identified from chronic or subchronic human epidemiology studies or animal exposure studies.

NO-OBSERVED-EFFECT-CONCENTRATION (NOEC) Highest tested concentration of an effluent or a toxicant at which no adverse effects are observed on the aquatic test organisms at a specific time of observation.

NO-OBSERVED-EFFECT-LEVEL (NOEL) Exposure level at which there are no statistically or biological significant differences in the frequency or severity of any effect in the exposed or control populations.

NOR'EASTER Strong low pressure system that affects the Mid-Atlantic and New England States; continuously strong northeasterly winds blowing in from the ocean ahead of the storm and over the coastal areas.

NORMAL Long-term average value of a meteorological parameter (i.e., temperature, humidity, etc.) for a certain area; central value (such as arithmetic average or median) of annual quantities for a 30-year period ending with an even 10-year, (i.e. 1921-50).

NORMAL DEPTH Depth of flow that would exist for a steady-uniform flow condition.

NORMAL LOADING CONDITIONS Loading conditions that occur or are anticipated to occur with some degree of regularity or frequency.

NORMAL POOL LEVEL See NORMAL WATER SURFACE ELEVATION.

NORMAL RESERVOIR LEVEL See NORMAL WATER SURFACE ELEVATION.

NORMAL WATER SURFACE Highest elevation at which water is normally stored, or that elevation which the reservoir should be operated for conservation purposes; elevation at the top of the active conservation capacity; maximum elevation to which the reservoir may rise under normal operating conditions exclusive of flood control capacity.

NORMAL WATER SURFACE ELEVATION For a reservoir with a fixed overflow sill, lowest crest level of that sill; for a reservoir whose outflow is controlled wholly or partly by moveable gates, siphons or other means, maximum level to which water may rise under normal operating conditions, exclusive of any provision for flood surcharge.

NORMAL YEAR Year during which the precipitation or stream flow approximates the average for a long period of record.

NORTH AMERICAN DATUM OF 1927 (NAD 27) Horizontal control datum that was defined by a location and azimuth on the Clarke spheroid of 1866.

NORTH AMERICAN ELECTRIC RELIABILITY COUNCIL (NERC) Principal organization for coordinating and promoting reliability for North America's electric utilities.

NORTH AMERICAN VERTICAL DATUM OF 1988 (NAVD 88) Fixed reference adopted as the official civilian vertical datum for elevations determined by federal surveying and mapping activities in the United States established in 1991 by minimum-constraint adjustment of the Canadian, Mexican, and United States first-order terrestrial leveling networks.

NORTH ATLANTIC OSCILLATION (NAO) Large-scale fluctuation in atmospheric pressure between the subtropical high pressure system located near the Azores in the Atlantic Ocean and the sub-polar low pressure system near Iceland and is quantified in the NAO Index.

NORTH PACIFIC HIGH Semi-permanent, subtropical area of high pressure in the North Pacific Ocean.

NORTH WALL North side boundary of the Gulf Stream generally extending northeast from Cape Hatteras where the Gulf Stream turns northeast.

NORTHERN LIGHTS See AURORA BOREALIS.

NOTICE OF DEFICIENCY Request to a facility owner or operator requesting additional information before a preliminary decision on a permit application can be made.

NOTICE OF INTENT TO CANCEL Notification sent to registrants when EPA decides to cancel registration of a product containing a pesticide.

NOTICE OF INTENT TO DENY Notification by EPA of its preliminary intent to deny a permit application.

NOTICE OF INTENT TO SUSPEND Notification sent to a pesticide registrant when EPA decides to suspend product sale and distribution because of failure to submit requested data in a timely and/or acceptable manner, or because of imminent hazard.

NOTIFICATION To inform appropriate individuals about an emergency condition so they can take appropriate action; third of five Early Warning System components consisting of communicating alerts and warnings about an emergency condition at a dam to appropriate local officials so they can take proper action(s).

NOWCAST Short-term weather forecast, generally out to six hours or less.

NR Near.

NRCS National Resources Conservation Services.

NRN Northern.

NRW Narrow.

NSDWR National Secondary Drinking Water Regulations.

NSSFC National Severe Storm Forecast Center.

NUCLEAR REACTORS AND SUPPORT FACILITIES Uranium mills, commercial power reactors, fuel reprocessing plants, and uranium enrichment facilities.

NUCLEAR WINTER Prediction by some scientists that smoke and debris rising from massive fires of a nuclear war could block sunlight for weeks or months, cooling the Earth's surface and producing climate changes.

NUCLIDE Atom characterized by the number of protons, neutrons, and energy in the nucleus.

NUISANCE SPECIES Undesirable plants and animals.

NUMERICAL EXPERIMENTS Varying the input data, or internal parameters, of a numerical model to ascertain the impact on the output.

NUMERICAL FORECASTING Computer forecast or prediction based on equations governing the motions and the forces affecting motion of fluids.

NUMERICAL METHODS Set of procedures used to solve equations in which the applicable partial differential equations are replaced by a set of algebraic equations written in terms of discrete values of state variables at discrete points in space and time.

NUMERICAL MODEL Mathematical model that uses numerical methods to solve the governing equations of the applicable problem.

NUMERICAL SOLUTION Approximate solution of a governing (partial) differential equation derived by replacing the continuous governing equation with a set of equations in discrete points of the model's time and space domains.

NUMERICAL WEATHER PREDICTION Computer forecast or prediction based on equations governing the motions and the forces affecting motion of fluids.

NUMEROUS National Weather Service convective precipitation descriptor for a 60 or 70 percent chance of measurable precipitation (0.01 inch).

NUTRIENT Any substance, element or compound essential for animal and plant growth; any inorganic or organic compound needed to sustain plant life.

NUTRIENT POLLUTION Contamination of water resources by excessive inputs of nutrients.

NW Northwest.

NWD Northward.

NWLY Northwesterly.

NWP Numerical Weather Prediction.

NWR NOAA WEATHER RADIO NOAA Weather Radio broadcasts National Weather Service warnings, watches, forecasts and other hazard information 24 hours a day.

NWRD Northwestward.

NWRN Northwestern.

NWS NATIONAL WEATHER SERVICE.

NWSH National Weather Service Headquarters.

NWSO National Weather Service Office.

NXT Next.

O

OAKUM Loosely woven hemp rope that has been treated with oil or other waterproofing agent; used to caulk joints in a bell and spigot pipe and fittings.

OBJECTIVE FUNCTION Mathematical expression that allows comparison between a calculated result and a specified goal.

OBLIGATE RIPARIAN SPECIES Species that depends completely upon habitat along a body of water.

OBS Observation.

OBSC Obscure.

OBSCURATION Any atmospheric phenomenon, except clouds, that restricts vertical visibility (e.g., dust, rain, snow, etc.).

OBSCURING PHENOMENA See OBSCURATION.

OBSERVATION WELL Non-pumping well (hole) used for observing the groundwater surface at atmospheric pressure within soil or rock.

OCCLUDED FRONT (OCFNT) Composite of two fronts, formed as a cold front overtakes a warm or quasi-stationary front.

OCCLUDED MESOCYCLONE Mesocyclone in which air from the rear-flank downdraft has completely enveloped the circulation at low levels, cutting off the inflow of warm unstable low-level air.

OCCUPATIONAL SAFETY AND HEALTH ACT OF 1970 (OSHA) Law designed to protect the health and safety of industrial workers.

OCCURRENCE AND DISTRIBUTION ASSESSMENT Characterization of the broad-scale spatial and temporal distributions of water-quality conditions in relation to major contaminant sources and background conditions for surface water and ground water.

OCEAN DISCHARGE WAIVER Variance from Clean Water Act requirements for discharges into marine waters.

OCEANOGRAPHY Study of the ocean, embracing and integrating all knowledge pertaining to the ocean's physical boundaries, the chemistry and physics of sea water, and marine biology.

OCNL Occasional.

ODOR THRESHOLD Minimum odor of a water or air sample that can just be detected after successive dilutions with odorless water.

OECD GUIDELINES Testing guidelines prepared by the Organization of Economic and Cooperative Development of the United Nations that assist in preparation of protocols for studies of toxicology, environmental fate, etc.

OFFICE OF EQUAL OPPORTUNITY AND DIVERSITY MANAGEMENT (OEODM) Advises and assists the Assistant Administrator in carrying out the National Weather Service's (NWS) responsibilities relative to Civil Rights laws, Executive Orders, regulatory guidelines, and other non-discrimination laws within the Federal Government; advises and assists the Assistant Administrator in carrying out the National Weather Service policy of diversity management by fostering an inclusive workforce, building an environment that respects the individual and offering opportunities for all employees to develop to their full potential.

OFFICE OF GLOBAL PROGRAMS (OGP) Sponsors focused scientific research, within approximately eleven research elements, aimed at understanding climate variability and its predictability.

OFFICE PAPER High grade papers such as copier paper, computer printout, and stationary almost entirely made of uncoated chemical pulp, although some ground wood is used.

OFF-PEAK ENERGY Electric energy supplied during periods of relatively low system demand.

OFFSETS Concept whereby emissions from proposed new or modified stationary sources are balanced by reductions from existing sources to stabilize total emissions.

OFFSHORE BREEZE Wind that blows from the land towards a body of water; land breeze.

OFFSHORE FLOW Occurs when air moves from land to sea and is usually associated with dry weather.

OFFSHORE WATERS That portion of the oceans, gulfs, and seas beyond the coastal waters extending to a specified distance from the coastline, to a specified depth contour, or covering an area defined by specific latitude and longitude points.

OFFSHORE WATERS FORECAST (OFF) National Weather Service marine forecast product for that portion of the oceans, gulfs, and seas beyond the coastal waters extending to a specified distance from the coastline, to a specified depth contour, or covering an area defined by specific latitude and longitude points.

OFF-SITE FACILITY Hazardous waste treatment, storage or disposal area that is located away from the generating site.

OFFSTREAM USE Water withdrawn from surface or groundwater sources for use at another place.

OFSHR Offshore.

OGEE Reverse curve, shaped like an elongated letter S.

OGEE CREST Shape of the concrete spillway crest that represents the lower profile of the undernappe of a jet of water flowing over a sharp-crested weir at a design depth.

OH (National Weather Service) Office of Hydrology,.

OHD Overhead.

OHM Unit of electrical resistance to current flow; resistance in a conductor in which one volt of potential difference produces a current of one ampere.

OIL AND GAS WASTE Gas and oil drilling muds, oil production brines, and other waste associated with exploration for, development and production of crude oil or natural gas.

OIL DESULFURIZATION Precombustion method for reducing sulfur dioxide emissions from oil-burning power plants where oil is treated with hydrogen, which removes some of the sulfur by forming hydrogen sulfide gas.

OIL FINGERPRINTING Method that identifies sources of oil and allows spills to be traced to their source.

OIL SPILL Accidental or intentional discharge of oil which reaches bodies of water.

OIL SPILL CONTINGENCY FUND
Revolving fund for spill control efforts has been authorized in cases where the federal government has taken over containment and cleanup operations.

OKTA Used for the measurement of total cloud cover where one okta of cloud cover is the equivalent of $1/8$ of the sky covered with cloud.

OLIGOTROPHIC Reservoirs and lakes that are nutrient poor, little organic matter (aquatic plant or animal life) and a high dissolved-oxygen level.

OLR Outgoing Longwave Radiation.

OMEGA Term used to describe vertical motion in the atmosphere determined by the amount of spin (or large scale rotation) and warm (or cold) advection present in the atmosphere.

OMEGA HIGH Warm high aloft which has become displaced and is on the polarward side of the jet stream.

OMNIVORE Animal that eats both vegetable and animal substances.

ONBOARD CONTROLS Devices placed on vehicles to capture gasoline vapor during refueling and route it to the engines when the vehicle is starting so that it can be efficiently burned.

ONCONOGENICITY Capacity to induce cancer.

ON-DISTRICT STORAGE Small water storage facilities located within the boundaries of an irrigation entity, including reregulating reservoirs, holding ponds, or other new storage methods that allow for efficient water use.

ONE HUNDRED YEAR FLOOD (100-YEAR FLOOD)
Statistic that indicates the magnitude of flood that can be expected to be equaled or exceeded during any year.

ONE HUNDRED YEAR FLOOD PLAIN (100-YEAR FLOOD PLAIN)
Flood plain that would be inundated in the event of a 100-year flood.

ONE PERCENT CHANCE FLOOD See ONE HUNDRED YEAR FLOOD.

ONE TWO THREE RULE (1-2-3 RULE) Means of avoiding winds associated with a tropical cyclone by taking into account the forecast track error of the National Weather Service over a 10 year period which is approximately 100 nm in 24 hours, 200 nm for 48 hours and 300 nm in 72 hours where the forecast track error is added to the 34 knot wind radii to compute the danger area.

ONE-HIT MODEL Mathematical model based on the biological theory that a single "hit" of some minimum critical amount of a carcinogen at a cellular target such as DNA can start an irreversible series events leading to a tumor.

ON-FARM Activities (especially growing crops and applying irrigation water) that occur within the legal boundaries of private property.

ON-FARM IRRIGATION EFFICIENCY Ratio of the volume of water used for consumptive use and leaching requirements in cropped areas to the volume of water delivered to a farm (applied water).

ON-PEAK ENERGY Electric energy supplied during periods of relatively high system demand.

ON-SCENE COORDINATOR (OSC) Federal official to coordinate and direct federal removal efforts at the scene of an oil or hazardous substance discharge.

ONSHORE BREEZE Wind that blows from a body of water towards the land; seabreeze.

ONSHORE FLOW Occurs when air moves from sea to land, and is usually associated with increased moisture.

ON-SITE FACILITY Hazardous waste treatment, storage or disposal area that is located on the generating site.

ONSITE RENEWABLE ENERGY GENERATION Electricity generated by renewable resources using a system or device located at the site where the power is used.

OPACITY Amount of light obscured by particulate pollution in the air; indicator of changes in performance of particulate control systems.

OPAQUE Condition where a material (e.g. cloud) blocks the passage of radiant energy, especially light.

OPC Ocean Prediction Center.

OPEN ACCESS SAME-TIME INFORMATION SYSTEM (OASIS) Electronic information system that allows users to instantly receive data on the current operating status and transmission capacity of a transmission provider.

OPEN BURNING Uncontrolled fires in an open dump.

OPEN DUMP Uncovered site used for disposal of waste without environmental controls.

OPEN INTERVAL Length of unscreened opening or of well screen through which water enters a well, in feet below land surface.

OPEN LAKES FORECAST (GLF) National Weather Service marine forecast product for the U.S. waters within a Great Lake not including the waters covered by an existing Nearshore Waters Forecast (NSH).

OPEN TRANSMISSION ACCESS Enables all participants in the wholesale market equal access to transmission service, as long as capacity is available, with the objective of creating a more competitive wholesale power market.

OPEN-CUT Method of excavation in which the working area is kept open to the sky.

OPEN-WORK MATERIALS Poorly-graded (uniform or gap-graded gradation) gravels, cobbles and boulders with few fines in the matrix, resulting in a deposit containing a large amount of interconnected void space through which seepage water (and soil particles) can easily move.

OPERABLE UNIT Term for each of a number of separate activities undertaken as part of a Superfund site cleanup.

OPERATING BASIS EARTHQUAKE (OBE) Earthquake that the structure must safely withstand with no damage.

OPERATING CONDITIONS Conditions specified in a permit that dictate how an incinerator must operate as it burns different waste types.

OPERATING LOG See LOGBOOK.

OPERATING POLICY See OPERATING RULE.

OPERATING RESERVE Generation capacity dedicated to maintaining adequate and reliable system operation during sudden reduction in system capacity.

OPERATING RULE Rule that specifies how water is managed throughout a water resource system.

OPERATIONAL LOSSES Losses of water resulting from evaporation, seepage, and spills.

OPERATIONAL PRODUCTS Product that has been fully tested and evaluated and is produced on a regular and ongoing basis.

OPERATIONAL WASTE Water that is lost or otherwise discarded from an irrigation system after having been diverted into it as part of normal operations.

OPERATIONS AND MAINTENANCE (O&M) Activities related to the performance of routine, preventive, predictive, scheduled, and unscheduled actions aimed at preventing equipment failure or decline with the goal of increasing efficiency, reliability, and safety.

Operations Manual Letter (OML) Serve as updates to policy and procedure for the National Weather Service Operations Manual (WSOM).

OPERATOR CERTIFICATION Certification of operators of community and non-transient non-community water systems, asbestos specialists, pesticide applicators, hazardous waste transporter, and other such specialists.

OPTIMAL CORROSION CONTROL TREATMENT Erosion control treatment that minimizes the lead and copper concentrations at users' taps while also ensuring that the treatment does not cause the water system to violate any national primary drinking water regulations.

OPTIMUM MOISTURE (WATER) CONTENT One water content (percent of dry weight of the total material) of a given soil and a given compactive effort that will result in a maximum dry density of the soil.

OPTION VALUE Value associated with people who know they can visit an area in the future if they so desire; reversible decision or an option to develop at some time in the future have option value.

ORAL TOXICITY Ability of a pesticide to cause injury when ingested.

ORE Rock or earth containing workable quantities of a mineral or minerals of commercial value.

ORGANIC Any chemical containing the element carbon; substances that come from animal or plant sources; referring to or derived from living organisms.

ORGANIC CARBON (OC) Measure of organic matter present in aqueous solution, suspension, or bottom sediment.

ORGANIC CHEMICALS/COMPOUNDS Naturally occurring (animal or plant-produced or synthetic) substances containing mainly carbon, hydrogen, nitrogen, and oxygen.

ORGANIC COMPOUND Chemicals that contain carbon.

ORGANIC CONTAMINANTS Carbon-based chemicals, such as solvents and pesticides, which can get into water through runoff from cropland or discharge from factories.

ORGANIC DETRITUS Any loose organic material in streams (i.e., leaves, bark, or twigs) removed and transported by mechanical means, such as disintegration or abrasion.

ORGANIC MASS (VOLATILE MASS) OF A LIVING SUBSTANCE Difference between the dry mass and ash mass and represents the actual mass of the living matter.

ORGANIC MATTER Plant and animal residues, or substances made by living organisms that are based upon carbon compounds.

ORGANIC SOIL Soil that contains more than 20 percent organic matter in the upper 16 inches.

ORGANISM Any form of animal or plant life.

ORGANISM COUNT/AREA Number of organisms collected and enumerated in a sample and adjusted to the number per area habitat.

ORGANISM COUNT/VOLUME Number of organisms collected and enumerated in a sample and adjusted to the number per sample volume.

ORGANOCHLORINE COMPOUND Synthetic organic compounds containing carbon and chlorine.

ORGANOCHLORINE INSECTICIDE Class of organic insecticides containing a high percentage of chlorine.

ORGANOCHLORINE PESTICIDE See ORGANOCHLORINE INSECTICIDE.

ORGANONITROGEN HERBICIDES Group of herbicides consisting of a nitrogen ring with associated functional groups and including such classes as triazines and acetanilides (e.g., atrazine, cyanazine, alachlor, and metolachlor).

ORGANOPHOSPHATE INSECTICIDES
Class of insecticides derived from phosphoric acid that tend to have high acute toxicity to vertebrates.

ORGANOPHOSPHATES Pesticides that contain phosphorus; short-lived, but some can be toxic when first applied.

ORGANOPHOSPHORUS INSECTICIDES
Insecticides derived from phosphoric acid and generally the most toxic of all pesticides to vertebrate animals.

ORGANOPHYLLIC Substance that easily combines with organic compounds.

ORGANOTINS Chemical compounds used in anti-foulant paints to protect the hulls of boats and ships, buoys, and pilings from marine organisms such as barnacles.

ORIENTATION EXERCISE (SEMINAR)
Activity designed to introduce, discuss, and update emergency planning documents, organization structure, or early warning system (EWS) component to familiarize key personnel with the emergency procedures and their responsibilities.

ORIFICE Opening with closed perimeter, usually sharp edged, and of regular form in a plate, wall, or partition through which water may flow, generally used for the purpose of measurement or control of water; end of a small tube, such as a Pitot tube, piezometer, etc.

ORIG Original.

ORIGINAL AHERA INSPECTION
Examination of school buildings arranged by Local Education Agencies to identify asbestos-containing-materials, evaluate their condition, and take samples of materials suspected to contain asbestos; performed by EPA-accredited inspectors.

ORIGINAL GENERATION POINT
Where regulated medical or other material first becomes waste.

ORIGINAL GROUND (SURFACE)
Surface of the Earth as it exists in an unaltered state (i.e. prior to any earthwork).

O-RING Rubber seal used around stems of some valves to prevent water from leaking past.

OROGRAPHIC Related to, or caused by, physical geography (such as mountains or sloping terrain) and their general effect on storm path and generation of rainfall.

OROGRAPHIC LIFTING Occurs when air is forced to rise and cool due to terrain features such as hills or mountains.

OROGRAPHIC PRECIPITATION
Precipitation which is caused by hills or mountain ranges deflecting the moisture-laden air masses upward, causing them to cool and precipitate their moisture.

OROGRAPHIC UPLIFT See OROGRAPHIC LIFTING.

OROGRAPHIC WAVES Wavelike airflow produced over and in the lee of a mountain barrier.

ORPHAN ANVIL Anvil from a dissipated thunderstorm, below which no other clouds remain.

OSCILLATION Shift in position of various high and low pressure systems that in climate terms is usually defined as an index.

OSHA Occupational Safety and Health Administration.

OSMOSIS Movement of water molecules through a thin membrane; passage of a liquid from a weak solution to a more concentrated solution across a semipermeable membrane that allows passage of the solvent (water) but not the dissolved solids.

OTHER FERROUS METALS Recyclable metals from strapping, furniture, and metal found in tires and consumer electronics.

OTHER GLASS Recyclable glass from furniture, appliances, and consumer electronics.

OTHER NON-FERROUS METALS Recyclable non-ferrous metals such as lead, copper, and zinc from appliances, consumer electronics, and non-packaging aluminum products.

OTHER PAPER Recyclable paper from books, third-class mail, commercial printing, paper towels, plates and cups, posters, photographic papers, cards and games, milk cartons, folding boxes, bags, wrapping paper, and paperboard.

OTHER PLASTICS Recyclable plastic from appliances, eating utensils, plates, containers, toys, and various kinds of equipment.

OTHER SOLID WASTE Recyclable non-hazardous solid wastes, other than municipal solid waste.

OTHER WOOD Recyclable wood from furniture, consumer electronics cabinets, and other non-packaging wood products.

OTLK Outlook.

OTR Other.

OTRW Otherwise.

OUTAGE Period during which a generating unit, transmission line, or other facility is out of service.

OUTBURST FLOOD See JOKULHLAUP.

OUTDOOR AIR SUPPLY Air brought into a building from outside.

OUTER CONVECTIVE BAND Bands in a hurricane that occur in advance of main rain shield and up to 300 miles from the eye of the hurricane.

OUTFALL Place where effluent (sewer, drain, or stream) is discharged into receiving waters; outlet or structure through which reclaimed water or treated effluent is finally discharged to a receiving water body.

OUTFLOW Amount of water passing a given point downstream of a structure; water flowing out of a body of water; air that flows outward from a thunderstorm.

OUTFLOW BOUNDARY Storm-scale or mesoscale boundary separating thunderstorm-cooled air (outflow) from the surrounding air.

OUTFLOW CHANNEL Natural stream channel which transports reservoir releases.

OUTGOING LONGWAVE RADIATION Polar satellite derived measurement of the radiative character of energy radiated from the warmer Earth surface to cooler space

OUTLET Opening through which water can be freely discharged from a reservoir.

OUTLET CAPACITY Amount of water that can be safely released through the outlet works.

OUTLET CHANNEL (EXIT CHANNEL) Channel downstream from terminal structure that conveys releases back to the "natural" stream or river.

OUTLET DISCHARGE STRUCTURE Protects the downstream end of the outlet pipe from erosion; often designed to slow down the velocity of released water to prevent erosion of the stream channel.

OUTLET GATE Gate controlling the flow of water through a reservoir outlet.

OUTLET WORKS Combination of structures and equipment required for the safe operation and control of water released from a reservoir to serve various purposes; series of components located in a dam through which normal releases from the reservoir are made; device to provide controlled releases from a reservoir; pipe that lets water out of a reservoir, mainly to supply downstream demands.

OUTLET WORKS TOWER Tower within a reservoir that contains the mechanisms to open the entrance to the outlet works.

OUTLOOK Broad discussion of the weather pattern expected across any given area, generally confined to forecast periods beyond 48 hours; used to indicate that a hazardous weather or hydrologic event may develop.

OUTPUT All information that is produced by the computer code.

OUTWASH Soil material washed down a hillside by rainwater and deposited upon more gently sloping land.

OVER CALIBRATION Achieving artificially low residuals by inappropriately adjusting model input parameters without field data to support the adjusted model parameter value.

OVERALL SAFETY OF DAMS CLASSIFICATION Classifications assigned to a dam following an onsite examination and subsequent analyses using available data and state-of-the-art knowledge: Satisfactory - no existing or potential dam safety deficiencies are recognized; Fair - no existing dam safety deficiencies are recognized for normal loading conditions; Conditionally Poor - Potential dam safety deficiency is recognized for unusual loading conditions that may realistically occur during the expected life of the structure; Poor - Potential dam safety deficiency is clearly recognized for normal loading conditions; Unsatisfactory - Dam safety deficiency exists for normal loading conditions.

OVERAMPING Exceeding the rated capacity of a system.

OVERBANK Surface area between the bank on the main channel and the limits of the floodplain.

OVERBREAK Moving or loosening of rock as a result of a blast, beyond the intended line of cut.

OVERBURDEN Rock and soil cleared away before mining.

OVERCAST (OVC) When the sky is completely covered by an aloft obscuring phenomenon.

OVERDRAFT Pumping of water from a groundwater basin or aquifer in excess of the supply flowing into the basin.

OVERDREDGING Additional depth dredged beyond the minimum dredging depth used to provide sufficient navigational depth, to minimize redredging, and to help compensate for the sloughing off and resettling of sediment after dredging occurs.

OVEREXCAVATION Excavation beyond specified or directed excavation; removing unsuitable foundation material.

OVERFIRE AIR Air forced into the top of an incinerator or boiler to fan the flames.

OVERFLOW DAM Dam designed to be overtopped.

OVERFLOW DAM SECTION Section or portion of a dam designed to be overtopped.

OVERFLOW RATE One of the guidelines for design of the settling tanks and clarifiers in a treatment plant; used by plant operators to determine if tanks and clarifiers are over or under-used.

OVERFLOW SPILLWAY (OGEE SPILLWAY) Spillway that has a control weir that is ogee-shaped (S-shaped) in profile; spillway on a dam that functions like a dam, but allows water to safely flow over it.

OVERHANG Projecting parts of a face or bank; radar term indicating a region of high reflectivity at middle and upper levels above an area of weak reflectivity at low levels.

OVERHAUL Movement of (earth) material far enough so that payment, in addition to excavation pay, is made for haulage; distance in excess of that given as the stated haul distance to haul excavated material.

OVERLAND FLOW Flow of rainwater or snowmelt over the land surface toward stream channels whereas after it enters a watercourse, it becomes runoff; land application technique that cleanses waste water by allowing it to flow over a sloped surface; part of surface runoff flowing over land surfaces toward stream channels.

OVERRUNNING Weather pattern in which a relatively warm air mass is in motion above another air mass of greater density at the surface.

OVERSHOOTING TOP (PENETRATING TOP) Dome-like protrusion above a thunderstorm anvil, representing a very strong updraft and hence a higher potential for severe weather with that storm.

OVERSIZED REGULATED MEDICAL WASTE Medical waste that is too large for plastic bags or standard containers.

OVERTOPPING Flow of water over the top of a dam or embankment.

OVERTURN Almost spontaneous mixing of all layers of water in a reservoir or lake when the water temperature becomes similar from top to bottom; one complete cycle of top to bottom mixing of previously stratified water masses.

OVERWINDING Rope or cable wound and attached so that it stretches from the top of a drum to the load.

OVNGT Overnight.

OVR Over.

OVRN Overrun.

OVRNGT Overnight.

OXBOW Bow-shaped lake formed in an abandoned meander of a river.

OXBOW CHANNEL Natural U-shaped channel in a river as viewed from above.

OXIDANT Collective term for some of the primary constituents of photochemical smog.

OXIDATION Chemical addition of oxygen to break down pollutants or organic waste (e.g., destruction of chemicals such as cyanides, phenols, and organic sulfur compounds in sewage by bacterial and chemical means).

OXIDATION POND Man-made (anthropogenic) body of water in which waste is consumed by bacteria, used most frequently with other waste-treatment processes; sewage lagoon.

OXIDATION-REDUCTION POTENTIAL Electric potential required to transfer electrons from one compound or element (the oxidant) to another compound (the reductant); used as a qualitative measure of the state of oxidation in water treatment systems.

OXYGEN DEMAND Need for molecular oxygen to meet the needs of biological and chemical processes in water.

OXYGENATED FUELS Gasoline which has been blended with alcohols or ethers that contain oxygen in order to reduce carbon monoxide and other emissions.

OXYGENATED SOLVENT Organic solvent containing oxygen as part of the molecular structure (e.g., alcohols and ketones).

OZONATION Application of ozone to water for disinfection or for taste and odor control.

OZONATOR Device for ozonation.

OZONE Gas that is bubbled through water to kill germs; form of oxygen; found in two layers of the atmosphere, the stratosphere and the troposphere and major component of photochemical smog.

OZONE ACTION DAY Message when ozone levels may reach dangerous levels the next day.

OZONE ADVISORY Issued when ozone levels reach 100 where levels above 100 are unhealthy for people with heat and/or respiratory ailments.

OZONE DEPLETION Destruction of the stratospheric ozone layer which shields the Earth from ultraviolet radiation harmful to life caused by the breakdown of certain chlorine and/or bromine containing compounds (chlorofluorocarbons or halons), which break down when they reach the stratosphere and then catalytically destroy ozone molecules.

OZONE HOLE Severe depletion of stratospheric ozone caused by a chemical reaction involving ozone and chlorine, primarily from human produced sources, cloud particles, and low temperatures; thinning break in the stratospheric ozone layer; designation is made when the detected amount of depletion exceeds fifty percent.

OZONE LAYER Protective layer in the atmosphere that contains a high proportion of oxygen that exists as ozone, about 15 miles above the ground, that absorbs some of the Sun's ultraviolet rays, thereby reducing the amount of potentially harmful radiation that reaches the Earth's surface.

P

P.T. Pipe Thread.

PAC Pacific.

PACIFIC DECADAL OSCILLATION (PDO) Pattern of climate variation on a timescale of decades that primarily affects weather patterns and sea surface temperatures in the Pacific Northwest, Alaska, and northern Pacific Islands.

PACIFIC FLYWAY Established air route of migratory birds along the west coast of the United States.

PACIFIC TSUNAMI WARNING CENTER (PTWC) Center has an international warning responsibility for the entire Pacific and a regional warning responsibility for the State of Hawaii.

PACKAGING Assembly of one or more containers and any other components necessary to ensure minimum compliance with a program's storage and shipment packaging requirements.

PACKED BED SCRUBBER Air pollution control device in which emissions pass through alkaline water to neutralize hydrogen chloride gas.

PACKED TOWER Pollution control device that forces dirty air through a tower packed with crushed rock or wood chips while liquid is sprayed over the packing material where the pollutants in the air stream either dissolve or chemically react with the liquid.

PACKER Mechanical device set immediately above the injection zone that seals the outside of the tubing to the inside of the long string casing; inflatable gland, or balloon, used to create a temporary seal in a borehole, probe hole, well, or drive casing made of rubber or non-reactive materials.

PALATABLE WATER Water, at a desirable temperature, that is free from objectionable tastes, odors, colors, and turbidity.

PALEO Studies of the geologic past.

PALEOFLOOD Peak discharges estimated using geology, fluvial, geomorphology, and stratigraphic records where geologic information is used to determine flood depths, carbon 14 dating techniques are typically used to determine the time frame when these depths were reached, and hydraulic models, such as step backwater techniques, are used to determine the associated flow given the depth.

PALEOFLOOD DATA Distinguished from both historical and systematic (conventional) flood data by lack of human observation, regardless of the time of occurrence and include two broad categories: fluvial geomorphic evidence and botanical evidence.

PALEOFLOOD HYDROLOGY Study of past or ancient floods which occurred prior to the time of human observation or direct measurement by modern hydrologic procedures; study of the movements of water and sediment in channels before the time of continuous hydrologic records or direct measurements.

PALEOHYDROLOGY Study of hydrologic processes and events, using geological, botanical, and cultural evidence, that occurred before the beginning of the systematic collection of hydrologic data and observations.

PALMER DROUGHT SEVERITY INDEX (PDSI) Index used to gauge the severity of drought conditions by using a water balance equation to track water supply and demand taking into account precipitation, potential and actual evapotranspiration, infiltration of water into the soil, and runoff.

PALUSTRINE HABITAT Marsh habitat; all non-tidal wetlands that are dominated by trees, shrubs, persistent emergents, emergent mosses, or lichens, and all such wetlands in tidal areas where salinity owing to ocean-derived salts is below 0.5 part per thousand.

PALUSTRINE WETLANDS Freshwater wetlands including open water bodies of less than 20 acres in which water is less than 2 meters deep; includes marshes, wet meadows, fens, playas, potholes, pocosins, bogs, swamps, and shallow ponds.

PAN EVAPORATION Evaporative water losses from a standardized pan sometimes used to estimate crop evapotranspiration and assist in irrigation scheduling.

PAN PAN Headline within National Weather Service high seas forecasts transmitted via the GMDSS to indicate that a hurricane or hurricane force winds are forecast.

PANCAKE ICE Circular flat pieces of ice with a raised rim where the shape and rim are due to repeated collisions.

PANDEMIC Widespread epidemic throughout an area, nation or the world.

PANHANDLE HOOK Low pressure systems that originate in the panhandle region of Texas and Oklahoma which initially move east and then "hook" or recurve more northeast toward the upper Midwest or Great Lakes region.

PAPER In the recycling business, refers to products and materials, including newspapers, magazines, office papers, corrugated containers, bags and some paperboard packaging that can be recycled into new paper products.

PAPER PROCESSOR Intermediate facility where recovered paper products and materials are sorted, decontaminated, and prepared for final recycling.

PARADOX GATE Similar to a ring follower gate except the gate leaf is supported along either side by endless trains of rollers. The gate seals by the rollers disengaging from support of the leaf when the gate is completely closed, allowing hydrostatic forces to seal the gate.

PARALLEL PATH FLOW Refers to the flow of electric power on an electric system's transmission facilities resulting from scheduled electric power transfers between two other electric systems.

PARAMETER Variable, measurable property whose value is a determinant of the characteristics of a system (e.g., temperature, pressure, and density); any of a set of physical properties which determine the characteristics or behavior of a system; variable, in a general model, whose value is adjusted to make the model specific to a given situation; numerical measure of the properties of the real-world system; subset of the group of evaluations that constitute each element of an observation.

PARAMETER ESTIMATION Selection of a parameter value based on the results of analysis and/or engineering judgment.

PARAMETER IDENTIFICATION MODEL (INVERSE MODEL) Computer code for determination of selected unknown parameters and stresses in a groundwater system, given that the response of the system to all stresses is known and that information is available regarding certain parameters and stresses.

PARAMETRIC DATA Data such as rating curves, unit hydrographs, and rainfall/runoff curves which define hydrologic variables in models.

PARAPET WALL Solid wall built along the top of a dam (upstream and/or downstream edge) used for ornamentation, for safety of vehicles and pedestrians, or to prevent overtopping caused by wave runup.

PARAQUAT Standard herbicide used to kill various types of crops (e.g. marijuana) that causes lung damage if smoke from the crop is inhaled.

PARCEL Volume of air small enough to contain uniform distribution of its meteorological properties and large enough to remain relatively self-contained and respond to all meteorological processes.

PARHELION Scientific name for Sun dogs; either of two colored luminous spots that appear at roughly 22 degrees on both sides of the Sun at the same elevation caused by the refraction of sunlight passing through ice crystals.

PARSHALL FLUME (IMPROVED VENTURI FLUME) Device used to measure the flow of water in an open channel; flume with a specially shaped open-channel flow section that may be installed in a drainage lateral or ditch to measure the rate of flow of water; calibrated device consisting of a broad and flat converging section, a narrow downward sloping throat, and a diverging upward sloping section developed to measure a wide range of flows in an open channel; calibrated device, based on the principle of critical flow, used to measure the flow of water in open channels.

PART PER MILLION (PPM) Unit of concentration equal to one milligram per kilogram or one milligram per liter.

PARTIAL BEAM FILLING Limitation of the rainfall estimation techniques where reflectivity and associated rainfall rates are underestimated.

PARTIAL-DURATION FLOOD SERIES
List of all flood peaks that exceed a chosen base stage or discharge, regardless of the number of peaks occurring in a year.

PARTIAL-RECORD STATION Site where discrete measurements of one or more hydrologic parameters are obtained over a period of time without continuous data being recorded or computed.

PARTICIPATION RATE Portion of population participating in a recycling program.

PARTICLE ACCELERATION Time rate of change of particle velocity.

PARTICLE COUNT Results of a microscopic examination of treated water with a special "particle counter" that classifies suspended particles by number and size.

PARTICLE DISPLACEMENT Difference between the initial position of a soil particle and any later temporary position during shaking.

PARTICLE SHAPE FACTOR Particle shape factor of a perfect sphere is 1.0 and can be as low as 0.1 for very irregular shapes.

PARTICLE SIZE Diameter of a particle determined by sieve or sedimentation methods; diameter of particles comprising a particular soil; linear dimension used to characterize the size of a particle: Clay 0.00024-0.004 millimeters (mm), Silt 0.004-0.062 mm, Sand 0.062-2.0 mm, and Gravel 2.0-64.0 mm.

PARTICLE TRAJECTORY MODEL
Computer sub-model that tracks the trajectories of multiple particles that are released into an atmospheric flow model.

PARTICLE VELOCITY Time rate of change of particle displacement.

PARTICULARLY DANGEROUS SITUATION (PDS) WATCH
Wording used in rare situations when long-lived, strong and violent tornadoes are possible.

PARTICULATE LOADING Mass of particulates per unit volume of air or water.

PARTICULATE MATTER Complex mixture of extremely small particles and liquid droplets made up of a number of components, including acids (such as nitrates and sulfates), organic chemicals, metals, and soil or dust particles.

PARTICULATES Fine liquid or solid particles (i.e., dust, smoke, mist, fumes, or smog) found in air or emissions; very small solids suspended in water.

PARTITION COEFFICIENT Measure of the sorption phenomenon, whereby a pesticide is divided between the soil and water phase.

PARTITIONING FUNCTION
Mathematical relation describing the distribution of a reactive solute between solution and other phases.

PARTLY CLOUDY Between 3/8 and 5/8 of the sky is covered by clouds.

PARTLY SUNNY Between 3/8 and 5/8 of the sky is covered by clouds during daylight hours.

PARTS PER BILLION (PPB) Number of "parts" by weight of a substance per billion parts of water; measurement of concentration on a weight or volume basis; equivalent to micrograms per liter.

PARTS PER MILLION (PPM) Number of "parts" by weight of a substance per million parts of water; measurement of concentration on a weight or volume basis; equivalent to milligrams per liter.

PASCAL (PA) Unit of pressure produced when one newton acts on one square meter.

PASS Working trip or passage of an excavating, grading, or compaction machine.

PASSIVE EARTH PRESSURE Maximum value of earth pressure that exists when a soil mass is compressed sufficiently to cause its internal shearing resistance along a potential failure surface to be completely mobilized.

PASSIVE SMOKING/SECONDHAND SMOKE Inhalation of others' tobacco smoke.

PASSIVE TREATMENT WALLS Technology in which a chemical reaction takes place when contaminated groundwater comes in contact with a barrier such as limestone or a wall containing iron filings.

PAT Pattern.

PATHOGEN Any living organism (e.g., bacteria, viruses, bacteria, fungi or parasites) that causes disease.

PATHWAY Physical course a chemical or pollutant takes from its source to the exposed organism.

PATTERN CRACKING Fine cracks in the form of a pattern on a concrete surface.

PAY FORMATION Layer or deposit of soil or rock whose value is sufficient to justify excavation.

PAY-AS-YOU-THROW/UNIT-BASED PRICING Systems under which residents pay for municipal waste management and disposal services by weight or volume collected, not a fixed fee.

PAYLINE Lines of excavation, backfill, compacted backfill or embankment which are described in the specifications or shown on the drawings which describe or show the limits to which earthwork is paid for.

PBL Probable.

PCPN Precipitation.

PCT Percent.

PD Period.

PDI Palmer Drought Index.

PDMT Predominant.

PDT Pacific Daylight Time.

PEA GRAVEL Uniformly graded gravel with a particle size of approximately 3/16″.

PEAK Highest elevation reached by a flood wave; maximum crest.

PEAK DEMAND Maximum electrical demand occurring within a specified period of time; maximum power used in a specific time period.

PEAK DISCHARGE Highest rate of discharge of a volume of water passing a given location during a given period of time.

PEAK ELECTRICITY DEMAND Maximum electricity used to meet the cooling load of a building or buildings in a given area.

PEAK FLOW (PEAK STAGE) Maximum instantaneous flow in a specified period of time at a given location; point of the hydrograph that has the highest flow.

PEAK GROUND ACCELERATION (PGA) Representing the peak acceleration of ground motion.

PEAK GUST Highest instantaneous wind speed observed or recorded.

PEAK LEVELS Levels of airborne pollutant contaminants much higher than average or occurring for short periods of time in response to sudden releases.

PEAK LOAD Maximum power load in a stated period of time.

PEAK LOAD PLANT Power plant that normally is operated to provide power during maximum load periods.

PEAK PULSE Amount of power transmitted by a radar during a given pulse.

PEAK STAGE Maximum height of a water surface above an established datum plane.

PEAK WIND SPEED Maximum instantaneous wind speed since the last observation that exceeded 25 knots.

PEAKEDNESS Describes the rate of rise and fall of a hydrograph.

PEAKING CAPACITY Capacity of generating equipment normally reserved for operation during the hours of highest daily, weekly, or seasonal loads.

PEAKING POWER Power plant capacity typically used to meet the highest levels of demand in a utility's load or demand profile.

PEAT Highly organic soil, composed of partially decomposed vegetable matter; soft light swamp soil consisting mostly of decayed vegetation.

PECELT NUMBER Relationship between the advective and diffusive components of solute transport expressed as the ratio of the product of the average interstitial velocity, times the characteristic length, divided by the coefficient of molecular diffusion; small values indicate diffusion is the dominant transport process, large values indicate advection dominance.

PEER-REVIEWED LITERATURE Referable, obtainable, published document that is reviewed by a minimum of two technical reviewers who are located external to the author's organization.

PEGMATITE Very coarse-grained intrusive igneous rock of granitic composition that typically fills fractures to form veins.

PENDANT ECHO Radar signature generally similar to a hook echo, except that the hook shape is not as well defined.

PENETRATING TOP See OVERSHOOTING TOP.

PENSTOCK Pipeline or conduit designed to withstand pressure surges leading from a forebay or reservoir to power-producing turbines, or pump units; conduit used to convey water under pressure to the turbines of a hydroelectric plant; pressurized pipeline or shaft between the reservoir and hydraulic machinery.

PENUMBRA Sunspot area that may surround the darker umbra or umbrae that consists of linear bright and dark elements radial from the sunspot umbra.

PER CAPITA USE Average amount of water used per person during a standard time period (e.g. per day).

PERCENT COMPOSITION Unit for expressing the ratio of a particular part of a sample or population to the total sample or population, in terms of types, numbers, weight, mass, or volume.

PERCENT OF TOTAL Unit for expressing the ratio of a particular part of a sample or population to the total sample or population, in terms of types, numbers, weight, mass, or volume.

PERCENT SATURATION Amount of a substance that is dissolved in a solution compared to the amount that could be dissolved in it.

PERCENT SHADING Measure of the amount of sunlight potentially reaching the stream.

PERCENT VOLATILE Percentage of a chemical that will evaporate at ordinary temperatures.

PERCHED Either a losing stream or an insulated stream that is separated from the underlying groundwater by a zone of aeration.

PERCHED GROUNDWATER Unconfined groundwater separated from an underlying body of groundwater by an unsaturated zone.

PERCHED WATER Zone of unpressurized water held above the water table by impermeable rock or sediment.

PERCHED WATER TABLE Water table of a relatively small groundwater body supported above the general groundwater body.

PERCOLATING WATER Water that passes through rocks or soil under the force of gravity.

PERCOLATION Downward movement of water through the soil or other porous media (rock); water soaking into the ground; flow through a porous substance; slow seepage of water through a filter; entrance of a portion of the streamflow into the channel materials to contribute to groundwater replenishment.

PERCOLATION PATH Course followed by water moving or percolating through any other permeable material, or under a dam which rests upon a permeable foundation.

PERCOLATION RATE Rate (velocity) at which water moves through porous media, such as soil; intake rate used for designing wastewater absorption systems.

PERENNIAL STREAM Stream that flows continually throughout the year.

PERENNIAL YIELD Maximum quantity of water that can be annually withdrawn from a groundwater basin over a long period of time (during which water supply conditions approximate average conditions) without developing an overdraft condition.

PERFLUOROCARBONS (PFCS) Potent greenhouse gases that accumulate in the atmosphere and remain there for thousands of years.

PERFORATED PIPE Pipe designed to discharge water through small, multiple, closely spaced orifices or nozzles, placed in a segment of its circumference for irrigation purposes.

PERFORMANCE BOND Cash or securities deposited before a landfill operating permit is issued, which are held to ensure that all requirements for operating ad subsequently closing the landfill are faithful performed.

PERFORMANCE DATA (FOR INCINERATORS)
Information collected, during a trial burn, on concentrations of designated organic compounds and pollutants found in incinerator emissions.

PERFORMANCE STANDARDS
Regulatory requirements limiting the concentrations of designated organic compounds, particulate matter, and hydrogen chloride in emissions from incinerators; operating standards established by EPA for various permitted pollution control systems, asbestos inspections, and various program operations and maintenance requirements.

PERFORMANCE-BASED METHODS SYSTEM
System that permits the use of any appropriate measurement methods that demonstrates the ability to meet established performance criteria and that complies with specified data-quality needs.

PERIGEE Closest distance between Moon and Earth or the Earth and Sun.

PERIHELION Point on the annual orbit of a body (about the Sun) that is closest to the Sun.

PERIODIC FACILITY REVIEW (PFR)
Field review performed on a high- or significant-hazard dam every 6 years that entails a thorough examination from both operation and maintenance and dam safety perspectives.

PERIODIC-RECORD STATION
Site where stage, discharge, sediment, chemical, physical, or other hydrologic measurements are made one or more times during a year but at a frequency insufficient to develop a daily record.

PERIPHYTON Microorganisms (e.g., bacteria, fungi, protozoa, rotifers, etc.) attached to and living upon submerged solid surfaces.

PERMAFROST Layer of soil, subsoil, surficial deposit, or bedrock at varying depths below the surface in which the temperature has remained below freezing continuously from a few to several thousands of years.

PERMANENT CONTROL Stream gaging control which is substantially unchanging and is not appreciably affected by scour, fill, or backwater.

PERMANENT MONUMENT
Fixed monuments placed away from the dam which allow movements in horizontal and vertical control points on the dam to be monitored by using accurate survey procedures.

PERMANENT WILTING POINT (PERCENTAGE)
Soil water content below which plants cannot readily obtain water and permanently wilt.

PERMEABILITY Ability of a material to transmit fluid through its pores when subjected to a difference in head; measure of the flow of water through soil under a gradient of force; ease (or measurable rate) with which gasses, liquids, or plant roots penetrate or pass through a layer of soil or porous media.

PERMEABILITY COEFFICIENT Rate of flow of a fluid through a cross section of a porous mass under a unit hydraulic gradient, at a temperature of 60 °F.

PERMEABLE Having pores or openings that permit liquids or gasses to pass through.

PERMEAMETER Laboratory instrument for determining permeability by measuring the discharge through a sample of material (e.g. soil) when a known hydraulic head is applied.

PERMEATE To penetrate and pass through.

PERMISSIBLE DOSE Dose of a chemical that may be received by an individual without the expectation of a significantly harmful result.

PERMISSIBLE EXPOSURE LIMIT (PEL) Federal limits for workplace exposure to contaminants as established by OSHA.

PERMISSIBLE EXPOSURE LIMITS (PEL) Guide to acceptable levels of chemical exposure.

PERMIT An authorization, license, or equivalent control document needed in order to operate.

PERSISTENCE Length of time a compound stays in the environment, once introduced; continuation of existing conditions.

PERSISTENCE FORECAST Forecast that the current weather condition will persist and that future weather will be the same as the present.

PERSISTENT PESTICIDES Pesticides that do not break down chemically or break down very slowly and remain in the environment after a growing season.

PERSISTENT POLLUTANT Not subject to decay, degradation, transformation, volatilization, hydrolysis, or photolysis.

PERSONAL AIR SAMPLES Air samples taken with a pump that is directly attached to the worker with the collecting filter and cassette placed in the worker's breathing zone.

PERSONAL MEASUREMENT Measurement collected from an individual's immediate environment.

PERSONAL PROTECTIVE EQUIPMENT Clothing and equipment worn by pesticide mixers, loaders and applicators and re-entry workers, hazmat emergency responders, workers cleaning up Superfund sites, et. al., which is worn to reduce their exposure to potentially hazardous chemicals and other pollutants.

PERTURBATION MODEL Computer model used to calculate air pollution concentrations.

PERVIOUS Permeable, having openings that allow water to pass through.

PERVIOUS ZONE Part of the cross section of an embankment dam comprising material of high permeability.

PEST Insect, rodent, nematode, fungus, weed or other form of terrestrial or aquatic plant or animal life that is injurious to health or the environment.

PEST CONTROL OPERATOR Person or company that applies pesticides as a business (e.g. exterminator).

PESTICIDE Substance used to control or kill undesirable organisms (e.g., plant or animal pests); major categories include herbicides and insecticides.

PESTICIDE REGULATION NOTICE Formal notice to pesticide registrants about important changes in regulatory policy, procedures, regulations.

PESTICIDE TOLERANCE Amount of pesticide residue allowed by law to remain in or on a harvested crop.

PETCOCK Small drain valve; small ground key type valve used with soft copper tubing.

PETROLEUM Crude oil or any fraction thereof that is liquid under normal conditions of temperature and pressure.

PETROLEUM DERIVATIVES Chemicals formed when gasoline breaks down in contact with ground water.

PH measure of the relative acidity or alkalinity of water or soil; symbol for the logarithm of the reciprocal of hydrogen ion concentration in gram atoms per liter; relative scale, from 0 to 14, of how acidic or basic (alkaline) a material is, where a pH of 7 is neutral, a pH less than 7 is acidic, and a pH greater than 7 is basic; indicator of acidity.

PHARMACOKINETICS Study of the way that drugs move through the body after they are swallowed or injected.

PHENOLPHTHALEIN ALKALINITY Alkalinity in a water sample measured by the amount of standard acid needed to lower the pH to a level of 8.3 as indicated by the change of color of the phenolphthalein from pink to clear.

PHENOLS Organic compounds that are byproducts of petroleum refining, tanning, and textile, dye, and resin manufacturing where low concentrations cause taste and odor problems in water and higher concentrations can kill aquatic life and humans.

PHENOMENOLOGICAL MODEL Computer model used to calculate air pollution concentrations.

PHFO Honolulu National Weather Service Forecast Office.

PHIEZOMETER Instrument used to measure pressure head; used in dams to measure the level of saturation.

PHILADELPHIA ROD Leveling rod in which the hundredths of feet, or eights of inches, are marked by alternate bars of color the width of the measurement.

PHOSPHATES Certain chemical compounds containing phosphorus.

PHOSPHOGYPSUM PILES (STACKS) Principal byproduct generated in production of phosphoric acid from phosphate rock.

PHOSPHORUS Essential chemical food element that can play a key role in stimulating aquatic growth in lakes and streams and contribute to the eutrophication of lakes and other water bodies.

PHOSPHORUS PLANTS Facilities using electric furnaces to produce elemental phosphorous for commercial use, such as high grade phosphoric acid, phosphate-based detergent, and organic chemicals use.

PHOTOCHEMICAL OXIDANTS Air pollutants formed by the action of sunlight on oxides of nitrogen and hydrocarbons.

PHOTOCHEMICAL SMOG Air pollution containing ozone and other reactive chemical compounds formed by the reaction of nitrogen oxides and hydrocarbons in the presence of sunlight.

PHOTOGRAMMETRY Techniques of obtaining precise measurements from images.

PHOTOSPHERE Intensely bright portion of the Sun visible to the unaided eye; "surface" of the Sun.

PHOTOSYNTHESIS Process that plant chlorophyll-containing cells manufacture carbohydrates and oxygen (chemical energy) from carbon dioxide in the presence of sunlight, forming organic compounds from inorganic compounds.

PHOTOVOLTAIC (PV) System that converts sunlight directly into electricity using cells made of silicon or other conductive materials.

PHREATIC SURFACE Free surface of groundwater at atmospheric pressure.

PHREATIC WATER Water within the Earth that supplies wells and springs; water in the zone of saturation where all openings in rocks and soil are filled, the upper surface of which forms the water table; groundwater.

PHREATIC ZONE Locus of points below the water table where soil pores are filled with water; zone of saturation.

PHREATOPHYTE Water-loving plant; plant that habitually obtains its water supply from the zone of saturation, either directly or through the capillary fringe; vegetation that consumes much water.

PHTHALATES Class of organic compounds containing phthalic acid esters and derivatives; used as plasticizers in plastics.

PHYSICAL AND CHEMICAL TREATMENT Processes generally used in large-scale wastewater treatment facilities where physical processes may include air-stripping or filtration and chemical treatment includes coagulation, chlorination, or ozonation.

PHYSIOGRAPHIC PROVINCE Region in which the landforms are distinctive and differ significantly from those of adjacent regions.

PHYSIOGRAPHY Description of nature or natural phenomenon in general; description of the surface features of the Earth, with an emphasis on the origin of landforms.

PHYTOPHAGOUS Plant eating.

PHYTOPLANKTON Portion of the plankton community comprised of small, usually microscopic, plants (e.g., algae and diatoms), found in lakes, reservoirs, and other bodies of water; plant part of the plankton.

PHYTOREMEDIATION Low-cost remediation option for sites with widely dispersed contamination at low concentrations.

PHYTOTOXIC Harmful to plants.

PHYTOTREATMENT Cultivation of specialized plants that absorb specific contaminants from the soil through their roots or foliage.

PICO Prefix meaning "one trillionth".

PICOCURIE (PC, PCI) One-trillionth of the amount of radioactive nuclide represented by a curie (Ci).

PICOCURIES PER LITER (PCI/L) Unit of measure for levels of radon gas; equal to Becquerels per cubic meter.

PIEZOMETER Instrument for measuring water levels or pore water pressures in embankments, foundations, abutments, soil, rock, concrete, tanks and soil; non-pumping well, generally of small diameter, for measuring the elevation of a water table; used in dams to measure the level of saturation.

PIEZOMETRIC LEVEL (SURFACE) Surface at which water will stand in a series of piezometers.

PIG Air manifold having a number of pipes which distribute compressed air coming through a single large line.

PILE Relatively slender structural element which is driven, or otherwise introduced, into the soil, usually for the purpose of providing vertical or lateral support.

PILOT BALLOON (PIBAL) Small helium-filled meteorological balloon that is tracked as it rises through the atmosphere to determine how wind speed and direction change with altitude.

PILOT REPORT (PIREP) Report of inflight weather by an aircraft pilot or crew member.

PILOT TESTS Testing a cleanup technology under actual site conditions to identify potential problems prior to full-scale implementation.

PINCH-OUT Location where a porous, permeable formation that is located between overlying and underlying confining formations thins to a zero thickness, and the confining formations are in contact with each other.

PINGO Large frost mound of more than one-year duration.

PIONEER PLANT Herbaceous annual and perennial seedling plants that colonize bare areas as a first stage in secondary succession.

PIONEERING First working over of rough or overgrown areas.

PIPE Circular conduit constructed of any one of a number of materials that conveys water by gravity or under pressure; cylindrical conduit or conductor with wall thickness sufficient to receive a standard pipe thread.

PIPE TRENCH Temporary excavation in which pipe is placed and eventually covered with soil material.

PIPING Progressive development of internal erosion by seepage, appearing downstream as a hole or seam discharging water that contains soil particles.

PIT Any mine, quarry, or excavation area worked by the open-cut method to obtain material of value.

PIT RUN GRAVEL Natural gravelly material taken from excavation; gravelly material that is not processed.

PITCHING Squared masonry, precast blocks, or embedded stones laid in regular fashion with dry or filled joints on the upstream slope of an embankment dam, reservoir shore or ,sides of a channel as protection against wave and ice action.

PITOT TUBE Device for measuring the velocity of flowing water using the velocity head of the stream as an index of velocity.

PIX Picture.

PK Peak.

PLACED To put in a particular place or in a particular state.

PLACER Surficial mineral deposit formed by mechanical concentration of mineral particles from weathered debris.

PLAGE Extended emission feature of an active region that exists from the emergence of the first magnetic flux until the widely scattered remnant magnetic fields merge with the background.

PLAGE CORRIDOR Space in chromospheric plage lacking plage intensity, coinciding with polarity inversion line.

PLAGENIL Spotless disc free of calcium plage.

Plan Position Indicates No Echoes (PPINE)
Refers to radar detects no precipitation within its range.

PLAN POSITION INDICATOR (PPI)
Displays radar data horizontally using a map projection.

PLANETARY BOUNDARY LAYER (PBL)
Layer within the atmosphere between 1 km and the Earth's surface where friction affects wind speed and wind direction.

PLANFORM Shape and size of channel and overbank features as viewed from directly above.

PLANIMETER Device that measures area.

PLANKTON Tiny plants and animals that live in water; community of suspended, floating or weakly swimming organisms at the mercy of the waves and currents that live in the open water of lakes and rivers; animals of the group are called zooplankton and the plants are called phytoplankton.

PLANNING BASIS Guidance concerning size of planning area (distance), time dependence of releases and flood characteristics of releases.

PLANNING STANDARDS Standards which emergency plans of the dam operating organization and local authorities should conform to.

PLANT Station where mechanical, chemical, and/or nuclear energy is converted into electric energy.

PLANT FACTOR Ratio of average load on the plant for period of time considered to be the aggregate rating of all the generating equipment installed in the plant.

PLANT-USE ELECTRICITY Electric energy used in the operation of a plant.

PLASMA Any ionized gas; that is, any gas containing ions and electrons.

PLASMA ARC REACTORS Devices that use an electric arc to thermally decompose organic and inorganic materials at ultra-high temperatures into gases and a vitrified slag residue.

PLASMID Circular piece of DNA that exists apart from the chromosome and replicates independently of it.

PLASTIC LIMIT Water content corresponding to an arbitrary limit between the plastic and the semisolid state of consistency of a soil; water content at which a soil will just begin to crumble when rolled into a thread approximately 1/8 inches in diameter; minimum amount of water in terms of percent of oven-dry weight of soil that will make the soil plastic.

PLASTIC SOIL Soil that exhibits plasticity.

PLASTICITY Property of a soil or rock which allows it to be deformed beyond the point of recovery without cracking or appreciable volume change.

PLASTICITY INDEX Numerical difference between the liquid limit and the plastic limit.

PLASTICS Non-metallic chemoreactive compounds molded into rigid or pliable construction materials, fabrics, etc.

PLASTICS PROCESSOR Intermediate facility where recovered plastic products and materials are sorted, decontaminated, and prepared for final recycling.

PLATE TOWER SCRUBBER Air pollution control device that neutralizes hydrogen chloride gas by bubbling alkaline water through holes in a series of metal plates.

PLATFORM Generic radar term often used to encompass the pedestal and antenna assembly.

PLAYA Intermittent shallow lake formed at the bottom of an undrained desert basin after heavy rains.

PLAYA LAKE Shallow, temporary lake in an arid or semiarid region, covering or occupying a playa in the wet season but drying up in summer; temporary lake that upon evaporation leaves or forms a playa.

PLOT Area of land that is studied or used for an experimental purpose, in which sample areas are often located.

PLOW WIND Strong, straight-line winds associated with the downdrafts spreading out in advance of squall lines and thunderstorms.

PLUG FLOW Type of flow the occurs in tanks, basins, or reactors when a slug of water moves through without ever dispersing or mixing with the rest of the water flowing through.

PLUGGING Act or process of stopping the flow of water, oil or gas into or out of a formation through a borehole or well penetrating that formation.

PLUMB Exactly vertical; at right angles to the horizontal.

PLUMB BOB Pointed weight hung from a string that is used for vertical alignment.

PLUMBLINES Measures the movement of a concrete dam due to applied reservoir water pressures and temperature changes.

PLUME Visible or measurable discharge of a contaminant from a given point of origin; area of radiation leaking from a damaged reactor; area downwind within which a release could be dangerous for those exposed to leaking fumes.

PLUME BLIGHT Visibility impairment caused by air pollution plumes aggregated from individual sources.

PLUME IMPINGEMENT Collision of a plume with topography that rises above the plume altitude.

PLUME-DOMINATED FIRE Fire whose behavior is governed primarily by the local wind circulation produced in response to the strong convection above the fire rather than by the general wind.

PLUNGE POOL Natural or artificially created pool that dissipates the energy of free falling water.

PLUTONIUM Radioactive metallic element chemically similar to uranium.

PLUVIAL Anything that is brought about directly by precipitation.

PLY One of several layers of fabric or of other strength-contributing material.

PM-10 Measure of particles in the atmosphere with a diameter of less than ten or equal to a nominal 10 micrometers.

PM-2.5 Measure of particles smaller than PM-10 in the atmosphere.

PMD Prognostic Discussion.

PMO Port Meteorological Officer.

PNA Pacific North American teleconnection.

PNEUMATIC Powered or inflated by compressed air.

PNEUMOCONIOSIS Health conditions characterized by permanent deposition of substantial amounts of particulate matter in the lungs and by the tissue reaction to its presence.

PNHDL Panhandle.

POCOSIN Local term along the Atlantic coastal plain, from Virginia south, for a shrub-scrub wetland located on a relatively flat terrain, commonly between streams.

POINT DISCHARGE Instantaneous rate of discharge, in contrast to the mean rate for an interval of time.

POINT OF INJECTION Last accessible sampling point prior to waste fluids being released into the subsurface environment through a Class V injection well.

POINT PRECIPITATION Precipitation at a particular site, in contrast to the mean precipitation over an area.

POINT SOURCE Stationery location or fixed facility from which pollutants are discharged or emitted; any single identifiable source of pollution; pollution that comes from a well-defined (i.e. discrete) source; any discernible, confined, or discrete conveyance from which pollutants are or may be discharged, including, but not limited to, any pipe, ditch, channel, tunnel, conduit, well, container, rolling stock, concentrated animal feeding operation, or vessel or other floating craft.

POINT-OF-CONTACT MEASUREMENT OF EXPOSURE Estimating exposure by measuring concentrations over time (while the exposure is taking place) at or near the place where it is occurring.

POINT-OF-DISINFECTANT APPLICATION Point where disinfectant is applied and water downstream of that point is not subject to recontamination by surface water runoff.

POINT-OF-USE TREATMENT DEVICE Treatment device applied to a single tap to reduce contaminants in the drinking water at the one faucet.

POINT-SOURCE CONTAMINANT Any substance that degrades water quality and originates from discrete locations such as discharge pipes, drainage ditches, wells, concentrated livestock operations, or floating craft.

POINT-SOURCE POLLUTION Pollution coming from a single point.

POISON Chemical that, in relatively small amounts, is able to produce injury by chemical action when it comes in contact with a susceptible tissue.

POLAR CAP ABSORPTION (PCA) Anomalous condition of the polar ionosphere whereby HF and VHF (3 - 300 MHz) radiowaves are absorbed, and LF and VLF (3 - 300 kHz) radiowaves are reflected at lower altitudes than normal.

POLAR FRONT Semi-permanent, semi-continuous front that separates tropical air masses from polar air masses.

POLAR JET Marked by a concentration of isotherms and strong vertical shear; boundary between the polar air and the subtropical air.

POLAR JET STREAM See POLAR JET.

POLAR ORBITING SATELLITE Weather satellite which travels over both poles each time it orbits the Earth.

POLARIZATION RADAR Radar which takes advantage of ways in which the transmitted waves' polarization affect the backscattering.

POLLEN Fertilizing element of flowering plants; background air pollutant.

POLLUTANT Any inorganic or organic substance that contaminates air, water or soil; any substance introduced into the environment that adversely affects the usefulness of a resource or the health of humans, animals, or ecosystems; any substance that, when present in a hydrologic system at sufficient concentration, degrades water quality in ways that are or could become harmful to human and/or ecological health or that impair the use of water for recreation, agriculture, industry, commerce, or domestic purposes.

POLLUTANT PATHWAYS Avenues for distribution of pollutants.

POLLUTANT STANDARD INDEX (PSI) Indicator of one or more pollutants that may be used to inform the public about the potential for adverse health effects from air pollution in major cities.

POLLUTANT, CONSERVATIVE Pollutants that do not readily degrade in the environment, and which are mitigated primarily by natural stream dilution after entering receiving bodies of waters.

POLLUTANT, NON-CONSERVATIVE Pollutants that are mitigated by natural biodegradation or other environmental decay or removal processes in the receiving stream after in-stream mixing and dilution have occurred.

POLLUTION Resource that is out of place; presence of matter or energy whose nature, location or quantity produces undesired environmental effects; man-made or man-induced alteration of the chemical, physical, biological and radiological integrity of water.

POLLUTION PREVENTION Identifying areas, processes, and activities which create excessive waste products or pollutants in order to reduce or prevent them through, alteration, or eliminating a process.

POLYCHLORINATED BIPHENYLS (PCBS) Industrial chemicals that are mixtures of chlorinated biphenyl compounds having various percentages of chlorine; synthetic, toxic, persistent chemicals used in electrical transformers and capacitors for insulating purposes, and in gas pipeline systems as lubricant.

POLYCHLORINATED NAPHTHALENES (PCNS) Industrial chemicals that are mixtures of chlorinated naphthalene compounds.

POLYCYCLIC AROMATIC HYDROCARBON (PAH) Class of organic compounds with a fused-ring aromatic structure.

POLYELECTROLYTES Synthetic chemicals that help solids to clump during sewage treatment.

POLYETHYLENE TEREPTHALATE (PETE) Thermoplastic material used in plastic soft drink and rigid containers.

POLYMER Natural or synthetic chemical structure where two or more like molecules are joined to form a more complex molecular structure.

POLYVINYL CHLORIDE (PVC) Type of tough, environmentally indestructible plastic used to make pipe and fittings for water distribution and irrigation that releases hydrochloric acid when burned.

POND Small lake; water impounded by a diversion dam.

PONDAGE Holding back of water for later release; storage capacity available for the use of such water.

PONDING In flat areas, runoff collects in depression and cannot drain out; method for compacting soil using water added after the soil is in place until free water stands on the surface.

POOL Deep area of a stream where current is slow; small part of a stream reach with little velocity; deep reach of a stream; reach of a stream between two riffles; elevation of the surface of a body of water such as a lake.

POOL HEIGHT Height of the water behind a dam.

POP Probability of Precipitation.

POPCORN CONVECTION Slang for showers and thunderstorms that form on a scattered basis with little or no apparent organization, usually during the afternoon in response to diurnal heating.

POPULATION Total of individuals occupying an area; group of interbreeding organisms occupying a particular space; number of humans or other living creatures in a designated area.

POPULATION AT RISK (PAR) Population subgroup that is more likely to be exposed to a chemical, or is more sensitive to the chemical, than is the general population; population potentially affected by flood waters as a result of large operational releases or dam failure; potential number of persons whose lives are at risk in the event of a dam failure.

POPULATION DENSITY Number per unit area of individuals of any given species at a given time.

PORE Small to minute opening or interstice in a rock or soil.

PORE PRESSURE Interstitial pressure of fluid (air or water) within a mass of soil, rock, or concrete.

PORE SPACE Open spaces in soil, sediments, and rocks that are filled by air or water.

PORE-WATER PRESSURE Internal hydrostatic pressure in an embankment caused by the level of water in the reservoir acting through pressure-transmitting paths between soil particles in the fill.

POROSITY Ratio of the volume of void space to the total volume of an undisturbed sample; degree to which soil, gravel, sediment, or rock is permeated with pores or cavities through which water or air can move; measure of the water-bearing capacity of subsurface rock; degree of perviousness.

PORTAL Opening into a tunnel.

PORTAL-OF-ENTRY EFFECT Local effect produced in the tissue or organ of first contact between a toxicant and the biological system.

POS Positive.

POSITIVE AREA Area on a sounding representing the layer in which a lifted parcel would be warmer than the environment; measure of the energy available for convection.

POSITIVE CLOUD TO GROUND LIGHTNING Lightning that delivers positive charge to the ground, as opposed to the more common negative charge.

POSITIVE PRESSURE Pressure within a pipe that is greater than atmospheric pressure.

POSITIVE VORTICITY ADVECTION (PVA) Advection of higher values of vorticity into an area.

POSITIVE-TILT TROUGH Upper level system which is tilted to the east with increasing latitude (i.e. from southwest to northeast).

POST-CHLORINATION Addition of chlorine to plant effluent for disinfectant purposes after the effluent has been treated.

POST-CLOSURE Time period following the shutdown of a waste management or manufacturing facility; for monitoring purposes, often considered to be 30 years.

POST-CONSUMER MATERIALS/WASTE Materials or finished products that have served their intended use and have been diverted or recovered from waste destined for disposal, having completed their lives as consumer items.

POST-CONSUMER RECYCLING Use of materials generated from residential and consumer waste for new or similar purposes.

POSTED OPERATING INSTRUCTIONS Operations and maintenance instructions taken from the Standing Operating Procedures (SOP) that pertain to the mechanical features in the immediate area of the dam and which are posted adjacent to those features.

POSTEMERGENCE HERBICIDE Herbicide applied to foliage after the crop has sprouted to kill or significantly retard the growth of weeds.

POST-FLARE LOOPS Loop prominence system often seen after a major two-ribbon flare, which bridges the ribbons.

POST-INJECTION SITE CARE Appropriate monitoring and other actions (including corrective action) needed following cessation of injection to assure that USDWs are not endangered.

POSTPROCESSING Using computer programs to assist in preparing data sets for use with generic simulation codes.

POST-STORM REPORT Report summarizing the impact of a storm on it's forecast area.

POTABLE WATER Water that is safe and satisfactory for drinking and cooking.

POTENTIAL DAM SAFETY DEFICIENCY Condition that currently does not significantly affect the safety of the dam, but is capable of becoming a dam safety deficiency.

POTENTIAL DOSE Amount of a compound contained in material swallowed, breathed, or applied to the skin.

POTENTIAL ENERGY Energy of a body with respect to the position of the body.

POTENTIAL EVAPOTRANSPIRATION Rate at which water, if available, would be removed from soil and plant surfaces by evapotranspiration.

POTENTIAL HABITAT Habitat that is suitable for, but currently unoccupied by, the species or community in question.

POTENTIAL NATURAL WATER LOSS Water loss during years when the annual precipitation greatly exceeds the average water loss.

POTENTIAL RATE OF EVAPORATION See EVAPORATIVITY.

POTENTIAL SPILL Accident or other circumstances which threatens to result in discharge of oil or other hazardous substances.

POTENTIAL TEMPERATURE Temperature a parcel of dry air would have if brought adiabatically (i.e., without transfer of heat or mass) to a standard pressure level of 1000 mb.

POTENTIALLY RESPONSIBLE PARTY (PRP) Any individual or company.

POTENTIATION Ability of one chemical to increase the effect of another chemical.

POTENTIOMETRIC SURFACE Surface to which water in an aquifer can rise by hydrostatic pressure;

imaginary surface that represents the total head in an aquifer.

POUNDS PER SQUARE INCH (PSI) Pressure designation for pounds per square inch that may be pounds per square inch gauge (psig) or absolute (psia) where psig measures pressure above the local atmospheric pressure and psia measures pressure with absolute vacuum as a reference.

POWDER FACTOR (EXPLOSIVES FACTOR) Term expressing the amount of explosives used to break a quantity of rock.

POWDER SNOW Dry, loose, unconsolidated snow.

POWER Electrical energy generated, transferred, or used.

POWER DEMAND Rate at which electric energy is required and delivered to or by a system over any designated period of time.

POWER FACTOR Ratio of real to total power.

POWER POOL Two or more interconnected electric systems which operate as a single system to supply power to meet combined load requirements.

POWERPLANT Structure that houses turbines, generators, and associated control equipment.

PPI Plan Position Indicator.

PRACTICAL QUANTIFICATION LIMIT (PQL) Lowest level that can be reliably achieved within specified limits of precision and accuracy during routine laboratory operating conditions.

PRAIRIE POTHOLE Shallow depression, generally containing wetlands, occurring in an outwash plain, a recessional moraine, or a till plain.

PRCP Precipitation.

PRECAST DAM Dam constructed mainly of large precast concrete blocks or sections.

PRECAUTIONARY PRINCIPLE When information about potential risks is incomplete, basing decisions about the best ways to manage or reduce risks on a preference for avoiding unnecessary health risks instead of on unnecessary economic expenditures.

PRECD Precede.

PRECHLORINATION Addition of chlorine at the headwork of a treatment plant prior to other treatment processes.

PRECIPITABLE WATER Measure of the depth of liquid water at the surface that would result after precipitating all of the water vapor in a vertical column over a given location, usually extending from the surface to 300 mb.

PRECIPITATE Substance separated from a solution or suspension by chemical or physical change.

PRECIPITATION Discharge of water, in a liquid or solid state, out of the atmosphere, generally onto a land or water surface; total measurable amount of water received in the form of snow, rain, drizzle, hail, and sleet; process by which atmospheric moisture falls onto a land or water surface as rain, snow, hail, or other forms of moisture; removal of hazardous solids from liquid waste to permit safe disposal; removal of particles from airborne emissions as in rain (e.g. acid precipitation); act or process of producing a solid phase within a liquid medium.

PRECIPITATION ATTENUATION Loss of energy that radar beam experiences as it passes through an area of precipitation.

PRECIPITATION MODE Standard, or default, operational mode of the WSR-88D.

PRECIPITATION PROCESSING SYSTEM WSR-88D system that generates 1-hour running, 3-hourly, and running storm total precipitation accumulations.

PRECIPITATOR Pollution control device that collects particles from an air stream.

PRECISION Accuracy with which a number can be represented (i.e. the number of digits used to represent a number).

PRE-CONSUMER MATERIALS/WASTE Materials generated in manufacturing and converting processes such as manufacturing scrap and trimmings and cuttings.

PRECURSOR Compound antecedent to a pollutant.

PREDEVELOPMENT HYDROLOGY Combination of runoff, infiltration, and evapotranspiration rates and volumes that typically existed on a site before human-induced land disturbance occurred (e.g., construction of infrastructure on undeveloped land such as meadows or forests).

PREDOMINANT PERIOD Period(s) at which maximum spectral energy is concentrated.

PREDOMINANT WIND Wind that prevails and generates the local component of the significant sea conditions across the forecast area; in marine forecast products this is the 10-meter wind; in nearshore marine zones this is the 3-meter wind.

PREEMERGENCE HERBICIDE Herbicide applied to bare ground after planting the crop but prior to the crop sprouting above ground to kill or significantly retard the growth of weed seedlings.

PREFERENCE Priority access to federal power by public bodies and cooperatives.

PREFERENCE CUSTOMERS Publicly owned systems and non-profit cooperatives which, by law, have preference over investor-owned systems for purchase of power from federal projects.

PRE-FRONTAL SQUALL LINE Line of thunderstorms that precedes an advancing cold front.

PRE-FRONTAL TROUGH Elongated area of relatively low pressure preceding a cold front that is usually associated with a shift in wind direction.

PRE-HARVEST INTERVAL Time between the last pesticide application and harvest of the treated crops.

PRE-HURRICANE SQUALL LINE First serious indication that a hurricane is approaching that is generally a straight line and resembles a squall-line that occurs with a mid-latitude cold front.

PRELABORATORY Methods that include all activities involved in collecting, preparing, and delivering a sample to the place of analysis.

PRELIMINARY ASSESSMENT Process of collecting and reviewing available information about a known or suspected waste site or release.

PRELIMINARY REPORT Report summarizing the life history and effects of a storm.

PRES Pressure.

PRESCRIBED ELEVATION Elevation as called out or dictated in notes, drawings or specifications paragraphs.

PRESCRIBED FIRE Management ignited or natural wildland fire that burns under specified conditions where the fire is confined to a predetermined area and produces the fire behavior and fire characteristics required to attain planned fire treatment and resource management objectives.

PRESCRIPTIVE Water rights which are acquired by diverting water and putting it to use in accordance with specified procedures.

PRESENT MOVEMENT Best estimate of the movement of the center of a tropical cyclone at a given time and given position.

PRESENT WEATHER Type of weather observed at the reporting time.

PRESSED WOOD PRODUCTS Materials used in building and furniture construction that are made from wood veneers, particles, or fibers bonded together with an adhesive under heat and pressure.

PRESSURE Load divided by the area over which it acts; exertion of force upon a surface by a fluid (e.g. atmosphere) in contact with it.

PRESSURE ALTIMETER Aneroid barometer calibrated to indicate altitude in feet instead of units of pressure.

PRESSURE ALTITUDE Altitude in standard atmosphere at which a given pressure will be observed.

PRESSURE CELL (GLOETZL CELL, CARLSON LOAD CELL) Instrument for measuring pressure within a mass of soil, rock, or concrete or at an interface between one and the other.

PRESSURE CHANGE Net difference between the barometric pressure at the beginning and ending of a specified interval of time.

PRESSURE CHARACTERISTIC Pattern of the pressure change during a specified period of time using three categories: falling, rising, or steady.

PRESSURE COUPLET Area where you have a high pressure area located adjacent to a low pressure area.

PRESSURE FALLING RAPIDLY Decrease in station pressure at a rate of 0.06 inch of mercury or more per hour which totals 0.02 inch or more.

PRESSURE GAUGE Device for registering the pressure of solids, liquids, or gases.

PRESSURE GRADIENT Amount of pressure change occurring over a given distance.

PRESSURE GRADIENT FORCE Three-dimensional force vector operating in the atmosphere that accelerates air parcels away from regions of high pressure and toward regions of low pressure in response to an air pressure gradient.

PRESSURE HEAD Amount of force or pressure created by a depth of one foot of water; energy contained by fluid because of its pressure; water pressure divided by the specific weight of water.

PRESSURE ICE Floating sea, river, or lake ice that has been deformed, altered, or forced upward in pressure ridges by the lateral stresses of any combination of wind, water currents, tides, waves, and surf.

PRESSURE INDUCED WAVE Rare type of wave that does not develop from wind or seismic activity but rather as a pressure perturbation moves over the water surface.

PRESSURE JUMP Sudden, sharp increase in atmospheric pressure, typically occurring along an active front and preceding a storm.

PRESSURE RELIEF PIPES Pipes used to relieve uplift or pore water pressure in a dam foundation or in the dam structure.

PRESSURE RISING RAPIDLY Increase in station pressure at a rate of 0.06 inch of mercury or more per hour which totals 0.02 inch or more.

PRESSURE SEWERS System of pipes in which water, wastewater, or other liquid is pumped to a higher elevation.

PRESSURE TENDENCY Character and amount of atmospheric pressure change during a specified period of time.

Pressure Tendency (PTDY) Sign (plus or minus) and the amount of pressure change (hPa) for a three hour period ending at the time of observation on a buoy report.

PRESSURE UNSTEADY Pressure that fluctuates by 0.03 inch of mercury or more from the mean pressure during the period of measurement.

PRESSURE, STATIC In flowing air, total pressure minus velocity pressure, pushing equally in all directions.

PRESSURE, TOTAL In flowing air, sum of the static and velocity pressures.

PRESSURE, VELOCITY In flowing air, pressure due to velocity and density of air.

PRESSURE-DRIVEN CHANNELING Channeling of wind in a valley by synoptic-scale pressure gradients superimposed along the valley's axis.

PRESTRESSED DAM Dam, the stability of which depends in part on the tension in steel wires, cables, or rods that pass through the dam and are anchored into the foundation rock.

PRETREATMENT Processes used to reduce, eliminate, or alter the nature of wastewater pollutants from non-domestic sources before they are discharged into publicly owned treatment works.

PREVAILING VISIBILITY Visibility that is considered representative of conditions at the station; greatest distance that can be seen throughout at least half the horizon circle, not necessarily continuous.

PREVAILING WESTERLIES Westerly winds that dominant in middle latitudes.

PREVAILING WINDS Wind that consistently blows from one direction more than from any other.

PREVALENT LEVEL SAMPLES Air samples taken under normal conditions (i.e. ambient background samples).

PREVALENT LEVELS Levels of airborne contaminant occurring under normal conditions.

PREVENTION OF SIGNIFICANT DETERIORATION Program, specified in the Clean Air Act, whose goal is to prevent air quality from deteriorating significantly in areas of the country that are presently meeting the ambient air quality standards.

PRICE CURRENT METER Current meter with a series of conical cups from which the velocity of the water may be computed.

PRIM Primary.

PRIMACY Having the primary responsibility for administering and enforcing regulations.

PRIMARY (PRINCIPAL) SPILLWAY Spillway which would be used first during normal inflow and flood flows.

PRIMARY AMBIENT AIR QUALITY STANDARDS Air quality standards designed to protect human health.

PRIMARY CONTROL TIDE STATION Tide station where continuous observations have been made for a minimum of 19 years.

PRIMARY DRINKING WATER REGULATION Applies to public water systems and specifies a contaminant level, which, in the judgment of the EPA Administrator, will not adversely affect human health.

PRIMARY EFFECT Effect where the stressor acts directly on the ecological component of interest, not on other parts of the ecosystem.

PRIMARY EXCAVATION Digging in undisturbed soil.

PRIMARY INFLOW POINTS Always at the upstream most end of a tributary or main stem segment.

PRIMARY POLLUTANT Substances that are pollutants immediately on entering the atmosphere.

PRIMARY PRODUCTION Creation of organic matter by photosynthesis.

PRIMARY PRODUCTIVITY Measure of the rate at which new organic matter is formed and accumulated through photosynthetic and chemosynthetic activity of producer organisms (chiefly, green plants); rate of primary production is estimated by measuring the amount of oxygen released (oxygen method) or the amount of carbon assimilated (carbon method) by the plants.

PRIMARY STANDARDS National ambient air quality standards designed to protect human health with an adequate margin for safety.

PRIMARY SWELL DIRECTION Prevailing direction of swell propagation.

PRIMARY TREATMENT Practice of removing some portion of the suspended solids and organic matter in a wastewater through sedimentation; first stage of wastewater treatment in which solids are removed by screening and settling.

PRIMARY TRIBUTARY Tributary that is directly connected to or that joins with the main river segment.

PRIME Action of filling a pump casing with water to remove the air.

PRIN Principal.

PRINCIPAL ORGANIC HAZARDOUS CONSTITUENTS (POHCS) Hazardous compounds monitored during an incinerator's trial burn, selected for high concentration in the waste feed and difficulty of combustion.

PRINCIPAL ORGANIZATIONS Federal, state, and local agencies or departments or executive offices having major or lead roles in emergency planning and preparedness for emergency incidents at dams.

PRIONS Microscopic particles made of protein that can cause disease.

PRIOR APPROPRIATION DOCTRINE Water law that allocates the right to use water on a first-come first-serve basis.

PRISTINE Earliest condition of the quality of a water body; unaffected by human activities.

PROBABILISTIC QUANTITATIVE PRECIPITATION FORECAST (PQPF) Form of Quantitative Precipitation Forecast QPF that includes an assigned probability of occurrence for each numerical value in the forecast product.

PROBABILITY Chance, or likelihood, that a certain event might happen.

PROBABILITY FORECAST Forecast of the probability that one or more of a mutually exclusive set of weather conditions will occur.

PROBABILITY OF DETECTION Likelihood, expressed as a percentage, that a test method will correctly identify a leaking tank.

PROBABILITY OF FAILURE Likelihood that, given the loading event and failure mode, the structure responds with the necessary adverse occurrences to ultimately result in uncontrolled reservoir release.

PROBABILITY OF HAIL (POH) Product from the NEXRAD hail detection algorithm that estimates the likelihood that hail is present in a storm.

PROBABILITY OF LOAD Likelihood that a certain loading event (static, seismic, hydrologic, other) occurs at a structure site.

PROBABILITY OF PRECIPITATION (POP) Likelihood that precipitation will be reported at a certain location during a specified period of time.

PROBABILITY OF THUNDERSTORMS Likelihood, based on climatology, that a thunderstorm will be reported at that location during a specified period of time.

PROBABILITY OF TROPICAL CYCLONE CONDITION
Probability, in percent, that the cyclone center will pass within 50 miles to the right or 75 miles to the left of the listed location within the indicated time period when looking at the coast in the direction of the cyclone's movement.

PROBABLE Likely to occur; reasonably expected; realistic.

PROBABLE MAXIMUM FLOOD (PMF)
See MAXIMUM PROBABLE FLOOD.

PROBABLE MAXIMUM PRECIPITATION (PMP)
Largest precipitation for which there is any reasonable expectancy in this climatic era.

PROCESS VARIABLE Physical or chemical quantity which is usually measured and controlled in the operation of a water treatment plant or industrial plant.

PROCESS VERIFICATION Verifying that process raw materials, water usage, waste treatment processes, production rate and other facts relative to quantity and quality of pollutants contained in discharges are substantially described in the permit application and the issued permit.

PROCESS WASTEWATER Any water that comes into contact with any raw material, product, byproduct, or waste.

PROCESS WEIGHT Total weight of all materials, including fuel, used in a manufacturing process.

PRODUCERS Plants that perform photosynthesis and provide food to consumers.

PRODUCT LEVEL Level of a product in a storage tank.

PRODUCT RESOLUTION Smallest spatial increment or data element that is distinguishable in a given Doppler radar product.

PRODUCT WATER Water that has passed through a water treatment plant and is ready to be delivered to consumers.

PRODUCTION Act or process of producing electrical energy from other forms of energy; amount of electrical energy produced expressed in kilowatt-hours (kWh).

PRODUCTION-BASED STANDARD
Discharge standard expressed in terms of pollutant mass allowed in a discharge per unit of product manufactured.

PRODUCTS OF INCOMPLETE COMBUSTION (PICS)
Organic compounds formed by combustion.

PROFILE Sectional view showing grades and distances, usually taken along the centerline; graph showing variation of elevation with distance along a traverse.

PROFILER Instrument designed to measure horizontal winds directly above its location, and thus measure the vertical wind profile.

PROFUNDAL Deepest part of the ocean or lake where light does not penetrate.

PROGRAMMATIC ENVIRONMENTAL IMPACT STATEMENT (PEIS)
Statement that addresses a proposal to

implement a specific policy, to adopt a plan for a group of related actions, or to implement a specific statutory program.

PROGRESSIVE DERECHO Derecho characterized by a short curved squall line oriented nearly perpendicular to the mean wind direction with a bulge in the general direction of the mean flow.

PROJECT Single financial entity which can be composed of several units or divisions, integrated projects, or participating projects.

PROJECT HAZARDOUS WASTE COORDINATOR (PHWC) Individual designated by the Area Manager to coordinate records and maintain proper management of hazardous wastes on each project; acts as primary contact with the Regional Hazardous Waste Management Coordinator (HWMC).

PROJECT XL EPA initiative to give states and the regulated community the flexibility to develop comprehensive strategies as alternatives to multiple current regulatory requirements in order to exceed compliance and increase overall environmental benefits.

PROJECT-USE ENERGY Primarily power and energy used for project operations such as pumping or other miscellaneous uses.

PROJECT-USE POWER Power for main conveyance pumping, designated drainage pumping and other designated miscellaneous electric loads directly associated with the operation of the project.

PROMINENCE Term identifying cloud-like features in the solar atmosphere.

PROPAGATION (PROPA) Movement; transmission of energy as waves through or along a medium.

PROPELLANT Liquid in a self-pressurized pesticide product that expels the active ingredient from its container.

PROPERTY PROTECTION Measures that are undertaken usually by property owners in order to prevent, or reduce flood damage.

PROPORTIONATE MORTALITY RATIO (PMR) Number of deaths from a specific cause in a specific period of time per 100 deaths from all causes in the same time period.

PROPOSED PERMIT Permit prepared after the close of the public comment period (and when applicable, any public hearing and administrative appeals) which is sent for review before final issuance.

PROPOSED PLAN Plan for a site cleanup that is available to the public for comment.

PRORATABLE WATER That portion of the total water supply available subject to proration in times of water shortage.

PROTECTIVE ACTION Action or measure taken to avoid or reduce exposure to a hazard.

PROTECTIVE ACTION DECISION MAKING (RECOMMENDATIONS) Process whereby local authorities select one of more actions to protect threatened populations.

PROTEINS Complex nitrogenous organic compounds of high molecular weight made of amino acids; essential for growth and repair of animal tissue.

PROTOCOL Series of formal steps for conducting a test.

PROTON Solar activity levels with at least one high energy event (Class X Flares).

PROTON EVENT Any flare producing significant fluxes of greater than 10 MeV (million electron volts) protons in the vicinity of the Earth.

PROTOPLAST Membrane-bound cell from which the outer wall has been partially or completely removed.

PROTOTYPE Full-sized structure, system process, or phenomenon being modeled.

PROTOZOA One-celled animals that are larger and more complex than bacteria.

PRST Persist.

PRVD Provide.

PSBL Possible.

PSBLY Possibly.

PSEUDO-COLD FRONT Boundary between a supercell's inflow region and the rear-flank downdraft.

PSEUDO-WARM FRONT Boundary between a supercell's inflow region and the forward-flank downdraft.

PSG Passage.

PST Pacific Standard Time.

PSYCHROMETER Instrument used to measure the water vapor content of the air; hygrometer consisting essentially of two similar thermometers with the bulb of one being kept wet so that the cooling that results from evaporation makes it register a lower temperature than the dry one and with the difference between the readings constituting a measure of the dryness of the atmosphere.

PTCHY Patchy.

PTCLDY Partly Cloudy.

PTLY Partly.

PTN Portion.

PTTN Pattern.

PTYPE Precipitation type.

PUBLIC AFFAIRS OFFICER Organization's designated person responsible for public affairs.

PUBLIC ALERT AND NOTIFICATION SYSTEM System for obtaining the attention of the public and providing appropriate emergency information.

PUBLIC COMMENT PERIOD Time allowed for the public to express its views and concerns regarding an action.

PUBLIC HEALTH APPROACH Regulatory and voluntary focus on effective and feasible risk management actions at the national and community level to reduce human exposures and risks, with priority given to reducing exposures with the biggest impacts in terms of the number affected and severity of effect.

PUBLIC HEALTH CONTEXT Incidence, prevalence, and severity of diseases in communities or populations and the factors that account for them, including infections, exposure to pollutants, and other exposures or activities.

PUBLIC HEARING Formal meeting wherein officials hear the public's views and concerns about an action or proposal.

PUBLIC INFORMATION OFFICER (PUBLIC RELATIONS OFFICER) Person on team who is in charge of public information affairs.

PUBLIC INVOLVEMENT Process of obtaining citizen input into each stage of development of planning documents.

PUBLIC INVOLVEMENT PLAN Document that presents the procedural plans to inform and gather information from project beneficiaries and the general public.

PUBLIC LAND ORDER Action on federal lands to withdraw it from public use for a specified purpose.

PUBLIC NOTICE Notification informing the public of actions.

PUBLIC NOTIFICATION Advisory that requires a water system to distribute to affected consumers when the system has violated MCLs or other regulations; advises consumers what precautions, if any, they should take to protect their health.

PUBLIC SUPPLY Water withdrawn by public and private water suppliers and delivered to users.

PUBLIC UTILITIES COMMISSION (PUBLIC SERVICE COMMISSION) Regulatory agency that governs retail utility rates and practices and, in many cases, issues approvals for the construction of new generation and transmission facilities.

PUBLIC UTILITY HOLDING COMPANY ACT (PUHCA) Federal legislation, enacted in 1935, which regulates the corporate structure and financial operations of certain utility holding companies.

PUBLIC UTILITY REGULATORY POLICIES ACT (PURPA) Bill aimed at expanding the use of cogeneration and renewable energy resources.

PUBLIC WATER SYSTEM (PWS) System that provides piped water for human consumption to at least 15 service connections or regularly serves 25 individuals for at least 60 days annually.

PUBLIC WATER USE Water supplied from a public-water supply and used for such purposes as firefighting, street washing, and municipal parks and swimming pools.

PUBLICLY OWNED TREATMENT WORKS (POTWS) Waste-treatment works owned by a state, unit of local government, or Indian tribe, usually designed to treat domestic wastewaters.

PUBLICS Any interested group or individual, including federal, state and local agencies, interest groups, ad hoc groups, and the general public.

PUBLIC-SUPPLY DELIVERIES Water provided to users through a public-supply distribution system.

PUBLIC-SUPPLY WITHDRAWALS Water withdrawn by public and private water suppliers for use within a general community.

PUDDLE Small pool of water; to compact loose soil by soaking it and allowing it to dry; act of compacting earth, soil clay, etc., by mixing them with water and rolling or tamping the mixture; compact mass of earth, soil, clay, or a mixture of material, which has been compacted through the addition of water, rolling and tamping.

PUDDLING Method of compacting soil with the soil deposited into a pool of water and then stirred or rodded.

PULL SHOVEL (DRAGSHOVEL, HOE) Shovel with a hinge-and-stick mounted bucket that digs while being pulled inward.

PULSE Short burst of electromagnetic energy that a radar sends out in a straight line to detect a precipitation target.

PULSE DURATION Time over which a radar pulse lasts.

PULSE LENGTH Linear distance in range occupied by an individual pulse from a radar.

PULSE RADAR Radar, designed to facilitate range (distance) measurements, where energy is transmitted in periodic, brief transmission.

PULSE REPETITION FREQUENCY (PRF) Amount of time between successive pulses, or bursts, of electromagnetic energy that is transmitted by a radar.

PULSE REPETITION TIME (PRT) Time elapsed between pulses by the radar.

PULSE RESOLUTION VOLUME Discrete radar sampling volume.

PULSE SEVERE THUNDERSTORMS Single cell thunderstorms which produce brief periods of severe weather (3/4 inch hail, wind gusts in the excess of 58 miles an hour, or a tornado).

PULSE STORM Thunderstorm within which a brief period (pulse) of strong updraft occurs, during and immediately after which the storm produces a short episode of severe weather.

PULSE WIDTH See PULSE LENGTH.

PUMP Machine that uses suction/pressure to move liquids or gases; to forcefully move a liquid or gas.

PUMP-BACK SYSTEM Return-flow system in which tailwater is pumped back to the head of an irrigation ditch for reuse.

PUMPED-STORAGE PLANT Power plant designed to generate electric energy for peak load use by releasing water previously pumped into an elevated storage reservoir, usually during off-peak periods.

PUMPING Mechanical transfer of liquids or gases.

PUMPING ENERGY Energy required to pump water from the lower reservoir to the upper reservoir of a pumped-storage plant.

PUMPING HEAD Energy given to a fluid by a pump.

PUMPING PLANT Facility that lifts water up and over hills.

PUMPING STATION Mechanical device installed in sewer or water system or other liquid-carrying pipelines to move the liquids to a higher level.

PUMPING TEST Test conducted to determine aquifer or well characteristics.

PUMPLIFT (PUMPING LIFT) Vertical distance that a pump will raise waters; distance water must be lifted in a well from the pumping level to the ground surface.

PURCHASED POWER Represents the purchase cost of energy for firming up the power supply.

PURGING Removing stagnant air or water from sampling zone or equipment prior to sample collection.

PURVEYOR Agency or person that supplies water.

PUTREFACTION Biological decomposition of organic matter that is associated with anaerobic conditions.

PUTRESCIBLE Able to rot quickly enough to cause odors and attract flies.

PVL Prevail.

PWO Public Severe Weather Outlook.

PYROCLASTIC Ash mixed with larger igneous rock fragments produced by explosive volcanic eruptions.

PYROLYSIS Decomposition of a chemical by extreme heat.

Q

QG Quasigeostrophic.

QN Question.

QPF (QUANTITATIVE PRECIPITATION FORECAST) Spatial and temporal precipitation forecast that will predict the potential amount of future precipitation for a specified region, or area.

QTR Quarter.

QUAD Quadrant.

QUADRAT Sampling area used for analyzing vegetation; square or rectangular plot of land marked off for the study of plants and animals.

QUADRATURE Component of the complex signal that is 90 degrees out of phase with the in-phase component.

QUALITATIVE Having to do with quality or qualities; descriptive of kind, type or direction, as opposed to size, magnitude or degree.

QUALITATIVE USE ASSESSMENT Report summarizing the major uses of a pesticide including percentage of crop treated, and amount of pesticide used on a site.

QUALITY ASSURANCE/QUALITY CONTROL System of procedures, checks, audits, and corrective actions to ensure that all research design and performance, environmental monitoring and sampling, and other technical and reporting activities are of the highest achievable quality.

QUALITY OF SNOW Amount of ice in a snow sample expressed as a percent of the weight of the sample.

QUANTITATIVE Having to do with quantity, capable of being measured; descriptive of size, magnitude or degree.

QUANTITATIVE PRECIPITATION FORECAST Spatial and temporal precipitation forecast that will predict the potential amount of future precipitation for a specified region, or area.

QUARRY Rock pit; open cut mine in rock chosen for physical rather than chemical characteristics.

QUARTZ Very hard, clear or translucent mineral composed of silica.

QUARTZITE Metamorphic rock composed of sand-sized quartz grains that have fused together by heat and pressure; hard, well-cemented quartz-rich sandstone.

QUASI-STATIONARY (QSTNRY) Describes a low or high pressure area or a front that is nearly stationary.

QUASI-STATIONARY FRONT Front which is nearly stationary or moves very little since the last synoptic position.

QUATERNARY Geologic period of the last 2,000,000 years and includes the Pleistocene and Holocene (last 10,000 years) epochs.

QUENCH TANK Water-filled tank used to cool incinerator residues or hot materials during industrial processes.

QUICKSAND Fine sand or silt that is prevented from settling firmly together by upward movement of ground water; any wet inorganic soil so unsubstantial that it will not support any load.

QUIESCENT PROMINENCE (FILAMENT) Long, sheet-like prominences nearly vertical to the solar surface.

QUIET Solar activity levels with less than one chromospheric event per day.

R

R Rain.

R VALUE Measure of insulation where a higher R-value indicates resistance to the transfer of heat.

RA Rain.

RADAR Acronym for RAdio Detection And Ranging; radio device or system for locating an object by means of ultrahigh-frequency radio waves reflected from the object and received, observed, and analyzed by the receiving part of the device in such a way that characteristics (as distance and direction) of the object may be determined.

RADAR BEAM Straight line that a radar pulse travels along that as the radar beam gets further away from the radar, it gets wider and wider.

RADAR CODED MESSAGE Alphanumeric coded message which will be used in preparation of a national radar summary chart.

RADAR CROSS SECTION Area of a fictitious, perfect reflector of electromagnetic waves (e.g. metal sphere) that would reflect the same amount of energy back to the radar as the actual target (e.g. lumpy snowflake).

RADAR DATA ACQUISITION (RDA) Hardware component of the NEXRAD system that consists of the radar antenna, transmitter, receiver, tower, and controlling computer that collects the unprocessed, analog voltages from the radar antenna and converts the signal to base reflectivity, base velocity, and spectrum width (in polar coordinate form).

RADAR METEOROLOGY Branch of meteorology that uses radars for weather observations and forecasts.

RADAR MOSAIC Radar product that combines information from multiple radars to give a regional or national view of reflectivity or precipitation.

RADAR PRODUCT GENERATOR (RPG) Computer in the NEXRAD system that receives polar-coordinate base radar data from the RDA and processes these data into end-user products.

RADAR RANGE Distance from the radar antenna.

RADAR REFLECTIVITY Sum of all backscattering cross-sections (e.g. precipitation particles) in a pulse resolution volume divided by that volume.

RADIAL Lines converging at a single center.

RADIAL GATE (TAINTER GATE) Pivoted crest gate, the face of which is usually a circular arc, with the center of curvature at the pivot about which the gate swings; gate with a curved upstream plate and radial arms hinged to piers or other supporting structure.

RADIAL VELOCITY Component of motion toward or away from a given location.

RADIANCE Measure of the intensity of the radiant energy flux emitted by a body in a given direction.

RADIATION Transmission of energy though space or any medium through electromagnetic waves.

RADIATION FOG Fog that forms when outgoing longwave radiation cools the near-surface air below its dew point temperature.

RADIATION LAWS Four physical laws which fundamentally describe the behavior of blackbody radiation: Kirchhoff's law, Planck's law, Stefan-Boltzmann law and Wien's displacement law.

RADIATION STANDARDS Regulations that set maximum exposure limits for protection of the public from radioactive materials.

RADIATIONAL COOLING Cooling of the Earth's surface (e.g. at night, the Earth suffers a net heat loss to space due to terrestrial cooling).

RADIATIONAL INVERSION Temperature inversion that develops during the night as a result of radiational cooling of the surface.

RADIO EMISSION Emissions of the Sun in radio wavelengths from centimeters to dekameters, under both quiet and disturbed conditions.

RADIO EVENT Flares with Centimetric Bursts and/or definite Ionospheric Event.

RADIO FREQUENCY RADIATION See NON-IONIZING ELECTROMAGNETIC RADIATION.

RADIOACTIVE DECAY Spontaneous change in an atom by emission of charged particles and/or gamma rays.

RADIOACTIVE SUBSTANCES Substances that emit ionizing radiation.

RADIOACTIVE WASTE Any waste that emits energy as rays, waves, streams or energetic particles.

RADIOFACSIMILE Broadcasting graphic weather maps and other graphic images via HF radio.

RADIOFAX Abbreviation for radio facsimile.

RADIOISOTOPE SNOW GAUGE Snow water equivalent gauge based on the absorption of gamma radiation by snow that can measure up to 55 inches water equivalent with a 2 to 5 percent error.

RADIOISOTOPES Chemical variants of radioactive elements with potentially oncogenic, teratogenic, and mutagenic effects on the human body; isotopic forms of elements that exhibit radioactivity.

RADIONUCLIDE Radioactive particle, man-made (anthropogenic) or natural, with a distinct atomic weight number; any man-made or natural element that emits radiation and that may cause cancer after many years of exposure through drinking water.

RADIOSONDE Instrument that is carried aloft by a balloon to send back information on atmospheric temperature, pressure and humidity by means of a small, expendable radio transmitter.

RADIUS Horizontal distance from the center of rotation.

RADIUS OF INFLUENCE Radial distance from the center of a wellbore to the point where there is no lowering of the water table or potentiometric surface (the edge of the cone of depression); radial distance from an extraction well that has adequate air flow for effective removal of contaminants when a vacuum is applied to the extraction well.

RADIUS OF MAXIMUM WINDS Distance from the center of a tropical cyclone to the location of the cyclone's maximum winds.

RADIUS OF VULNERABILITY ZONE Maximum distance from the point of release of a hazardous substance in which the airborne concentration could reach the level of concern under specified weather conditions.

RADON Naturally occurring, colorless, odorless, radioactive gas formed by radioactive decay of radium atoms in soil or rocks that is damaging to human lungs when inhaled.

RADON DAUGHTERS/RADON PROGENY Short-lived radioactive decay products of radon that decay into longer-lived lead isotopes that can attach themselves to airborne dust and other particles and, if inhaled, damage the linings of the lungs.

RADON DECAY PRODUCTS Immediate products of the radon decay chain including Po-218, Pb-214, Bi-214, and Po-214, which have an average combined half-life of about 30 minutes.

RAFC Regional Area Forecast Center.

RAFFINATE PONDS Liquid resulting from extractions with a solvent.

RAFS Regional Analysis and Forecasting System.

RAIN Liquid precipitation; precipitation that falls to Earth in drops more than 0.5 mm in diameter.

RAIN CHAINS Water feature that is used as an alternative to a downspout.

RAIN FOOT Horizontal bulging near the surface in a precipitation shaft, forming a foot-shaped prominence.

RAIN FOREST Forest which grows in a region of heavy annual precipitation with two major types: tropical and temperate.

RAIN GARDEN Depressed area of the ground planted with vegetation, allowing runoff from impervious surfaces such as parking lots and roofs the opportunity to be collected and infiltrated into the groundwater supply or returned to the atmosphere through evaporation and evapotranspiration.

RAIN GAUGE Instrument for measuring the quantity of rain that has fallen.

RAIN INDUCED FOG When warm rain falls through cooler air, water evaporates from the warm rain that subsequently condenses in the cool air forming fog.

RAIN SHADOW Fry region on the lee side of a topographic obstacle, usually a mountain range, where rainfall is noticeably less than on the windward side.

RAIN SHIELD Solid or nearly solid area of rain that typically becomes heavier as one approaches the eye of a hurricane.

RAINBOW Luminous arc featuring all colors of the visible light spectrum (red, orange, yellow, green, blue, indigo, and violet) created by refraction, total reflection, and the dispersion of light.

RAINBOW REPORT Comprehensive document giving the status of all pesticides now or ever in registration or special reviews.

RAINFALL Quantity of water that falls as rain only.

RAINFALL ESTIMATES Series of NEXRAD products that employ a Z-R relationship to produce accumulations of surface rainfall from observed reflectivity.

RAINFALL EXCESS Volume of rainfall available for direct runoff that is equal to the total rainfall minus interception, depression storage, and absorption.

RAINFALL, EXCESSIVE Rainfall in which the rate of fall is greater than certain adopted limits, chosen with regard to the normal precipitation (excluding snow) of a given place or area.

RAIN-FREE BASE Dark, horizontal cloud base with no visible precipitation beneath it.

RAMP FLUME Flume calibrated to measure the flow of liquid in an open channel.

RAMP RATE Rate of change in instantaneous output from a power plant.

RANDOM EARTHQUAKE Earthquakes not associated with known geological structures.

RANGE Geographic region in which a given plant or animal normally lives or grows; distance from the radar antenna.

RANGE FOLDING Occurs when the radar receives a signal return from a pulse other than the most recent pulse.

RANGE GATE Discrete point in range along a single radial of radar data at which the received signal is sampled.

RANGE HEIGHT INDICATOR (RHI)
Radar display in which the radar scans vertically, with the antenna pointing at a specific azimuth or radial.

RANGE NORMALIZATION Receiver gain function in the radar which compensates for the effect of range (distance) on the received power for an equivalent reflectivity.

RANGE RESOLUTION Ability of the radar to distinguish two targets along the same radial but at different ranges.

RANGE UNFOLDING Process of removing range ambiguity in apparent range of a multi-trip target on the radar.

RANKINE VORTEX Velocity profile for a symmetric circulation in which the inner core is in solid rotation and tangential winds outside the core vary inversely with radial distance from the center.

RAOB Radiosonde Observation (Upper-Air Observation).

RAPID Section of a river where the current is very fast moving, caused by a steep descent in the riverbed through a constriction of the main channel.

RAPID DEEPENING Decrease in the minimum sea-level pressure of a tropical cyclone of 1.75 mb/hr or 42 mb for 24 hours.

RAPID FLOW See SUPERCRITICAL FLOW.

RAPID UPDATE CYCLE (RUC) MODEL Numerical model run by NCEP that focuses on short-term forecasts out to 12 hours.

RAPIDLY INTENSIFYING Any maritime cyclone whose central pressure is dropping, or is expected to drop, at a rate of 1 MB per hour for 24 hours.

RAPTORS Birds of prey.

RASP Machine that grinds waste into a manageable material and helps prevent odor.

RATED CAPACITY Capacity which a hydrogenerator can deliver without exceeding mechanical safety factors or a nominal temperature rise; nameplate rating except where turbine power under maximum head is insufficient to deliver the nameplate rating of the generator.

RATED HEAD Water depth for which a hydroelectric generator and turbines were designed.

RATING CURVE Graph showing the relationship between the (gauge height) stage, usually plotted vertically (Y-axis) and the discharge, usually plotted horizontally (X-axis) at a given gaging station.

RATING TABLE Table of stage values and the corresponding discharge for a river gaging site.

RATIO OF REDUCTION Relationship between the maximum size of the stone which will enter a crusher, and the size of its product.

RAW AGRICULTURAL COMMODITY Unprocessed human food or animal feed crop (e.g., raw carrots, apples, corn, or eggs).

RAW SEWAGE Untreated wastewater and its contents.

RAW WATER Water in its natural state, prior to any treatment for drinking or use.

RAWINSONDE Radiosonde that is tracked to measure winds.

RAWINSONDE OBSERVATION Radiosonde observation which includes wind data.

RAWS Remote Automated Weather Stations.

RAYLEIGH SCATTERING Changes in directions of electromagnetic energy by particles whose diameters are 1/16 wavelength or less that is responsible for the sky being blue.

RCKY Rocky Mountains.

RCMD Recommend.

RCV Receive.

RDG Ridge.

RDS Radius.

REACH Continuous part of a stream between two specified points chosen to represent a uniform set of physical (e.g., discharge, depth, area, and slope), chemical, and biological conditions within a segment; section of river or stream between an upstream and downstream location, for which the stage or flow measured at a point somewhere along the section (e.g., gaging station or forecast point) is representative of conditions in that section of river or stream; any specified length of stream, channel, or other water conveyance; area of a canal or lateral between check structures.

REACTION PATH MODELING Simulation approach to studying the chemical evolution of a (natural) system.

REACTIVE POWER Portion of power that is produced by load inductances or capacitances; time average of the instantaneous product of the voltage and current, with current phase shifted 90 degrees.

REACTIVITY Those hazardous wastes that are normally unstable and readily undergo violent chemical change but do not explode.

READY RESERVE Generation capacity that can be synchronized and ready to serve load in a short time period, usually 10 minutes or less.

REAERATION Replenishment of oxygen in water from which oxygen has been removed; introduction of air into the lower layers of a reservoir.

REAL-TIME Refers to the rapid retrieval, processing and transmission of data.

REAL-TIME DATA Data collected by automated instrumentation and telemetered and analyzed quickly enough to influence a decision that affects the monitored system.

REAL-TIME MONITORING Monitoring and measuring environmental developments with technology and communications systems that provide time-relevant information to the public in an easily understood format people can use in day-to-day decision-making about their health and the environment.

REAR FLANK DOWNDRAFT Region of dry air subsiding on the back side of, and wrapping around, a mesocyclone.

REASONABLE FURTHER PROGRESS Annual incremental reductions in air pollutant emissions as reflected in a State Implementation Plan that EPA deems sufficient to provide for the attainment of the applicable national ambient air quality standards by the statutory deadline.

REASONABLE MAXIMUM EXPOSURE Maximum exposure reasonably expected to occur in a population.

REASONABLE WORST CASE Estimate of the individual dose, exposure, or risk level received by an individual in a defined population that is greater than the 90th percentile but less than that received by anyone in the 98th percentile in the same population.

REASONABLY AVAILABLE CONTROL MEASURES (RACM) Broadly defined term referring to technological and other measures for pollution control.

REASONABLY AVAILABLE CONTROL TECHNOLOGY (RACT) Control technology that is reasonably available, and both technologically and economically feasible.

REATTACHMENT DEPOSIT Sand deposit located where downstream flow meets the channel bank at the downstream end of a recirculating zone.

REBAR Reinforcing steel bar.

RECARBONIZATION Process in which carbon dioxide is bubbled into water being treated to lower the pH.

RECEIVER Electronic device which detects the backscattered radiation, amplifies it and converts it to a low-frequency signal which is related to the properties of the target.

RECEIVING WATER River, lake, ocean, stream or other watercourse into which wastewater or treated effluent is discharged.

RECEPTOR Ecological entity exposed to a stressor.

RECESSION CONSTANT Constant used to reduce the Antecedent Precipitation Index (API) value daily in the API method of estimating runoff.

RECESSION CURVE Portion of the hydrograph where runoff is predominantly produced from basin storage (subsurface and small land depressions); hydrograph showing the decreasing rate of runoff following a period of rain or snowmelt; separated from the falling limb of the hydrograph by an inflection point.

RECESSIONAL MORAINE End moraine built during a temporary but significant pause in the final retreat of a glacier.

RECHARGE Increases in groundwater storage due to precipitation, infiltration from streams, or human activity; water that infiltrates the ground and reaches the zone of saturation; amount of water added to an aquifer.

RECHARGE AREA Land area in which water infiltrates the ground and reaches the zone of saturation.

RECHARGE RATE Quantity of water per unit time that replenishes or refills an aquifer.

RECIRCULATION ZONE Area of flow composed of one or more eddies immediately downstream from a constriction in the channel, such as a debris fan or rock outcrop.

RECLAIMED WASTEWATER Treated wastewater that can be used for beneficial purposes.

RECLAMATION REFORM ACT OF 1982 (RRA) Limits the amount of owned land that is eligible to receive project (Reclamation) irrigation water and addresses the rate paid for such water delivered to owned and leased land.

RECOMBINANT BACTERIA Microorganism whose genetic makeup has been altered by deliberate introduction of new genetic elements.

RECOMBINANT DNA New DNA that is formed by combining pieces of DNA from different organisms or cells.

RECOMMENDED MAXIMUM CONTAMINANT LEVEL (RMCL) Maximum level of a contaminant in drinking water at which no known or anticipated adverse effect on human health would occur, and that includes an adequate margin of safety.

RECONNAISSANCE CODE Aircraft weather reconnaissance code that has come to refer primarily to in-flight tropical weather observations, but actually signifies any detailed weather observation or investigation from an aircraft in flight.

RECONSTRUCTED SOURCE Facility in which components are replaced to such an extent that the fixed capital cost of the new components exceeds 50 percent of the capital cost of constructing a comparable brand-new facility.

RECONSTRUCTION OF DOSE Estimating exposure after it has occurred by using evidence within an organism such as chemical levels in tissue or fluids.

RECORD EVENT REPORT Issued by the National Weather Service to report meteorological and hydrological events that equal or exceed existing records.

RECORD OF DECISION (ROD) Public document that explains which cleanup alternative(s) will be used at National Priorities List sites where, under CERCLA, Trust Funds pay for the cleanup.

RECOVERABLE Amount of a given constituent that is in solution after a representative water sample has been extracted or digested.

RECOVERY PHASE That period when the depressed northward field component returns to normal levels.

RECOVERY RATE Percentage of usable recycled materials that have been removed from the total amount of municipal solid waste generated in a specific area or by a specific business.

RECREATION REPORT National Weather Service product that reports on conditions for resorts and recreational areas and/or events.

RECREATIONAL BENEFIT Value of recreational activity to the recreationist, usually measured in dollars above the cost of participating in the recreational activity (travel, entrance fees, etc.).

RECRUITMENT Survival of young plants and animals from birth to a life stage less vulnerable to environmental change.

RECTANGULAR WEIR Contracted or suppressed weir with a horizontal crest, rectangular in shape, having vertical sides.

RECURRENCE INTERVAL (RETURN PERIOD) Average period in years between storm events equal to or larger than a given amount; reciprocal of the probability of that storm event being equaled or exceeded in any year.

RECYCLE/REUSE Minimizing waste generation by recovering and reprocessing usable products that might otherwise become waste (i.e., recycling of aluminum cans, paper, and bottles, etc.).

RECYCLED WATER Water that is used more than one time before it passes back into the natural hydrologic system.

RECYCLING ECONOMIC DEVELOPMENT ADVOCATES Individuals hired by state or tribal economic development offices to focus financial, marketing, and permitting resources on creating recycling businesses.

RECYCLING MILL Facility where recovered materials are remanufactured into new products.

RECYCLING TECHNICAL ASSISTANCE PARTNERSHIP NATIONAL NETWORK National information-sharing resource designed to help businesses and manufacturers increase their use of recovered materials.

RED BAG WASTE See INFECTIOUS WASTE.

RED FLAG Fire weather program which highlights the onset of critical weather conditions conducive to extensive wildfire occurrences.

RED FLAG WARNING Term used by fire-weather forecasters to call attention to limited weather conditions of particular importance that may result in extreme burning conditions.

RED TIDE Proliferation of a marine plankton toxic and often fatal to fish, perhaps stimulated by the addition of nutrients; tide can be red, green, or brown, depending on the coloration of the plankton.

RED WATCH (RED BOX) Slang for Tornado Watch.

REDD Nest that a spawning female salmon digs in gravel to deposit her eggs; depression in riverbed or lakebed dug by fish to deposit eggs.

REDEMPTION PROGRAM Program in which consumers are monetarily compensated for the collection of recyclable materials, generally through prepaid deposits or taxes on beverage containers.

REDUCTION Addition of hydrogen, removal of oxygen, or addition of electrons to an element or compound.

RE-ENTRY Air exhausted from a building that is immediately brought back into the system through the air intake and other openings.

RE-ENTRY INTERVAL Period of time immediately following the application of a pesticide during which unprotected workers should not enter a field.

REF Reference.

REFERENCE AMBIENT CONCENTRATION (RAC) Concentration of a chemical in water which will not cause adverse impacts to human health expressed in units of mg/l.

REFERENCE CONDITIONS Characteristics of water body segments least impaired by human activities.

REFERENCE DOSE (RFD) Estimate of the daily exposure to human population that is not likely to cause harmful effects during a lifetime.

REFERENCE MARK Relatively permanent point of known elevation that is tied to a benchmark.

REFERENCE SITE NAWQA sampling site selected for its relatively undisturbed conditions.

REFERENCE TISSUE CONCENTRATION (RTC) Concentration of a chemical in edible fish or shellfish tissue which will not cause adverse impacts to human health when ingested expressed in units of mg/kg.

REFERENCE VALUE/CONDITIONS Single measurement or set of selected measurements of unimpaired water bodies characteristic of an ecoregion and (or) habitat; chemical, physical, or biological quality or condition that is exhibited at either a single site or an aggregation of sites that represent the least impacted or reasonably attainable condition at the least impacted reference sites.

REFILL Material which had previously been excavated as a result of construction activities and again placed to the lines as shown on the drawings.

REFLECTION Process whereby radiation (or other waves) incident upon a surface is directed back into the medium through which it traveled.

REFLECTIVITY Measure of the efficiency of a radar target in intercepting and returning Electro Magnetic Energy; sum of all backscattering cross-sections (e.g. precipitation particles) in a pulse resolution volume divided by that volume.

REFLECTIVITY CROSS SECTION WSR-88D radar product displays a vertical cross section of reflectivity on a grid with heights up to 70,000 feet on the vertical axis and distance up to 124 nm on the horizontal axis.

REFLECTIVITY FACTOR Result of a mathematical equation (called the Weather Radar Equation) that converts the analog power (in Watts) received by the radar antenna into a more usable quantity.

REFORMULATED GASOLINE Gasoline with a different composition from conventional gasoline (e.g. lower aromatics content) that cuts air pollutants.

REFRACTION Changes in the direction of energy propagation as a result of density changes within the propagating medium; part of the wave advancing in shallower water moves more slowly than that part still advancing in deeper water, causing the wave crest to bend toward alignment with the underwater contours.

REFRACTIVE INDEX Measure of the amount of refraction equal to the ratio of wave velocity in a vacuum to a wave speed in the medium.

REFUELING EMISSIONS Emissions released during vehicle refueling.

REFUGE COMPATIBILITY REQUIREMENTS Requirements under the Refuge Administration Act that all uses of a national wildlife refuge must be compatible with the purpose for which the refuge was established.

REFUSE See SOLID WASTE.

REFUSE RECLAMATION Conversion of solid waste into useful products.

REGENERATION Manipulation of cells to cause them to develop into whole plants.

REGIME THEORY Theory of the forming of channels in material carried by the streams.

REGIMEN OF A STREAM System or order characteristic of a stream.

REGIONAL HAZARDOUS WASTE MANAGEMENT COORDINATOR (HWMC) Individual designated by the Regional Director to assess and coordinate records and maintain proper management of hazardous waste activities in each region.

REGIONAL HAZE Haze that is mixed uniformly between the surface and the top of a convective boundary layer.

REGIONAL RESPONSE TEAM (RRT)
Representatives of federal, local, and state agencies who may assist in coordination of activities at the request of the On-Scene Coordinator before and during a significant pollution incident such as an oil spill, major chemical release, or Superfund response.

REGIONAL TRANSMISSION GROUP (RTG)
Voluntary organization of transmission owners, transmission users, and other entities interested in coordinating transmission planning, expansion, operation, and use within a region.

REGISTRANT Any manufacturer or formulator who obtains registration for a pesticide active ingredient or product.

REGISTRATION Formal listing with EPA of a new pesticide before it can be sold or distributed.

REGISTRATION STANDARDS
Published documents which include summary reviews of the data available on a pesticide's active ingredient, data gaps, and the Agency's existing regulatory position on the pesticide.

REGOLITH Layer or mantle of fragmented and unconsolidated rock material, residual or transported, that nearly everywhere forms the surface of the land and overlies or covers the bedrock.

REGULATED ASBESTOS-CONTAINING MATERIAL (RACM)
Friable asbestos material or nonfriable ACM that will be or has been subjected to sanding, grinding, cutting, or abrading or has crumbled, or been pulverized or reduced to powder in the course of demolition or renovation operations.

REGULATED MEDICAL WASTE
Any solid waste generated in the diagnosis, treatment, or immunization of human beings or animals, in research pertaining thereto, or in the production or testing of biologicals.

REGULATING DAM (REREGULATING DAM)
Dam impounding a reservoir from which water is released to regulate the flow downstream.

REGULATING GATE (OPERATING GATE, REGULATING VALVE)
Gate used to regulate the rate of flow through an outlet works or spillway; gate or valve that operates under full pressure flow conditions to regulate the rate of discharge.

REGULATING SLEEVE VALVE
Valve for regulating high pressure outlets and ensuring energy dissipation.

REGULATION OF A STREAM
Artificial manipulation of the flow of a stream.

REGULATORY FLOODWAY Area where construction regulations require special provisions to account for this extra hazard.

REHABILITATION Process of renovating a facility or system whose performance is failing to meet the original criteria and needs of the project.

REIMBURSABLE Costs of constructing, operating, or maintaining a project that are repaid by some other individual, entity, or organization.

RELATIVE ABUNDANCE Number of organisms of a particular kind present in a sample relative to the total number of organisms in the sample.

RELATIVE DENSITY Ratio of the difference between the void ratio of a cohesionless soil in the loosest state and any given void ratio to the difference between its void ratios in the loosest and in the densest states.

RELATIVE ECOLOGICAL SUSTAINABILITY Ability of an ecosystem to maintain relative ecological integrity indefinitely.

RELATIVE HUMIDITY (RH) Dimensionless ratio, expressed in percent, of the amount of moisture in the air to the maximum amount of moisture the air could hold under the same conditions.

RELATIVE IONOSPHERIC OPACITY METER (RIOMETER) Radio receiver specially designed for continuous monitoring of cosmic noise.

RELATIVE PERMEABILITY Permeability of a rock to gas, NAIL, or water, when any two or more are present.

RELATIVE RISK ASSESSMENT Estimating the risks associated with different stressors or management actions.

RELATIVE VORTICITY Sum of the rotation of an air parcel about the axis of the pressure system and the rotation of the parcel about its own axis.

RELATIVE WIND Wind with reference to a moving point.

RELBL Reliable.

RELEASE Amount of water released after use; any spilling, leaking, pumping, pouring, emitting, emptying, discharging, injecting, escaping, leaching, dumping, or disposing into the environment of a hazardous or toxic chemical or extremely hazardous substance.

RELIABILITY Probability that a device will function without failure over a specified time period or amount of usage.

RELICT Species, population, etc., that is a survivor of a nearly extinct group; any species surviving in a small local area and widely separated from closely related species.

RELIEF Path of least resistance through which energy from explosions can be released.

RELIEF VALVE Valve which will allow air or fluid to escape if its pressure becomes higher than the valve setting; safety device that automatically provides protection against excessive temperatures, excessive pressures, or both.

RELIEF WELLS See DRAINAGE WELLS.

REMEDIAL ACTION (RA) Actual construction or implementation phase of a Superfund site cleanup that follows remedial design.

REMEDIAL DESIGN Phase of remedial action that follows the remedial investigation/feasibility study and includes development of engineering drawings and specifications for a site cleanup.

REMEDIAL INVESTIGATION In-depth study designed to: gather data needed to determine the nature and extent of contamination at a Superfund site, establish site cleanup criteria, identify preliminary alternatives for remedial action, and support technical and cost analyses of alternatives.

REMEDIAL PROJECT MANAGER (RPM) Official responsible for overseeing on-site remedial action.

REMEDIAL RESPONSE Long-term action that stops or substantially reduces a release or threat of a release of hazardous substances that is serious but not an immediate threat to public health.

REMEDIATION Cleanup or other methods used to remove or contain a toxic spill or hazardous materials from a Superfund site; abatement methods including evaluation, repair, enclosure, encapsulation, or removal of greater than 3 linear feet or square feet of asbestos-containing materials from a building.

REMOTE OBSERVATION SYSTEM AUTOMATION (ROSA) Type of automated data transmitter used by National Weather Service Cooperative Program observers.

REMOTE OBSERVING SYSTEM AUTOMATION (ROSA) Type of automated data transmitter used by National Weather Service Cooperative Program observers.

REMOTE OPERATION Operation of mechanical features from an on-site location other than at the feature.

REMOTE SENSING Method for determining characteristics of an object, organism or community from afar; collection and interpretation of information about an object without physical contact with the object (e.g., satellite imaging, aerial photography, and open path measurements).

REMOVAL ACTION Short-term immediate actions taken to address releases of hazardous substances that require expedited response.

RENEWABLE ENERGY CERTIFICATES (RECS) Represent the technology and environmental attributes of electricity generated from renewable sources that are referred to as Green Tags.

RENEWABLE ENERGY CREDITS See RENEWABLE ENERGY CERTIFICATES.

RENEWABLE ENERGY PRODUCTION INCENTIVE (REPI) Incentive established for renewable energy power projects owned by a state or local government or non-profit electric cooperative.

RENEWABLE RESOURCES Naturally replenishable, but flow limited that are virtually inexhaustible in duration but limited in the amount of energy that is available per unit of time (e.g., biomass, hydro, geothermal, solar and wind).

REP Represent; representative.

REPEAT COMPLIANCE PERIOD Any subsequent compliance period after the initial one.

REPLICATE SAMPLES Group of samples collected in a manner such that the samples are thought to be essentially identical in composition.

REPORT Statement of the actual conditions observed at a specific time at a specific site.

REPORTABLE DISCHARGE OF OIL Harmful quantities of oil discharged or released to navigable waters of the United States are any amount that violates state water quality standards and/or causes a sheen or film (as little as 10 parts per million of oil can cause a sheen), upon or discoloration of the surface of the water or adjoining shorelines or cause a sludge or emulsion to be deposited beneath the surface of the water or upon adjoining shorelines.

REPORTABLE QUANTITIES Quantities of hazardous substances which require notification.

REPOWERING Rebuilding and replacing major components of a power plant instead of building a new one.

REPRESENTATIVE SAMPLE Portion of material or water that is as nearly identical in content and consistency as possible to that in the larger body of material or water being sampled.

REREGISTRATION Reevaluation and relicensing of existing pesticides originally registered prior to current scientific and regulatory standards.

RE-REGULATING RESERVOIRS Reservoir for reducing diurnal fluctuations resulting from the operation of an upstream reservoir for power production.

REREGULATION DAM See AFTERBAY DAM.

RESERVATION (FEDERAL INDIAN RESERVATION, INDIAN RESERVATION) All land within the limits of any Indian reservation under the jurisdiction of the United States Government, notwithstanding the issuance of any patent, and including rights-of-way running through the reservation.

RESERVE CAPACITY Extra treatment capacity built into solid waste and wastewater treatment plants and interceptor sewers to accommodate flow increases due to future population growth.

RESERVE GENERATING CAPACITY Extra generating capacity available to meet unanticipated capacity demand for power in the event of generation loss due to scheduled or unscheduled outages of regularly used generating capacity.

RESERVOIR Man-made facility for the storage, regulation and controlled release of water; pond, lake, or basin, either natural or artificial, for the storage, regulation, and control of water.

RESERVOIR CAPACITY Sum of the dead and live storage of the reservoir.

RESERVOIR INFLOW Amount of water entering a reservoir.

RESERVOIR REGULATION (OPERATING) PROCEDURE (RULE CURVE) Operating procedures that govern reservoir storage and releases; compilation of operating criteria, guidelines, and specifications that govern the storage and release function of a reservoir.

RESERVOIR RIM Boundary of the reservoir including all areas along the valley sides above and below the water surface elevation associated with the routing of the IDF.

RESERVOIR SURFACE AREA Area covered by a reservoir when filled to a specified level.

RESERVOIR VOLUME Volume of a reservoir when filled to a specified level.

RESIDENTIAL USE Pesticide application in and around houses, office buildings, apartment buildings, motels, and other living or working areas.

RESIDENTIAL WASTE Waste generated in single and multi-family homes (i.e., newspapers, clothing, disposable tableware, food packaging, cans, bottles, food scraps, etc.).

RESIDENTIAL WATER USE See DOMESTIC WATER USE.

RESIDUAL Amount of a pollutant remaining in the environment after a natural or technological process has taken place; difference between the model-computed and field-measured values of a variable (e.g., hydraulic head, groundwater flow rate) at a specific time and location.

RESIDUAL LAYER Elevated portion of a convective boundary layer that remains after a stable boundary layer develops at the ground (usually in late afternoon or early evening) and cuts off convection.

RESIDUAL MOISTURE Atmospheric moisture which lingers over an area after the main weather system has departed.

RESIDUAL RISK Extent of health risk from air pollutants remaining after application of the Maximum Achievable Control Technology (MACT).

RESIDUAL SATURATION Saturation level below which fluid drainage will not occur.

RESIDUAL-MASS CURVE Graph of the cumulative departures from a given reference such as the arithmetic average plotted against time or date.

RESIDUE Dry solids remaining after the evaporation of a sample of water or sludge.

RESILIENCE Ability of any system to resist or to recover from stress or hardship.

RESISTANCE Ability for plants and animals to withstand poor environmental conditions or attacks by chemicals or disease.

RESONANCE State of a system in which an abnormally large vibration is produced in response to an external stimulus, occurring when the frequency of the stimulus is the same, or nearly the same, as the natural vibration frequency of the system.

RESOURCE CONSERVATION AND RECOVERY ACT (RCRA) Federal statute which establishes a framework for proper management and disposal of all hazardous wastes (i.e., generation, transportation, storage, treatment, and disposal).

RESOURCE MANAGEMENT PLAN Written plan that addresses the existing resources of an area and provides future objectives, goals, and management direction.

RESOURCE RECOVERY Process of obtaining matter or energy from materials formerly discarded.

RESPONSE ACTION Actions taken in response to actual or potential health-threatening environmental events such as spills, sudden releases, and asbestos abatement/management problems; action involving either a short-term removal action or a long-term removal response; any of the following actions taken in school buildings to reduce the risk of exposure to asbestos: removal, encapsulation, enclosure, repair, and operations and maintenance.

RESPONSE INDICATOR Environmental indicator measured to provide evidence of the biological condition of a resource at the organism, population, community, or ecosystem level of organization (EMAP).

RESPONSE LEVEL I First, and least serious, of three response levels that the dam operating organization will declare after analyzing a potentially threatening event that may be perceived as an emergency or may be of general interest to the public, but does NOT pose a hazard, either at the dam or to downstream populations at risk when observed.

RESPONSE LEVEL II Second of three response levels the dam operating organization will declare after analyzing a threatening event that means emergency conditions are such that populations at risk should prepare to leave predetermined inundation areas for higher ground and safe shelter.

RESPONSE LEVEL III Third, and most serious, response level the dam operating organization will declare after analyzing threatening events that indicates life-threatening flood waters, as a result of high operational releases or dam failure, present imminent danger to the public located downstream from a dam.

RESPONSE SPECTRUM Plot of the maximum values of acceleration, velocity, and/or displacement response of an infinite series of single degree of freedom systems subjected to a time history of earthquake ground motion.

RESPONSE TIME Amount of time in which it will take a watershed to react to a given rainfall event.

RESPONSIVENESS SUMMARY Summary of oral and/or written public comments received during a comment period on documents and the response to those comments.

RESTORATION Measures taken to return a site to pre-violation conditions.

RESTRICTED ENTRY INTERVAL Time after a pesticide application during which entry into the treated area is restricted.

RESTRICTED USE Pesticide may be classified (under FIFRA regulations) for restricted use if it requires special handling because of its toxicity, and, if so, it may be applied only by trained, certified applicators or those under their direct supervision.

RESTRICTION ENZYMES Enzymes that recognize specific regions of a long DNA molecule and cut it at those points.

RETAINING WALL Wall separating two levels.

RETARDATION FACTOR Used to simulate the resistance of the contamination to move through the groundwater aquifer.

RETARDING (DETENTION) RESERVOIR
Ungated reservoir for temporary storage of flood water.

RETENTION BASIN Water in storage is permanently obstructed from flowing downstream.

RETROFIT Addition of device on an existing facility without making major changes to the facility.

RETROGRESSION (RETROGRADE MOTION)
Movement of a weather system in a direction opposite to that of the basic flow in which it is embedded.

RETROSPECTIVE ANALYSIS Review and analysis of existing data in order to address objectives and to aid in the design of studies.

RETURN FLOW Drainage water from irrigated farmlands that re-enters the water system to be used further downstream; water that reaches a ground- or surface-water source after release from the point of use and thus becomes available for further use; irrigation water that is applied to an area and which is not consumed in evaporation or transpiration and returns to a surface stream or aquifer; south winds on the back (west) side of an eastward-moving surface high pressure system.

RETURN PERIOD See RECURRENCE INTERVAL.

RETURN STROKE Electrical discharge that propagates upward along a lightning channel from the ground to the cloud.

RETURN-FLOW SYSTEM (REUSE SYSTEM)
System of pipelines or ditches to collect and convey surface or subsurface runoff from an irrigated field for reuse.

REUSE Using a product or component of municipal solid waste in its original form more than once.

REVERSE OSMOSIS (RO) Treatment process used in water systems by adding pressure to force water through a semi-permeable membrane; process of removing salts (desalination) from water using a membrane; advanced method of water or wastewater treatment that relies on a semi-permeable membrane to separate waters from pollutants.

REVERSE OSMOSIS REJECT WATER
Waste water released from the reverse osmosis process.

REVERSIBLE EFFECT Effect which is not permanent.

REVETMENT Embankment or wall of sandbags, earth, etc., constructed to restrain material from being transported away; facing of stone, cement, sandbags, etc., to protect a wall or embankment.

REVIEW OF OPERATION AND MAINTENANCE (RO&M) Periodic evaluation of operation and maintenance activities at a particular facility.

REWIND Act of putting new copper insulated wire in the armature windings of a generator; replacement of an armature (stator) winding of a synchronous generator.

REX BLOCK Blocking pattern where there is an upper level high located directly north of a closed low.

RFD Reference Dose.

RGD Ragged.

RGN Region.

RHI Range-Height Indicator.

RHYOLITE Light-colored volcanic igneous rock that is the extrusive equivalent of granite.

RIBBON LIGHTNING Succession of strokes, each blown a bit to the side of the previous strokes by wind, but striking so fast that all the strokes are seen at once as a ribbon-like flash.

RIBONUCLEIC ACID (RNA) Molecule that carries the genetic message from DNA to a cellular protein-producing mechanism.

RICHTER SCALE Scale of numerical values to describe the magnitude of an earthquake, ranging from 1 to 9, by measurements made in well-defined conditions and with a given type of seismograph where a larger value indicates a larger earthquake.

RIDGE Line or wall of broken ice forced up by pressure; elongated area of relatively high atmospheric pressure; opposite of trough.

RIDGE ICE Ice piled haphazardly one piece over another in the form of ridges or walls.

RIFFLE Stretch of choppy water caused by an underlying rock shoal or sandbar; rapid in a stream.

RIFFLE AND POOL COMPLEX Water habitat composed of riffles (water flowing rapidly over a coarse substrate) and pools (deeper areas of water associated with riffles).

RIGHT ABUTMENT See ABUTMENT.

RIGHT ASCENSION Celestial longitude of the Sun where the value is 0 at the vernal equinox, 90 at the summer solstice, 180 at the autumnal equinox and 270 at the winter solstice.

RIGHT ENTRANCE REGION See RIGHT REAR QUADRANT.

RIGHT MOVER Thunderstorm that moves appreciably to the right relative to the main steering winds and to other nearby thunderstorms.

RIGHT OVERBANK See OVERBANK.

RIGHT REAR QUADRANT (RRQ) Area upstream from and to the right of an upper-level jet max.

RIGHT-OF-WAY Land on and immediately surrounding a structure that is owned by the entity holding title to the structure.

RIGID PIPE Pipe designed to transmit the backfill load to the foundation beneath the pipe.

RILL Small channel eroded into the soil surface by runoff; small grooves, furrows, or channels in soil made by water flowing down over its surface; small stream.

RIME ICE Opaque coating of tiny, white, granular ice particles caused by the rapid freezing of supercooled water droplets on impact with an object.

RING FOLLOWER GATE Gate consisting of a rectangular leaf and an opening equal in diameter to that of the conduit that forms an unobstructed passageway when the leaf is in the raised or open position.

RING GATE Ring- or annular-shaped steel drum operating in a recess or gate chamber in a spillway crest and controlled in a manner similar to a drum gate.

RING SEAL GATE Sealing by a moveable seal that is extended by water pressure when the gate is closed.

RINGLEMANN CHART Series of shaded illustrations used to measure the opacity of air pollution emissions, ranging from light grey through black.

RIOGD Rio Grande.

RIP CURRENT Relatively small-scale surf-zone current moving away from the beach.

RIP TIDE See RIP CURRENT.

RIPARIAN Areas adjacent to rivers and streams with a high density, diversity, and productivity of plant and animal species relative to nearby uplands; living on or adjacent to a water supply such as a riverbank, lake, or pond; of, on, or pertaining to the bank of a river, pond, or lake.

RIPARIAN AREAS Geographically delineable areas with distinctive resource values and characteristics that compose the aquatic and riparian ecosystems.

RIPARIAN DEPENDENT RESOURCES Resources that owe their existence to a riparian area .

RIPARIAN ECOSYSTEMS Transition between the aquatic ecosystem and the adjacent terrestrial ecosystem.

RIPARIAN HABITAT Transition zone between aquatic and upland habitat; areas adjacent to rivers and streams with a differing density, diversity, and productivity of plant and animal species relative to nearby uplands.

RIPARIAN RIGHTS Entitlement of a land owner to certain uses of water on or bordering the property.

RIPARIAN WATER RIGHTS See RIPARIAN RIGHTS.

RIPARIAN ZONE Stream and all the vegetation on its banks.

RIPPLE Small triangular-shaped bed forms; small water waves caused by disturbance (i.e. pebble thrown into pond).

RIPRAP Layer of large uncoursed stone, precast blocks, bags of cement, or other suitable material, generally placed on the slope of an embankment or along a watercourse as protection against wave action, erosion, or scour.

RISING LIMB Portion of the hydrograph where runoff is increasing.

RISK Relationship between the consequences resulting from an adverse event and its probability of occurrence; ability to describe potential outcomes using historic probability;

likelihood or chance of an unacceptable event occurring.

RISK ANALYSIS Procedure to identify and quantify risks by establishing potential failure modes, providing numerical estimates of the likelihood of an event in a specified time period, and estimating the magnitude of the consequences.

RISK ASSESSMENT Process of identifying the likelihood and consequences of failure (qualitative and quantitative evaluation) to provide the basis for informed decisions; process of deciding whether existing risks are tolerable and present risk control measures are adequate and, if not, whether alternative risk control measures are justified; incorporates the risk analysis and risk evaluation phases.

RISK CHARACTERIZATION Last phase of the risk assessment process that estimates the potential for adverse effects to occur and evaluates the uncertainty involved.

RISK COMMUNICATION Exchange of information about risks among risk assessors and managers, the general public, news media, interest groups, etc.

RISK ESTIMATE Description of the probability to develop an adverse response.

RISK FACTOR Characteristics (e.g., race, sex, age, obesity) or variables (e.g., smoking, occupational exposure level) associated with increased probability of an adverse effect.

RISK FOR NON-ENDANGERED SPECIES Risk to species if anticipated pesticide residue levels are equal to or greater than LC50.

RISK MANAGEMENT Process of evaluating and selecting alternative regulatory and non-regulatory responses to risk.

RISK-BASED TARGETING Direction of resources to those areas that have been identified as having the highest potential or actual adverse effect on human health and/or the environment.

RISK-SPECIFIC DOSE Dose associated with a specified risk level.

RIVER Natural stream of water of considerable volume, larger than a brook or creek.

RIVER BASIN Land area drained by a river and its tributaries; drainage area for a river above a particular point.

RIVER CORRIDOR River and the strips of land adjacent to it, including the talus slopes at the bases of cliffs, but not the cliffs themselves.

RIVER FLOOD STATEMENT Product is used by the local National Weather Service Forecast Office (NWFO) to update and expand the information in the River Flood Warning.

RIVER FLOOD WARNING Product is issued by the local National Weather Service Forecast Office (NWFO) when forecast points (those that have formal gaging sites and established flood stages) at specific communities or areas along rivers where flooding has been forecasted, is imminent, or is in progress.

RIVER FLOODING Rise of a river to an elevation such that the river overflows its natural banks causing or threatening damage.

RIVER FORECAST (RVF) Contains stage and/or flow forecasts for specific locations based on existing, and forecasted hydrometeorologic conditions.

RIVER FORECAST CENTER (RFC) Centers that serve groups of Weather Service Forecast offices and Weather Forecast offices, in providing hydrologic guidance and is the first echelon office for the preparation of river and flood forecasts and warnings.

RIVER GAUGE Device for measuring the river stage.

RIVER GAUGE DATUM Arbitrary zero datum elevation which all stage measurements are made from.

RIVER ICE STATEMENT (RVI) Public product containing narrative and numeric information on river ice conditions.

RIVER MILE (RM) Unit of measurement (in miles) used on rivers.

RIVER OBSERVING STATION Established location along a river designated for observing and measuring properties of the river.

RIVER OUTLET WORKS (ROW) See OUTLET WORKS.

RIVER REACH River or stream segment of a specific length.

RIVER RECREATION STATEMENT (RVR) Statement released to inform river users of current and forecast river and lake conditions.

RIVER SEGMENT See STREAM SEGMENT.

RIVER STATEMENT (RVS) Product issued to communicate notable hydrologic conditions which do not involve flooding.

RIVER SUMMARY (RVA) Summary of river and/or crest stages for selected forecast points along the river.

RIVER SYSTEM All of the streams and channels draining a river basin.

RIVER TRASH WALL Walls constructed to deflect heavy floating debris away from the upper ends of a fishway.

RIVERINE Riparian; pertaining to a riverbank.

RIVERINE HABITAT All wetlands and deep-water habitats within a channel, with two exceptions - wetlands dominated by trees, shrubs, persistent emergents, emergent mosses, or lichens and habitats with water that contains ocean-derived salt in excess of 0.5 part per thousand.

RIVERINE WETLANDS Wetlands within river and stream channels; ocean-derived salinity is less than 0.5 part per thousand.

RLS Release.

RLTV Relative.

RMN Remain.

RMTN Regional Meteorological Telecommunications Network.

RMV Remove.

RNFL Rainfall.

ROCK Any naturally formed, consolidated or unconsolidated material (but not soil) consisting of two or more minerals; hard, firm and stable parts of the Earth's crust.

ROCK ANCHOR Steel rod or cable placed in a hole drilled in rock, held in position by grout, mechanical means, or both.

ROCK BOLT See ROCK ANCHOR.

ROCK EXCAVATION Hard and firm parts of the Earth's crust which is dug out and removed from a particular site or area; boulders or detached pieces of solid rock more than 1 cubic yard in volume.

ROCK FRAGMENT Detached pieces of rock which generally are not rounded.

ROCK REINFORCEMENT Placement of rock bolts, untensioned rock dowels, prestressed rock anchors, or wire tendons in a rock mass to reinforce and mobilize the rock's natural competency to support itself.

ROCKETSONDE Type of radiosonde that is shot into the atmosphere by a rocket, allowing it to collect data during its parachute descent from a higher position in the atmosphere than a balloon could reach.

ROCKFILL DAM An embankment dam in which more than 50 percent of the total volume is comprised of compacted or dumped cobbles, boulders, rock fragments, or quarried rock generally larger than 3-inch size; embankment dam of earth or rock in which the material is placed in layers and compacted by using rollers or rolling equipment.

ROD Graduated staff used in determining the difference in elevation between two points.

RODENTICIDE Chemical or agent used to destroy rats or other rodent pests, or to prevent them from damaging food, crops, etc.

ROGUE WAVE Unexpected wave of much greater height or steepness than other waves in the prevailing sea or swell system.

ROLL CLOUD Low, horizontal tube-shaped arcus cloud associated with a thunderstorm gust front (or sometimes with a cold front).

ROLLED FILLED DAM Embankment dam of earth or rock in which the material is placed in layers and compacted by the use of rollers or rolling equipment.

ROLLER COMPACTED CONCRETE (RCC) Mixture of cement, water, and aggregate compacted by rolling.

ROLLER COMPACTED CONCRETE DAM Concrete gravity dam constructed by the use of a dry mix concrete transported by conventional construction equipment and compacted by rolling, usually with vibratory rollers.

ROLLER DRUM GATE See DRUM GATE.

ROLLER GATE (STONEY GATE) Gate rolled up or down inclined supporting rails by a hoist through sprocket chains around the ends of a cylinder.

ROLLING Method of compacting soil using a sheepsfoot roller or a smooth drum roller.

ROOT ZONE That depth of soil which plant roots readily penetrate and in which the predominant root activity occurs; area where a low-angle thrust fault steepens and descends into the crust.

ROPE CLOUD Narrow, rope-like band of clouds sometimes seen on satellite images along a front or other boundary.

ROPE FUNNEL Narrow, often contorted condensation funnel usually associated with the decaying stage of a tornado.

ROPE STAGE Dissipating stage of a tornado, characterized by thinning and shrinking of the condensation funnel into a rope (or rope funnel).

ROSSBY WAVES Series of troughs and ridges on quasi-horizontal surfaces in the major belt of upper tropospheric westerlies where waves are thousands of kilometers long and have significant latitudinal amplitude.

ROTARY KILN INCINERATOR Incinerator with a rotating combustion chamber that keeps waste moving, thereby allowing it to vaporize for easier burning.

ROTATION Spinning of a body (e.g. Earth) about its axis.

ROTG Rotating.

ROTOR CLOUD Turbulent altocumulus cloud formation found in the lee of some mountain barriers when winds cross the barrier at high speed.

ROTTEN ICE Ice in an advanced stage of disintegration.

ROUGH FISH Fish not prized for sport or eating.

ROUGH SEAS Sea conditions associated with regionally defined wind thresholds over bays, inlets, harbors, inland waters, and estuaries where larger waves are forming with whitecaps and spray everywhere.

ROUTE ALERTING Method for alerting people in areas not covered by the primary system or if the primary system fails.

ROUTE OF EXPOSURE Avenue by which a chemical comes into contact with an organism (e.g., inhalation, ingestion, dermal contact, injection).

ROUTING Methods of predicting the attenuation of a flood wave as it moves down the course of a river.

RPD Rapid.

RPLC Replace.

RPRT Report.

RQR Require.

RSG Rising.

RSN Reason.

RTE Route.

RTRD Retard.

RTRN Return.

RTVS Real Time Verification System.

RUBBISH Solid waste, excluding food waste and ashes, from homes, institutions, and workplaces.

RUBBLE DAM Stone masonry dam in which the stones are unshaped or uncoursed.

RULE AUTHORIZATION Allowing an owner or operator to construct or operate as long as the requirements are followed.

RUN Relatively shallow part of a stream with moderate velocity and little or no surface turbulence; seasonal upstream migration of anadromous fish; one or more lengths of pipe that continue in a straight line.

RUNNING AND QUICK-START CAPABILITY Net capability of generating units that carry load or have quick-start capability.

RUNNING LOSSES Evaporation of motor vehicle fuel from the fuel tank while the vehicle is in use.

RUNOFF That part of water yield (e.g., precipitation, snow melt, or irrigation) that flows by overland flow, tile drains, or groundwater toward the streams; composed of baseflow and surface runoff; quantity of water that is discharged from a drainage basin during a given time period.

RUN-OF-RIVER PLANTS Regulated inflow of one power plant is equal to the outflow from a power plant upstream.

RUNUP Vertical distance above the setup that the rush of water reaches when a wave breaks.

RUNWAY VISUAL RANGE (RVR) Maximum distance at which the runway, or the specified lights or markers delineating it, can be seen from a position above a specified point on its center line.

RURAL AREA Predominantly agricultural, prairie, forest, range, or undeveloped land where the population is small.

RURAL ELECTRIC COOPERATIVE (COOPERATIVELY-OWNED ELECTRIC UTILITY) Customer-owned electric utility that was created to transmit and distribute power in rural areas.

RURAL WATER USE Water used in suburban or farm areas for domestic and livestock needs.

RURAL WITHDRAWAL See RURAL WATER USE.

RVI River Ice Statement.

S

S South; snow.

S1 AND S2 CURVES Represent steep sloping water surface profiles.

SAB Satellite Analysis Branch.

SACRED SITE Any specific, discrete, narrowly delineated location on federal land that is identified by an Indian Tribe, or Indian individual determined to be an appropriately authoritative representative of an Indian religion, as sacred.

SACRIFICAL ANODE Easily corroded material deliberately installed in a pipe or intake to give it up (sacrifice it) to corrosion while the rest of the water supply facility remains relatively corrosion-free.

SADDLE DAM (DIKE) Subsidiary dam of any type constructed across a saddle or low point on the perimeter of a reservoir.

SAFE Condition under which there is a practical certainty that no harm will result.

SAFE DRINKING WATER ACT (SDWA) Main federal law that ensures the quality of Americans' drinking water.

SAFE WATER Water that does not contain harmful bacteria, toxic materials, or chemicals, and is considered safe for drinking even if it may have taste, odor, color, and certain mineral problems.

SAFE YIELD Annual amount of water that can be taken from a source of supply over a period of years without depleting that source beyond its ability to be replenished naturally in "wet years."

SAFENER Chemical added to a pesticide to keep it from injuring plants.

SAFETY EVALUATION EARTHQUAKE (SEE) Earthquake expressed in terms of magnitude and closest distance from the dam site or in terms of the characteristics of the time history of free-field ground motions for which the safety of the dam and critical structures associated with the dam are to be evaluated.

SAFETY EVALUATION FLOOD (SEF) Largest flood for which the safety of a dam and appurtenant structure(s) are to be evaluated.

SAFETY EVALUATION OF EXISTING DAMS (SEED) EXAMINATION Onsite examination performed initially and at predetermined intervals where the design, construction, operation, performance, and existing condition of all features are evaluated in accordance with state-of-the-art criteria.

SAFETY EVALUATION OF EXISTING DAMS (SEED) REPORT Compilation of independent technical reports that evaluate the design, construction, and performance of a dam for its structural and hydraulic integrity using available data; identify existing or potential dam safety deficiencies; and recommend future actions appropriate for the safety of the dam.

SAFFIR-SIMPSON HURRICANE WIND SCALE 1 to 5 categorization based on the hurricane's intensity at the indicated time.

SALINA Area where deposits of crystalline salt are formed (e.g. salt flat); a body of saline water (e.g., saline playa or salt marsh).

SALINE Condition of containing dissolved or soluble salts.

SALINE FORMATIONS Deep and geographically extensive sedimentary rock layers saturated with waters or brines that have a high total dissolved solids (TDS) content (i.e. over 10,000 mg/L TDS).

SALINE INTRUSION BARRIER WELLS Class V wells that are used to inject fluids to prevent the intrusion of salt water into an aquifer.

SALINE SODIC LANDV Soil that contains soluble salts in amounts that impair plant growth but not an excess of exchangeable sodium.

SALINE WATER Water that is considered unsuitable for human consumption or for irrigation because of its contains significant amounts of dissolved solids; water that contains more than 1,000 milligrams per liter of dissolved solids.

SALINITY (SAL) Saltiness; percentage of salt in water; relative concentration of dissolved salts, usually sodium chloride, in a given water supply; measure of the concentration of dissolved mineral substances in water.

SALMONIDS Family of fish that includes salmon and steelhead.

SALT WATER INTRUSION Invasion of fresh surface or groundwater by salt water.

SALTATION Movement of sand or fine sediment by short jumps above a streambed under the influence of a water current too weak to keep it permanently suspended in the moving water.

SALTS Minerals that water picks up as it passes through the air, over and under the ground, or from households and industry.

SALVAGE Utilization of waste materials.

SAMEX Storm and Mesoscale Ensemble Experiment.

SAMPLE Water that is analyzed for the presence of EPA-regulated drinking water contaminants; portion thought to represent whole.

SAMPLE ERROR Random variation reflecting the inherent variability within a population being counted.

SAMPLING FREQUENCY Interval between the collection of successive samples; rate at which sensor data is read.

SANCTIONS Actions taken for failure to provide or implement a plan.

SAND Mineral grains whose particle size vary from a No. 4 sieve to a No. 200 sieve; loose soil composed of particles between 1/16 mm and 2 mm in diameter.

SAND BACKFILL Material which has a particle size which varies from a No. 4 sieve to a No. 200 sieve and is used for refilling an excavation.

SAND BOIL Seepage characterized by a boiling action at the surface surrounded by a cone of material from deposition of foundation and/or embankment material carried by the seepage; swirling upheaval of sand or soil on the surface of or downstream from an embankment caused by water leaking through the embankment; ejection of sand and water resulting from piping.

SAND FILTERS Devices that remove some suspended solids from sewage; air and bacteria decompose additional wastes filtering through the sand so that cleaner water drains from the bed.

SANDSTONE Sedimentary rock composed of sand-sized grains (usually quartz) cemented together.

SANDSTORM Particles of sand carried aloft by strong wind.

SANITARY LANDFILL See LANDFILL.

SANITARY SEWER Pipe or conduit (sewer) intended to carry wastewater or water-borne wastes from homes, businesses, and industries to the publicly owned treatment works.

SANITARY SEWER OVERFLOWS (SSO) Untreated or partially treated sewage overflows from a sanitary sewer collection system.

SANITARY SEWERS Underground pipes that carry off only domestic or industrial waste, not storm water.

SANITARY SURVEY On-site review of the water sources, facilities, equipment, operation, and maintenance of a public water systems for the purpose of evaluating the adequacy of the facilities for producing and distributing safe drinking water.

SANITARY WASTE Liquid or solid wastes originating solely from humans and human activities, such as wastes collected from toilets, showers, wash basins, sinks used for cleaning domestic areas, sinks used for food preparation, clothes washing operations, and sinks or washing machines where food and beverage serving dishes, glasses, and utensils are cleaned.

SANITARY WATER See GRAY WATER.

SANITARY WATER USE Domestic water uses, such as faucets, toilets, and showers.

SANITATION Control of physical factors in the human environment that could harm development, health, or survival.

SANTA ANA WIND Weather condition in southern California, in which strong, hot, dust-bearing winds descend to the Pacific Coast around Los Angeles from inland desert regions.

SAPROLITE Soft, clay-rich, thoroughly decomposed rock formed in place by chemical weathering of igneous or metamorphic rock.

SAPROPHYTES Organisms living on dead or decaying organic matter that help natural decomposition of organic matter in water.

SARA Superfund Amendments and Reauthorization Act of 1986.

SASTRUGI Ridges of snow formed on a snow field by the action of the wind.

SATL Satellite.

SATURATED UNIT WEIGHT Wet unit weight of soil when saturated.

SATURATED ZONE Zone in which all voids are filled with water; area below the water table where all open spaces are filled with water under pressure equal to or greater than that of the atmosphere.

SATURATION Condition of a liquid when it has taken into solution the maximum possible quantity of a given substance at a given temperature and pressure; degree to which voids in soil are filled with water; method of compacting soil using water added to soil and internal vibrators (such as a concrete vibrator) are worked down through the depth of soil placed; condition of being filled to capacity.

SATURATION VAPOR PRESSURE
Vapor pressure of a system, at a given temperature, wherein the vapor of a substance is in equilibrium with a plane surface of that substance's pure liquid or solid phase.

SATURATION ZONE Portion of the soil profile where available water storage is completely filled.

SBCAPE Surface Based CAPE; CAPE calculated using a Surface based parcel.

SBND Southbound.

SBSD Subside.

SC Stratocumulus.

SCA Small Craft Advisory.

SCALE Ratio of map distance to Earth distance.

SCALING Prying loose pieces of rock off a face or roof to avoid danger of their falling unexpectedly; adjustment to an earthquake time history or response spectrum where the amplitude of acceleration, velocity, and/or displacement is increased or decreased, usually without change to the frequency content of the ground motion.

SCATTER Concentration of artifacts.

SCATTERED (SCT) Area coverage of convective weather affecting 30 percent to 50 percent of a forecast zone.

SCATTERING Process in which a beam of light is diffused or deflected by collisions with particles suspended in the atmosphere.

SCHEDULED OUTAGE Shutdown of a generating unit, or other facility, for inspection or maintenance, in accordance with an advance schedule.

SCHIST Metamorphic rock composed of platy mica minerals aligned in the same direction.

SCIENCE ADVISORY BOARD (SAB)
Group of external scientists who advise on science and policy.

SCOPE 1 GHG EMISSIONS Direct GHG emissions from sources that are owned or controlled by the reporting entity that include emissions from fossil fuels burned on site, emissions from agency-owned or agency-leased vehicles, and other direct sources.

SCOPE 2 GHG EMISSIONS Indirect GHG emissions resulting from the generation of electricity, heat, or steam generated off site but purchased by the reporting agency.

SCOPE 3 GHG EMISSIONS Indirect GHG emissions from sources not owned or directly controlled by the reporting agency but related to the agencies activities such as vendor supply chains, delivery services, outsourced activities, and employee travel and commuting.

SCOUR Erosion in a stream bed, particularly if caused or increased by channel changes; enlargement of a flow section by the removal of bed material through the action of moving water.

SCRAP Materials discarded from manufacturing operations that may be suitable for reprocessing.

SCRAP METAL PROCESSOR Intermediate operating facility where recovered metal is sorted, cleaned of contaminants, and prepared for recycling.

SCRAPE Nest made from scratching in the ground.

SCREEN Mesh or bar surface used for separating pieces or particles of different sizes; filter.

SCREENED INTERVAL Length of unscreened opening or of well screen through which water enters a well, in feet below land surface.

SCREENING Use of screens to remove coarse floating and suspended solids from sewage.

SCREENING RISK ASSESSMENT Risk assessment performed with few data and many assumptions to identify exposures that should be evaluated more carefully for potential risk.

SCRUBBER Air pollution device that uses a spray of water or reactant or a dry process to trap pollutants in emissions.

SCS Soil Conservation Service; now known as the National Resources Conservation Services (NRCS).

SCS CURVE NUMBER Empirically derived relationship between location, soil-type, land use, antecedent moisture conditions and runoff.

SCT Scattered.

SCUD Small, ragged, low cloud fragments that are unattached to a larger cloud base and often seen with and behind cold fronts and thunderstorm gust fronts.

SDM Station Duty Manual.

SE Southeast.

SEA BREEZE Thermally produced wind blowing during the day from a cool ocean surface onto the adjoining warm land, caused by the difference in the rates of heating of the surfaces of the ocean and of the land.

SEA BREEZE CONVERGENCE ZONE Zone at the leading edge of a sea breeze where winds converge.

SEA BREEZE FRONT Leading edge of a sea breeze, whose passage is often accompanied by showers, a wind shift, or a sudden drop in temperature.

SEA FOG Advection fog caused by transport of moist air over a cold body of water.

SEA ICE Any form of ice found at sea which has originated from the freezing of sea water.

SEA LEVEL Long-term average position of the sea surface commonly referenced to two national vertical datums (NGVD 1929 or NAVD 1988).

SEA LEVEL PRESSURE Atmospheric pressure at sea level at a given location.

SEA SURFACE TEMPERATURES Mean temperature of the ocean in the upper few meters.

SEA WATER See SALINE WATER.

SEAM Layer of rock, coal, or ore.

SEAS Combination of both wind waves and swell.

SEASONAL STREAM See INTERMITTENT STREAM.

SECCHI DEPTH Measure of water clarity.

SECONDARY AMBIENT AIR QUALITY STANDARDS Air quality standards designed to protect human welfare, including the effects on vegetation and fauna, visibility and structures.

SECONDARY CURRENTS (OR FLOW) Movement of water particles on a cross section normal to the longitudinal direction of the channel.

SECONDARY DRINKING WATER REGULATIONS Non-enforceable federal regulations applying to public water systems and specifying the maximum contamination levels that are required to protect the public welfare.

SECONDARY DRINKING WATER STANDARDS Non-enforceable federal guidelines regarding cosmetic effects (i.e., tooth or skin discoloration) or aesthetic effects (i.e., taste, odor, or color) of drinking water.

SECONDARY EFFECT Action of a stressor on supporting components of the ecosystem, which in turn impact the ecological component of concern.

SECONDARY INDUSTRY CATEGORY Any industry category which is not a primary industry category.

SECONDARY MATERIALS Materials that have been manufactured and used at least once and are to be used again.

SECONDARY MAXIMUM CONTAMINANT LEVEL (SMCL) Maximum level of a contaminant or undesirable constituent in public water systems that is required to protect the public welfare.

SECONDARY POLLUTANT Pollutants generated by chemical reactions occurring within the atmosphere.

SECONDARY STANDARDS National ambient air quality standards designed to protect welfare, damage to property, transportation hazards, economic values and personal comfort/well-being.

SECONDARY WASTEWATER TREATMENT Second step in most publicly owned waste treatment systems in which bacteria consume the organic parts of the waste; treatment (following primary wastewater treatment) involving the biological process of reducing suspended, colloidal, and dissolved organic matter in effluent from primary treatment systems and which generally removes 80 to 95 percent of the Biochemical Oxygen Demand (BOD) and suspended matter.

SECOND-DAY FEET Volume of water represented by a flow of one cubic foot per second for 24 hours; equal to 84,000 ft^3.

SECOND-FOOT Volume of water represented by a flow of one cubic foot per second.

SECTION Area equal to 640 acres or 1 square mile.

SECTION 304(A) CRITERIA Developed by EPA under authority of section 304(a) of the Act based on the latest scientific information on the relationship that the effect of a constituent concentration has on particular aquatic species and/or human health.

SECTOR BOUNDARY In the solar wind, the area of demarcation between sectors, which are large-scale features distinguished by the predominant direction of the interplanetary magnetic field, toward or away from the Sun.

SECTOR VISIBILITY Visibility in a specific direction that represents at least a 45° arc of a horizontal circle.

SECTORIZED HYBRID SCAN Single reflectivity scan composed of data from the lowest four elevation scans.

SECURE MAXIMUM CONTAMINANT LEVEL Maximum permissible level of a contaminant in water delivered to the free flowing outlet of the ultimate user, or of contamination resulting from corrosion of piping and plumbing caused by water quality.

SECURITE Headline within National Weather Service high seas forecasts transmitted via the GMDSS to indicate that no hurricane or hurricane force winds are forecast.

SEDIMENT Particles derived from rocks or biological materials that have been transported by a fluid; solid material (sludges) suspended in or settled from water; collective term meaning an accumulation of soil, rock and mineral particles transported or deposited by flowing water; all kinds of deposits from the waters of streams, lakes, or seas.

SEDIMENT CONCENTRATION Quantity of sediment relative to the quantity of transporting fluid, or fluid-sediment mixture.

SEDIMENT DISCHARGE Rate at which sediment passes a stream cross-section in a given period of time.

SEDIMENT GUIDELINE Threshold concentration above which there is a high probability of adverse effects on aquatic life from sediment contamination, determined using modified U.S. Environmental Protection Agency USEPA (1996) procedures.

SEDIMENT LOAD Material in suspension and/or in transport; amount of sediment that is being moved by a stream; mass of sediment passing through a stream cross section in a specified period of time.

SEDIMENT PARTICLE Fragments of mineral or organic material in either a singular or aggregate state.

SEDIMENT QUALITY GUIDELINE Threshold concentration above which there is a high probability of adverse effects on aquatic life from sediment contamination.

SEDIMENT STORAGE CAPACITY Volume of a reservoir planned for the deposition of sediment.

SEDIMENT TRANSPORT (RATE) See SEDIMENT DISCHARGE.

SEDIMENT TRANSPORT FUNCTION Formula or algorithm for calculating the sediment transport rate given the hydraulics and bed material at a cross section.

SEDIMENT TRANSPORT ROUTING Computation of sediment movement for a selected length of stream (reach) for a period of time with varying flows.

SEDIMENT TRAP EFFICIENCY See TRAP EFFICIENCY.

SEDIMENT YIELD Quantity of sediment arriving at a specific location; amount of mineral or organic soil material that is in suspension, is being transported, or has been moved from its site of origin; portion of eroded material that does travel through the drainage network to a downstream measuring or control point; dry weight of sediment per unit volume of water-sediment mixture in place; ratio of the dry weight of sediment to the total weight of water-sediment mixture in a sample or a unit volume of the mixture.

SEDIMENT YIELD RATE Sediment yield per unit of drainage area.

SEDIMENTARY Rock resulting from the consolidation of loose eroded sediment, remains of organisms, or crystals forming directly from water.

SEDIMENTARY ROCKS Rocks formed by the consolidation of loose sediment that has accumulated in layers; rocks formed of sediment (e.g., sandstone and shale, formed of fragments of other rock transported from their sources and deposited in water); rocks formed by or from secretions of organisms (e.g. most limestone).

SEDIMENTATION Deposition of waterborne sediments due to a decrease in velocity and corresponding reduction in the size and amount of sediment which can be carried; act or process of forming or accumulating sediment in layers; letting solids settle out of wastewater by gravity during treatment.

SEDIMENTATION DIAMETER Diameter of a sphere of the same specific weight and the same terminal settling velocity as the given particle in the same fluid.

SEDIMENTATION TANKS Wastewater tanks in which floating wastes are skimmed off and settled solids are removed for disposal.

SEDIMENT-DISCHARGE RELATIONSHIP Tables that relate inflowing sediment loads to water discharge for the upstream ends of the main stem, tributaries, and local inflows.

SEDIMENTS Soil, sand, and minerals washed from land into water.

SEED PROTECTANT Chemical applied before planting to protect seeds and seedlings from disease or insects.

SEEP Small area where groundwater percolates slowly to the land surface, usually forming a pool.

SEEPAGE Slow movement or percolation of water through soil or rock; movement of water through soil without formation of definite channels; movement of water into and through the soil from unlined canals, ditches, watercourses, and water storage facilities; slow movement or percolation of water through small cracks, pores, interstices, etc., from an embankment, abutment, or foundation; interstitial movement of water that may take place through a dam, its foundation, or abutments.

SEEPAGE COLLAR Projecting collar of concrete built around the outside of a tunnel or conduit, within an embankment dam, to reduce seepage along the outer surface of the conduit.

SEEPAGE FACE Physical boundary segment of a groundwater system along which groundwater discharges and which is present when a water table surface ends at the downstream external boundary of a flow domain.

SEEPAGE FORCE Force transmitted to the soil or rock grains by seepage.

SEEPAGE LOSS Water loss by capillary action and slow percolation.

SEEPAGE VELOCITY Rate of discharge of seepage water through a porous medium per unit area of void space perpendicular to the direction of the flow.

SEICHE Oscillating wave of water caused by a landslide, earthquake or meteorological event (e.g., wind or large barometric pressure gradient).

SEISMIC Of or related to movement in the Earth's crust caused by natural relief of rock stresses.

SEISMIC EVALUATION CRITERIA Guideline for determining which faults or seismic sources need to be assigned maximum credible earthquake (MCE).

SEISMIC INTENSITY Subjective measurement of the degree of shaking at a specified place by an experienced observer using a descriptive scale.

SEISMIC MOMENT (MO) Measure of the earthquake size containing information on the rigidity of the elastic medium in the source region, average dislocation, and area of faulting.

SEISMIC PARAMETERS Descriptors of earthquake loading or earthquake size, such as magnitude, peak acceleration, location (distance and focal depth), spectrum intensity, or any of many other parameters useful in characterizing earthquake loadings.

SEISMO Pertains to earthquakes.

SEISMOTECTONIC Of, relating to, or designating structural features of the Earth which are associated with or revealed by earthquakes.

SEISMOTECTONIC PROVINCE Geographic area characterized by a combination of geology and seismic history.

SEISMOTECTONIC SOURCE AREA Area or areas of known or potential seismic activity that may lack a specific identifiable seismotectonic structure.

SEISMOTECTONIC STRUCTURE Identifiable dislocation or distortion within the Earth's crust resulting from recent tectonic activity or revealed by seismologic or geologic evidence.

SEIZE To bind wire rope with soft wire; to prevent rope from unraveling when cut.

SEL Watch cancellation statement issued to terminate a watch before its original expiration time.

SELECT MATERIAL Backfill materials specially selected and segregated from excavated materials.

SELECTION CRITERIA Set of statements that describe suitable indicators or a rationale for selecting indicators.

SELECTIVE HERBICIDE Compound that kills or significantly retards growth of an unwanted plant species without significantly damaging desired plant species.

SELECTIVE PESTICIDE Chemical designed to affect only certain types of pests, leaving other plants and animals unharmed.

SELF-MONITORING Sampling and analyses performed by a facility to determine compliance with a permit or other regulatory requirements.

SELF-SUPPLIED WATER Water withdrawn from a surface- or groundwater source by a user rather than being obtained from a public supply.

SELS Severe Local Storm.

SELY Southeasterly.

SEMI-ANALYTICAL MODEL Mathematical model in which complex analytical solutions are evaluated using approximate techniques, resulting in a solution discrete in either the space or time domain.

SEMI-CONFINED AQUIFER Aquifer partially confined by soil layers of low permeability through which recharge and discharge can still occur.

SEMIPERMEABLE MEMBRANE DEVICE (SPMD) Long strip of low-density, polyethylene tubing filled with a thin film of purified lipid such as triolein that simulates the exposure to and passive uptake of highly lipid-soluble organic compounds by biological membranes.

SEMIPERVIOUS ZONE See TRANSITION ZONE.

SEMIVOLATILE ORGANIC COMPOUND (SVOC) Organic compounds that volatilize slowly at standard temperature (20 °C and 1 atm pressure).

SENESCENCE Aging process; describes lakes or other bodies of water in advanced stages of eutrophication.

SENSIBLE HEAT FLUX (LATENT SENSIBLE HEAT-FLUX DENSITY) Amount of heat energy that moves by turbulent transport through the air across a specified cross-sectional area per unit time and goes to heating (cooling) the air; flux of heat from the Earth's surface to the atmosphere that is not associated with phase changes of water; component of the surface energy budget.

SENSITIVE SPECIES Plant and animal species for which population viability is a concern; species whose populations are small and widely dispersed or restricted to a few localities; species whose numbers are declining so rapidly that official listing may be necessary; species not yet officially listed but undergoing status review for listing on the U.S. Fish and Wildlife Service's (FWS) official threatened and endangered list.

SENSITIVITY Variation in the value of one or more output variables or quantities calculated from the output variables due to changes in the value of one or more inputs to the model.

SENSITIVITY ANALYSIS Analysis in which the relative importance of one or more of the variables thought to have an influence on the phenomenon under consideration is determined.

SEPARATION DEPOSIT Sand deposit located at the upstream end of a recirculation zone, where downstream flow becomes separated from the channel bank.

SEPARATION EDDY Eddy that forms near the ground on the windward or leeward side of a bluff object or steeply rising hillside.

SEPTIC SYSTEM On-site system designed to treat and dispose of domestic sewage.

SEPTIC TANK Underground storage tank for wastes from homes not connected to a sewer line; tank used to detain domestic wastes to allow the settling of solids prior to distribution to a leach field for soil absorption.

SERC State Emergency Response Commission.

SERIAL DERECHO Derecho that consists of an extensive squall line which is oriented such that the angle between the mean wind direction and the squall line axis is small.

SERN Southeastern.

SERVICE CONNECTOR Pipe that carries tap water from a public water main to a building.

SERVICE HYDROLOGIST Designated expert of the hydrology program at a WFO.

SERVICE LINE SAMPLE One-liter sample of water that has been standing for at least 6 hours in a service pipeline and is collected according to federal regulations.

SERVICE PIPE Pipeline extending from the water main to the building served or to the consumer's system.

SERVICE SPILLWAY (PRIMARY SPILLWAY) Structure located on or adjacent to a storage or detention dam over or through which surplus or floodwaters which cannot be contained in the allotted storage space are passed, and at diversion dams to bypass flows exceeding those which are turned into the diversion system.

SERVO LOOP Hardware needed to remotely control the motion of the antenna dish.

SET Direction towards which a current is headed.

SET-BACK Setting a thermometer to a lower temperature when the building is unoccupied to reduce consumption of heating energy; setting the thermometer to a higher temperature during unoccupied periods in the cooling season.

SETTLEABLE SOLIDS Material heavy enough to sink to the bottom of a wastewater treatment tank.

SETTLEMENT Vertical downward movement of a structure or its foundation; sinking of land surfaces because of subsurface compaction, usually occurring when moisture, added deliberately or by nature, causes a reduction in void volumes.

SETTLEMENT SENSORS (PNEUMATIC AND VIBRATING-WIRE)
Monitors the difference in elevation between the sensor unit and its reservoir.

SETTLING CHAMBER Series of screens placed in the way of flue gases to slow the stream of air, thus helping gravity to pull particles into a collection device.

SETTLING POND Open lagoon into which wastewater contaminated with solid pollutants is placed and allowed to stand where solid pollutants suspended in the water sink to the bottom of the lagoon and the liquid is allowed to overflow out of the enclosure.

SETTLING TANK Holding area for wastewater, where heavier particles sink to the bottom for removal and disposal.

SETTLING VELOCITY See FALL VELOCITY.

SET-UP Vertical rise in the stillwater level at the upstream face of a dam caused by wind stresses on the water surface.

SEVEN-DAY, 10-YEAR LOW FLOW (7Q10)
Lowest stream flow for seven consecutive days that would be expected to occur once in ten years; seven-day, consecutive low flow with a ten year return frequency.

7Q10 See SEVEN-DAY, 10-YEAR LOW FLOW.

SEVERE GEOMAGNETIC STORM
Storm for which the Ap index was 100 or more.

SEVERE ICING Rate of ice accumulation on an aircraft is such that de-icing/anti-icing equipment fails to reduce or control the hazard.

SEVERE LOCAL STORM Convective storm that usually covers a relatively small geographic area, or moves in a narrow path, and is sufficiently intense to threaten life and/or property.

SEVERE LOCAL STORM WATCH
Alert issued by the National Weather Service for the contiguous U.S. and its adjacent waters of the potential for severe thunderstorms or tornadoes.

SEVERE THUNDERSTORM
Thunderstorm that produces a tornado, winds of at least 58 mph (50 knots), and/or hail at least $3/4''$ in diameter.

SEVERE THUNDERSTORM WARNING
Warning issued by the National Weather Service when either a severe thunderstorm is indicated by the WSR-88D radar or a spotter reports a thunderstorm producing hail $3/4$ inch or larger in diameter and/or winds equal or exceed 58 miles an hour.

SEVERE THUNDERSTORM WATCH
Issued by the National Weather Service when conditions are favorable for the development of severe thun-

derstorms in and close to the watch area.

SEVERE WEATHER POTENTIAL STATEMENT Statement issued by the National Weather Service that is designed to alert the public and state/local agencies to the potential for severe weather up to 24 hours in advance.

SEVERE WEATHER PROBABILITY WSR-88D radar product algorithm displays numerical values proportional to the probability that a storm will produce severe weather within 30 minutes.

SEVERE WEATHER STATEMENT Statement issued by the National Weather Service that provides follow up information on severe weather conditions which have occurred or are currently occurring.

SEVERE WEATHER THREAT (SWEAT) INDEX Stability index developed by the Air Force which incorporates instability, wind shear, and wind speeds.

SEWAGE Waste and wastewater produced by residential and commercial sources and discharged into sewers.

SEWAGE LAGOON See LAGOON.

SEWAGE SLUDGE Sludge produced at a publicly owned treatment works (POTWs) where disposal is regulated under the Clean Water Act.

SEWAGE TREATMENT EFFLUENT WELLS Class V wells that are used by privately or publicly owned treatment works (POTWs) to inject treated or untreated domestic sewage through a vertical well or a leach field.

SEWAGE TREATMENT PLANT Facility designed to receive the wastewater from domestic sources and to remove materials that damage water quality and threaten public health and safety when discharged into receiving streams or bodies of water.

SEWD Southeastward.

SEWER System of underground pipes, channel or conduit that collect and deliver wastewater and storm-water runoff from the source to a treatment plant or receiving stream.

SEWER TILE Glazed waterproof clay pipe with bell joints.

SEWERAGE Entire system of sewage collection, treatment, and disposal.

SFC Surface.

SFERIC Transient electric or magnetic field generated by any feature of lightning discharge (entire flash).

SG Snow grains.

SGFNT Significant.

SHADING COEFFICIENT Amount of the Sun's heat transmitted through a given window compared with that of a standard $^{1}/_{8}$ inch thick single pane of glass under the same conditions.

SHAFT Round bar that rotates or provides an axis of revolution; vertical or steeply inclined tunnel.

SHAFT (MORNING GLORY) SPILLWAY Vertical or inclined shaft into which flood water spills and then is conducted through, under, or around a dam by means of a conduit or tunnel.

SHAFT HOUSE House at top of the access shaft to gate chamber that secures access to the gates.

SHALE Rock formed of consolidated mud; fine-grained sedimentary rock formed by the consolidation of clay, silt, or mud.

SHALLOW FOG Fog in which the visibility at 6 feet above ground level is 5/8ths statute mile or more and the apparent visibility in the fog layer is less than 5/8ths statute mile.

SHALLOWS Shallow place or area in a body of water.

SHAPE FACTOR See PARTICLE SHAPE FACTOR.

SHARPS Hypodermic needles, syringes (with or without the attached needle), Pasteur pipettes, scalpel blades, blood vials, needles with attached tubing, and culture dishes used in animal or human patient care or treatment, or in medical, research or industrial laboratories.

SHARS Subtle Heavy Rainfall Signature.

SHEAR Variation in wind speed (speed shear) and/or direction (directional shear) over a short distance within the atmosphere; structural break where differential movement has occurred along a surface or zone of failure.

SHEAR INTENSITY Dimensionless number that is taken from Einstein's bed load function; inverse of Shield's parameter.

SHEAR STRENGTH Maximum resistance of a soil or rock to shearing stresses.

SHEAR STRESS Frictional force per unit of bed area exerted on the bed by the flowing water.

SHEAR WALL Vertical lateral-force-resisting element in a structure designed to resist wind or earthquake generated lateral forces.

SHEAR ZONE Area where the rock mass has moved along the plane of contact which often becomes a channel for ground water.

SHEEPSFOOT ROLLER Tamping roller having lugs with feet extending at their outer tips.

SHEET (OVERLAND) FLOW Flow that occurs overland in places where there are no defined channels, the flood water spreads out over a large area at a uniform depth.

SHEET ICE Ice formed by the freezing of liquid precipitation or the freezing of melted solid precipitation.

SHEET PILING Steel strips shaped to interlock with each other when driven into the ground.

SHEET WASH Flow of rainwater that covers the entire ground surface with a thin film and is not concentrated into streams.

SHEFPARS Software decoder for SHEF Data.

SHELF CLOUD Low, horizontal wedge-shaped arcus cloud, associated with a thunderstorm gust front.

SHELL See SHOULDER.

SHELTERBELT Natural or planned barrier of trees or shrubs to reduce erosion and provide shelter from winds or storms.

SHELTERING (IN-PLACE) Protective action that involves taking cover in upper levels of a building that is able to withstand high flood levels.

SHELVES Streambank features extending nearly horizontally from the flood plain to the lower limit of persistent woody vegetation.

SHFT Shift.

SHIELD'S DETERMINISTIC CURVE Curve of the dimensionless tractive force plotted against the grain Reynolds number that is used to help determine the critical tractive force.

SHIELD'S PARAMETER Dimensionless number referred to as a dimensionless shear stress where the beginning of motion of bed material is a function of this dimensionless number.

SHIFTING CONTROL See CONTROL.

SHLW Shallow.

SHOAL Relatively shallow place in a stream, lake, or sea.

SHOCK LOAD Arrival at a water treatment plant of raw water containing unusual amounts of algae, colloidal matter, color, suspended solids, turbidity, or other pollutants.

SHORE ICE Ice sheet in the form of a long border attached to the bank or shore; border ice.

SHORING Temporary bracing to hold the sides of an excavation from caving.

SHORT TERM FORECAST Product used to convey information regarding weather or hydrologic events in the next few hours.

SHORT WAVE FADE (SWF) Particular ionospheric solar flare effect under the broad category of sudden ionospheric disturbances (SIDs) whereby short-wavelength radio transmissions, VLF, through HF, are absorbed for a period of minutes to hours.

SHORT-CIRCUITING When some of the water in tanks or basins flows faster than the rest; may result in shorter contact, reaction, or settling times than calculated or presumed.

SHORT-FUSE WARNING Warning issued by the National Weather Service for a local weather hazard of relatively short duration.

SHORT-THROATED FLUMES Flumes that control and produce curvilinear flow (e.g. Parshall flume).

SHORTWAVE (S/W, S/WV, SHRTWV) Disturbance in the mid or upper part of the atmosphere which induces upward motion ahead of it.

SHORTWAVE RADIATION Radiant energy emitted by the Sun in the visible and near-ultraviolet wavelengths (between about 0.1 and 2 micrometers).

SHORTWAVE TROUGH See SHORTWAVE.

SHORT-WAVE TROUGH (METEORLOGICAL) Wave of low atmospheric pressure in the form of a trough that has a wave length of 600 to 1,500 miles and moves progressively through the lower troposphere in the same direction as that of the prevailing current of air motion.

SHOULDER (SHELL) Upstream and downstream parts of the cross section of an embankment dam on each side of the core or core wall.

SHOWALTER INDEX (SWI) Stability index used to determine thunderstorm potential.

SHOWER (SH) Precipitation characterized by the suddenness with which they start and stop, by the rapid changes of intensity, and usually by rapid changes in the appearance of the sky.

SHRA Rain showers.

SHRAS Showers.

SHRINKAGE Loss of bulk of soil when compacted in a fill.

SHRINKAGE INDEX (SI) Numerical difference between the plastic and shrinkage limits.

SHRINKAGE LIMIT (SL) Maximum water content at which a reduction in water content will not cause a decrease in volume of the soil mass.

SHRINKAGE RATIO (R) Ratio of a given volume change, expressed as a percentage of the dry volume, to the corresponding change in moisture content above the shrinkage limit, expressed as a percentage of the weight of the oven-dried soil.

SHRT Short.

SHRUBLAND Land covered predominantly with shrubs.

SHSN Snow showers.

SHUT-OFF DEVICE Valve coupled with a control device which closes the valve when a set pressure or flow value is exceeded.

SHWR Shower.

SICK BUILDING SYNDROME Building whose occupants experience acute health and/or comfort effects that appear to be linked to time spent therein, but where no specific illness or cause can be identified.

SIDE CHANNEL SPILLWAY Spillway whose crest is roughly parallel to the channel immediately downstream of the spillway.

SIDELOBE Secondary energy maximum located outside the main radar beam.

SIDESLOPE GRADIENT Representative change in elevation in a given horizontal distance perpendicular to a stream; valley slope along a line perpendicular to the stream; layers of rock.

SIEVE DIAMETER Smallest standard sieve opening size through which a given particle of sediment will pass.

SIGMET Significant Meteorological Advisory.

SIGNAL Volume or product-level change produced by a leak in a tank.

SIGNAL WORDS Words used on a pesticide label (e.g., Danger, Warning, Caution) to indicate level of toxicity.

SIGNAL-TO-NOISE RATIO Ratio that measures the comprehensibility of data, usually expressed as the signal power divided by the noise power, usually expressed in decibels (dB).

SIGNIFICANT DETERIORATION Pollution resulting from a new source in previously "clean" areas.

SIGNIFICANT HAZARD Hazard classification in which lives are in jeopardy and appreciable economic loss (rural area with notable agriculture, industry, or worksites, or outstanding natural resources) would occur as a result of failure.

SIGNIFICANT INDUSTRIAL USER (SIU) Indirect discharger that is the focus of control efforts under the national pretreatment program.

SIGNIFICANT MUNICIPAL FACILITIES Those publicly owned sewage treatment plants that discharge a million gallons per day or more and are therefore considered by states to have the potential to substantially affect the quality of receiving waters.

SIGNIFICANT NON-COMPLIANCE See SIGNIFICANT VIOLATIONS.

SIGNIFICANT POTENTIAL SOURCE OF CONTAMINATION Facility or activity that stores, uses, or produces compounds with potential for significant contaminating impact if released into the source water of a public water supply.

SIGNIFICANT VIOLATIONS Violations by point source dischargers of sufficient magnitude or duration to be a regulatory priority.

SIGNIFICANT WAVE HEIGHT (H_s or WVHT) Average height of the highest one-third of the waves during a 20 minute sampling period.

SIGNIFICANT WEATHER OUTLOOK Narrative statement issued by the National Weather Service to provide information regarding the potential of significant weather expected during the next 1 to 5 days.

SIGWX Significant Weather.

SILICICLASTIC ROCKS Rocks such as shale and sandstone that are formed by the compaction and cementation of quartz-rich mineral grains.

SILL Submerged structure across a river to control the water level upstream; crest of a spillway; horizontal gate seating, made of wood, stone, concrete, or metal at the invert of any opening or gap in a structure.

SILT (ROCK FLOUR) Sedimentary materials composed of fine or intermediate-sized mineral particles; heavy soil intermediate between clay and sand; fine-grained portion of soil that is non-plastic or very slightly plastic and that exhibits little or no strength when air dry; non-plastic soil which passes a No. 200 United States Standard sieve; soil composed of particles between $1/256$ mm and $1/16$ mm in diameter.

SILTATION Deposition or accumulation of silt (or small-grained material) in a body of water.

SILTING Filling with soil or mud deposited by water.

SILTSTONE Fine-grained sedimentary rock composed mainly of silt-sized particles; indurated silt having the texture and composition of shale but lacking its fine lamination.

SILVICULTURE Management of forest land for timber; cultivation of forest trees.

SIMPLEX TELETYPE OVER RADIO (SITOR) Means of transmitting text broadcasts over radio.

SIMULATE To express a physical system in mathematical terms.

SIMULATION One complete execution of a modeling computer program, including input and output.

SIMULID Group of two-winged flying insects who live their larval stage underwater and emerge to fly about as adults.

SINGLE CELL THUNDERSTORM Thunderstorm that develops in weak vertical wind shear environments.

SINGLE-BREATH CANISTER Small one-liter canister designed to capture a single breath.

SINGLE-PASS COOLING To remove heat load, water is circulated once through a piece of equipment and then disposed down the drain.

SINGLE-STAGE PUMP Pump that has only one impeller.

SINK Depression in the land surface, especially one having a central playa or saline lake with no outlet; place in the environment where a compound or material collects.

SINKHOLE Depression in the Earth's surface caused by dissolving of underlying limestone, salt, or gypsum; steep-sided depression formed when removal of subsurface embankment or foundation material causes overlying material to collapse into the resulting void.

SINKING Controlling oil spills by using an agent to trap the oil and sink it to the bottom of the body of water where the agent and the oil are biodegraded.

SINUOSITY Ratio of the length of a river's thalweg to the length of the valley proper; ratio of the channel length between two points on a channel to the straight-line distance between the same two points; measure of a river's meandering.

SIP CALL EPA action requiring a state to resubmit all or part of its State Implementation Plan to demonstrate attainment of the require national ambient air quality standards within the statutory deadline.

SIPHON Pipe connecting two canals; tube or pipe through which water flows over a high point by gravity.

SIPHON SPILLWAY Spillway with one or more siphons built at crest level.

SIPHON TUBE Relatively short, light-weight, curved tube used to convey water over ditch banks to irrigate furrows or borders.

SIPHONAGE Partial vacuum created by the flow of liquids in pipes.

SITE Any location of past human activity; area or place within jurisdiction.

SITE ASSESSMENT PROGRAM Means of evaluating hazardous waste sites through preliminary assessments and site inspections to develop a Hazard Ranking System score.

SITE CHARACTERIZATION General term applied to the investigation activities at a specific location that examines natural phenomena and human-induced conditions important to the resolution of environmental, safety and water resources issues; means the program of exploration and research, both in the laboratory and in the field, undertaken to establish the geologic conditions are the ranges of those parameters of a particular site relevant to the program.

SITE CLOSURE Point/time, as determined by the Director, at which the owner or operator of the site has completed their post-injection site care responsibilities.

SITE INSPECTION Collection of information from a site to determine the extent and severity of hazards posed by the site.

SITE SAFETY PLAN Crucial element in all removal actions, it includes information on equipment being used, precautions to be taken, and steps to take in the event of an on-site emergency.

SITE SPECIFIC HYDROLOGIC PREDICTION SYSTEM (SSHP) WFO hydrologic forecast model for small rivers and streams that uses RFC soil moisture state variables, stage and precipitation data.

SITE-SPECIFIC Term used to convey the fact that the report is for an individual location as opposed to a general area.

SITE-SPECIFIC AQUATIC LIFE CRITERION Water quality criterion for aquatic life that has been derived to be specifically appropriate to the water quality characteristics and/or species composition at a particular location.

SITING Process of choosing a location for a facility.

SKEWED On a horizontal angle, or in an oblique course or direction.

SKEWNESS Numerical measure of the lack of symmetry of an asymmetrical frequency distribution.

SKIMMER GATE Gate at the spillway crest whose prime purpose is to control the release of debris and logs with a limited amount of water.

SKIMMING Using a machine to remove oil or scum from the surface of the water; diversion of water from a stream or conduit by a shallow overflow used to avoid diversion of sand, silt, or other debris carried as bottom load.

SKIP Non-digging bucket or tray that hoists material.

SKIVING To dig in thin layers.

SKY CONDITION Used in a forecast to describes the predominant/average sky condition based upon octants (eighths) of the sky covered by opaque (not transparent) clouds defined as Sky Condition - Cloud Coverage: Clear / Sunny - $^{0}/_{8}$, Mostly Clear / Mostly Sunny - $^{1}/_{8}$ to $^{2}/_{8}$, Partly Cloudy / Partly Sunny - $^{3}/_{8}$ to $^{4}/_{8}$, Mostly Cloudy / Considerable Cloudiness - $^{5}/_{8}$ to $^{7}/_{8}$, Cloudy - $^{8}/_{8}$, Fair (mainly for night) - Less than $^{4}/_{10}$ opaque clouds (no precipitation, no extremes of visibility/temperature/wind).

SKYWARN Nationwide network of volunteer weather spotters who report to and are trained by the National Weather Service.

SL Sea Level.

SLAB Deck or floor of a concrete bridge.

SLAB AND BUTTRESS DAM Buttress dam with buttresses which support the flat slab of reinforced concrete which forms the upstream face.

SLAKING Process of breaking up or sloughing when an indurated soil is immersed in water.

SLATE Fine-grained metamorphic rock formed by "baking" and recrystallizing shale or mudstone and which splits easily along flat, parallel planes.

SLD Solid.

SLEET Precipitation that consists of clear pellets of ice; pellets of ice composed of frozen or mostly frozen raindrops or refrozen partially melted snowflakes.

SLEET WARNING Issued when accumulation of sleet in excess of $1/2''$ is expected.

SLGT Slight.

SLIDE Small landslide.

SLIDE GATE (SLUICE GATE) Gate that can be opened or closed by sliding in supporting guides; steel gate that upon opening or closing slides on its bearings in edge guide slots.

SLIGHT CHANCE In probability of precipitation statements, usually equivalent to a 20 percent chance.

SLIGHT RISK OF SEVERE THUNDERSTORMS Severe thunderstorms are expected to affect between 2 and 5 percent of the area.

SLING PSYCHROMETER Instrument used to measure the water vapor content of the atmosphere in which wet and dry bulb thermometers are mounted on a frame connected to a handle at one end by means of a bearing or a length of chain.

SLO Slow.

SLOPE Inclination from the horizontal; inclined surface usually defined by the ratio of the horizontal distance to the vertical distance; change in elevation per unit of horizontal distance; side of a hill or a mountain; inclined face of a cut, canal, or embankment; inclination from the horizontal.

SLOPE PROTECTION Protection of a slope against wave action or erosion.

SLOSH (SEA, LAKE AND OVERLAND SURGES FROM HURRICANES) Computer model run by the National Hurricane Center (NHC) to estimate storm surge heights resulting from historical, hypothetical, or predicted hurricanes by taking into account pressure, size, forward speed, track, and winds.

SLOUGH Movement of a soil mass downward along a slope because of a slope angle too great to support the soil, wetness reducing internal friction among particles, or seismic activity; small marshy tract lying in a swale or other local shallow, undrained depression; sluggish creek or channel in a wetland.

SLOW SAND FILTRATION Passage of raw water through a bed of sand at low velocity, resulting in substantial removal of chemical and biological contaminants.

SLP Sea Level Pressure.

SLUDGE Semi-solid residue from any of a number of air or water treatment processes.

SLUDGE DIGESTER Tank in which complex organic substances like sewage sludges are biologically dredged.

SLUG Large initial amount; unit of mass which will undergo an acceleration of 1 foot per second squared when a force of 1 pound is applied to it.

SLUICE Opening for releasing water from below the static head elevation.

SLUICE GATE Gate that can be opened or closed by sliding in supporting guides.

SLUICEWAY Opening in a diversion dam used to discharge heavy floating debris safely past the dam.

SLUICING Method of compacting soil where the soil is washed into placed with a high velocity stream of water.

SLURRY Watery mixture of insoluble matter which is pumped beneath a dam to form an impervious barrier; watery mixture of insoluble matter resulting from some pollution control techniques; cement grout.

SLURRY TRENCH Narrow excavation whose sides are supported by a mud slurry filling the excavation.

SLY Southerly.

SM Statute Miles; sum total for month.

SMA Soil Moisture Accounting Model.

SMALL CONCENTRATED ANIMAL FEEDING OPERATION (SMALL CAFO) Animal feeding operation that is designated as a CAFO and is not a Medium CAFO.

SMALL CONSTRUCTION ACTIVITY Clearing, grading, and excavating resulting in a land disturbance that will disturb equal to or more than one acre and fewer than five acres of total land area but is part of a larger common plan of development or sale that will ultimately disturb equal to or fewer than five acres.

SMALL CRAFT ADVISORY (SCA) Advisory issued by coastal and Great Lakes Weather Forecast Offices (WFO) for areas included in the Coastal Waters Forecast or Nearshore Marine Forecast (NSH) products where any vessel that may be adversely affected.

SMALL CRAFT ADVISORY FOR HAZARDOUS SEAS (SCAHS) Advisory for wind speeds lower than small craft advisory criteria, yet waves or seas are potentially hazardous due to wave height, wave period, steepness, or swell direction.

SMALL CRAFT ADVISORY FOR ROUGH BAR (SCARB) Advisory for specialized areas near harbor or river entrances known as bars.

SMALL CRAFT ADVISORY FOR WINDS (SCAW) Advisory for wave heights lower than small craft advisory criteria, yet wind speeds are potentially hazardous.

SMALL CRAFT SHOULD EXERCISE CAUTION Precautionary statement issued to alert mariners with small, weather sensitive boats.

SMALL HAIL Snow pellets or graupel.

SMALL POWER PRODUCER (SPP) Facility (or small power producer) generates electricity using waste, renewable (water, wind and solar), or geothermal energy as a primary energy source.

SMALL QUANTITY GENERATOR (SQG) Persons or enterprises that produce 220-2200 pounds per month of hazardous waste.

SMALL STREAM FLOODING Flooding of small creeks, streams, or runs.

SMELTER Facility that melts or fuses ore, often with an accompanying chemical change, to separate its metal content.

SMOG Mixture of smoke and fog; air pollution typically associated with oxidants; air that has restricted visibility due to pollution or pollution formed in the presence of sunlight (e.g. photochemical smog).

SMOKE (K) Particles suspended in air after incomplete combustion.

SMOKE DISPERSAL Ability of the atmosphere to ventilate smoke.

SMOKE MANAGEMENT Use of meteorology, fuel moisture, fuel loading, fire suppression and burn techniques to keep smoke impacts from prescribed fires within acceptable limits.

SMOLTS Adolescent salmon 3 to 7 inches long.

SMOOTH RESPONSE SPECTRUM Response spectrum devoid of sharp peaks and valleys that specifies the amplitude of the spectral acceleration, velocity, and/or displacement to be used in the analyses of the structure.

SMOOTHED SUNSPOT NUMBER Average of 13 monthly RI numbers, centered on the month of concern.

SMPDBK Simplified Dam Break (DAMBRK) Model.

SNOW Precipitation that consists of frozen flakes formed when water vapor accumulates on ice crystals, going directly to the ice phase.

SNOW ACCUMULATION AND ABLATION MODEL Model which simulates snow pack accumulation, heat exchange at the air-snow interface, areal extent of snow cover, heat storage within the snow pack, liquid water retention, and transmission and heat exchange at the ground-snow interface.

SNOW ADVISORY Issued by the National Weather Service when a low pressure system produces snow that may cause significant inconveniences, but do not meet warning criteria and if caution is not exercised could lead to life threatening situations.

SNOW CORE Sample of either freshly fallen snow, or the combined old and new snow on the ground obtained by pushing a cylinder down through the snow layer and extracting it.

SNOW CORNICE Mass of snow or ice projecting over a mountain ridge.

SNOW COURSE Line or series of connecting lines along which snow samples are taken at regularly spaced points.

SNOW DENSITY Mass of snow per unit volume which is equal to the water content of the snow divided by its depth.

SNOW DEPTH Combined total depth of both the old and new snow on the ground.

SNOW FLURRIES Intermittent light snowfall of short duration with no measurable accumulation.

SNOW GRAINS Precipitation consisting of white, opaque ice particles usually less than 1 mm in diameter.

SNOW PACK Combined layers of snow and ice on the ground at any one time.

SNOW PELLETS Precipitation, usually of brief duration, consisting of crisp, white, opaque ice particles, round or conical in shape and about 2 to 5 mm in diameter.

SNOW PILLOW Instrument used to measure snow water equivalents; window of snow deposited in the immediate lee of a snow fence or ridge.

SNOW SHOWER Short duration of moderate snowfall where some accumulation is possible.

SNOW SQUALL Intense, but limited duration, period of moderate to heavy snowfall, accompanied by strong, gusty surface winds and possibly lightning (generally moderate to heavy snow showers) where snow accumulation may be significant.

SNOW STAKE 1-3/4 inch square, semi-permanent stake, marked in inch increments to measure snow depth.

SNOW STICK Portable rod used to measure snow depth.

SNOW TELEMETRY (SNOTEL) Automated network of snowpack data collection sites.

SNOW WATER EQUIVALENT Water content obtained from melting accumulated snow.

Snow Water Equivalent (SWE) Amount of water content in a snowpack or snowfall.

SNOWBOARD Flat, solid, white material, approximately two feet square, which is laid on the ground, or snow surface by weather observers to obtain more accurate measurements of snowfall and water content.

SNOWCOVER Combined layers of snow and ice on the ground at any one time.

SNOWFLAKE Agglomeration of snow crystals falling as a unit.

SNOWLINE General altitude to which the continuous snow cover of high mountains retreats in summer, chiefly controlled by the depth of the winter snowfall and by the temperature of the summer.

SNOWLINE, TEMPORARY Line sometimes drawn on a weather map during the winter showing the southern limit of the snow cover.

SNOWMELT FLOODING Flooding caused primarily by the melting of snow.

SNOWPACK Total snow and ice on the ground, including the new snow, the previous snow and ice which has not melted.

SNR Signal-to-Noise Ratio.

SNW Snow.

SNWFL Snowfall.

SOCIAL VALUE (PSYCHOLOGICAL VALUE) Concept that the existence of wilderness provides a condition that could allow an individual to achieve control over stressful conditions, thus contributing to the psychological health of many off-site users.

SODIC SOIL Contains sufficient exchangeable sodium to interfere with the growth of most crop plants.

SODICITY Exchangeable-sodium content of the soil; high sodium content.

SODIUM ADSORPTION RATIO (SAR) Expression of relative activity of sodium ions in exchange reactions within soil and is an index of sodium or alkali hazard to the soil.

SODIUM HAZARD IN WATER Index that can be used to evaluate the suitability of water for irrigating crops.

SOFT DETERGENTS Cleaning agents that break down in nature.

SOFT WATER Any water that does not contain a significant amount of dissolved minerals such as salts of calcium or magnesium.

SOIL Sediments or other unconsolidated accumulations of solid particles produced by the physical and chemical disintegration of rocks, and which may or may not contain organic matter; components may consist of clay, silt, sand, or gravel; loose surface material of the Earth's crust; layer of material at the land surface that supports plant growth.

SOIL ADSORPTION FIELD Subsurface area containing a trench or bed with clean stones and a system of piping through which treated sewage may seep into the surrounding soil for further treatment and disposal.

SOIL AND WATER CONSERVATION PRACTICES Control measures consisting of managerial, vegetative, and structural practices to reduce the loss of soil and water.

SOIL BULK DENSITY Mass of dry soil per unit bulk soil.

SOIL CEMENT Mixture of water, cement, and natural soil, usually processed in a tumble and mixed to a specific consistency, then placed in lifts and rolled to compact to provide slope protection; mixture of Portland cement and pulverized soil placed in layers on the upstream face of a dam to provide slope protection; tightly compacted mixture of pulverized soil, Portland cement, and water that, as the cement hydrates, forms a hard, durable, low-cost paving material.

SOIL CLASSIFICATION Systematic arrangement of soils into classes of one or more categories or levels of classification for a specific objective where broad groupings are made on the basis of general characteristics and subdivisions are made on the basis of more detailed differences in specific properties.

SOIL CONDITIONER Organic material like humus or compost that helps soil absorb water, build a bacterial community, and take up mineral nutrients.

SOIL CONSERVATION Protection of soil against physical loss by erosion and chemical deterioration by the application of management and land-use methods that safeguard the soil against all natural and human-induced factors.

SOIL CONSERVATION SERVICE Former name of a branch of the United States Department of Agriculture, renamed the Natural Resources Conservation Service (NRCS) that has responsibilities in soil and water conservation, and flood prevention.

SOIL ERODIBILITY Indicator of a soil's susceptibility to raindrop impact, runoff, and other erosive processes.

SOIL GAS Gaseous elements and compounds in the small spaces between particles of the earth and soil.

SOIL HEAT FLUX (SOIL HEATFLUX DENSITY) Amount of heat energy that moves by conduction across a specified cross-sectional area of soil per unit time and goes to heating (or cooling) the soil.

SOIL HORIZON Layer of soil that is distinguishable from adjacent layers by characteristic physical and chemical properties.

SOIL MOISTURE (SOIL WATER) Water stored in soils; water contained in the upper regions near the Earth's surface; water contained in the pore space of the unsaturated zone; upper part of the zone of aeration from which water is discharged by the transpiration of plants or by soil evaporation.

SOIL MOISTURE ACCOUNTING (SMA) Modeling process that accounts for continuous fluxes to and from the soil profile where models can be event-based or continuous.

SOIL PROFILE Uppermost layers of the ground down to bedrock; portion of the ground subject to infiltration, evaporation and percolation fluxes.

SOIL STERILANT Chemical that temporarily or permanently prevents the growth of all plants and animals.

SOIL SUBSIDENCE Lowering of the normal level of the ground, usually due to over pumping of water from wells.

SOIL-CEMENT BEDDING Mixture of soil, portland cement, and water placed for pipe bedding.

SOIL-WATER CONTENT Water lost from the soil upon drying to constant mass at 105 °C that is expressed either as mass of water per unit mass of dry soil or as the volume of water per unit bulk volume of soil.

SOLAR COORDINATES Angular distance in solar longitude measured from the central meridian.

SOLAR CYCLE Approximately 11-year quasi-periodic variation in frequency or number of solar active events.

SOLAR MAXIMUM Month(s) during the solar cycle when the 12-month mean of monthly average sunspot numbers reaches a maximum.

SOLAR MINIMUM Month(s) during the solar cycle when the 12-month mean of monthly average sunspot numbers reaches a minimum.

SOLAR NOON Time of day at which the Sun is the highest in the sky.

SOLAR SECTOR BOUNDARY (SSB) Apparent solar origin, or base, of the interplanetary sector boundary marked by the larger-scale polarity inversion lines.

SOLAR WIND Outward flux of solar particles and magnetic fields from the Sun where velocities are typically near 350 km/s.

SOLDER Metallic compound used to seal joints between pipes.

SOLE SOURCE AQUIFER Aquifer that supplies 50 percent or more of the drinking water to a particular human population (e.g. area).

SOLID HEAD BUTTRESS DAM Buttress dam in which the upstream end of each buttress is enlarged to span the gap between buttresses.

SOLID WASTE Non-liquid, non-soluble materials ranging from municipal garbage to industrial wastes (e.g., sewage sludge, agricultural refuse, demolition wastes, and mining residues) that contain complex and sometimes hazardous substances.

SOLID WASTE DISPOSAL Final placement of refuse that is not salvaged or recycled.

SOLID WASTE MANAGEMENT Supervised handling of waste materials from their source through recovery processes to disposal.

SOLIDIFICATION AND STABILIZATION Removal of wastewater from a waste or changing it chemically to make it less permeable and susceptible to transport by water.

SOLID-PHASE EXTRACTION Procedure to isolate specific organic compounds onto a bonded silica extraction column.

SOLSTICE Either of the two times per year when the Sun is at its greatest angular distance from the celestial equator: about June 21 (the Northern Hemisphere summer solstice), when the Sun reaches its northernmost point on the celestial sphere, or about December 22 (the Northern Hemisphere winter solstice), when it reaches its southernmost point.

SOLUBILITY Amount of mass of a compound that will dissolve in a unit volume of solution; total amount of solute species that will remain indefinitely in a solution maintained at constant temperature and pressure in contact with the solid crystals from which the solutes were derived.

SOLUBILITY IN WATER Indicator of the amount of a chemical that can be dissolved in water, shown as a percentage or as a description.

SOLUTE Substance that is dissolved in another substance, thus forming a solution.

SOLUTE CONCENTRATION Concentration of a chemical species dissolved in groundwater.

SOLUTE TRANSPORT MODEL Application of a model to represent the movement of chemical species dissolved in groundwater.

SOLUTION Mixture of a solvent and a solute; formed when a solid, gas, or another liquid in contact with a liquid becomes dispersed homogeneously throughout the liquid.

SOLUTION MINING WELLS Class V wells that inject leaching solutions (lixiviants) to remove mineral ores from their original geological settings.

SOLVENT Substance that dissolves other substances, thus forming a solution (e.g. water is known as the "universal solvent.").

SOO Science and Operations Officer.

SOOT Carbon dust formed by incomplete combustion.

SORB To take up and hold either by absorption or adsorption.

SORPTION Process of absorption and adsorption; interaction (binding or association) of a solute ion or molecule with a solid; all processes which remove solutes from the fluid phase and concentrate them on the solid phase of the medium.

SORTING Dynamic process by which sedimentary particles having some particular characteristic (such as similarity of size, shape, or specific gravity) are naturally selected and separated from associated but dissimilar particles by the agents of transportation.

SOUNDING Set of data measuring the vertical structure of a parameter (depth, temperature, humidity, pressure, winds, etc.) at a given time.

SOURCE Process, or a feature from which, water, vapor NAPL, solute or heat is added to the groundwater or vadose zone flow system.

SOURCE AREA Location of liquid hydrocarbons or the zone of highest soil or groundwater concentrations, or both, of the chemical of concern.

SOURCE CHARACTERIZATION MEASUREMENTS Measurements made to estimate the rate of release of pollutants into the environment from a source such as an incinerator, landfill, etc.

SOURCE LOADING Rate at which a contaminant is entering the groundwater system at a specific source.

SOURCE OF CONTAMINANTS Physical location (and spatial extent) of the source contaminating the aquifer.

SOURCE REDUCTION Reducing the amount of materials entering the waste stream from a specific source by redesigning products or patterns of production or consumption.

SOURCE ROCKS Rocks from which fragments and other detached pieces have been derived to form a different rock.

SOURCE SEPARATION Segregating various wastes at the point of generation.

SOURCE WATER Water in its natural state, prior to any treatment for drinking.

SOURCE-WATER PROTECTION AREA Area delineated by a state for a Public Water Supply or including numerous such suppliers, whether the source is groundwater or surface water or both.

SOUTHERN OSCILLATION (SO) "See-saw" in surface pressure in the tropical Pacific characterized by simultaneously opposite sea level pressure anomalies at Tahiti, in the eastern tropical Pacific and Darwin, on the northwest coast of Australia.

SPACE ENVIRONMENT CENTER (SEC) Provides real-time monitoring and forecasting of solar and geophysical events, conducts research in solar-terrestrial physics, and develops techniques for forecasting solar and geophysical disturbances.

SPACING In blasting, the distance between holes in a row.

SPALLING (SPALL) Loss of surface concrete usually caused by impact, abrasion, or compression; to break off from a surface in sheets or pieces.

SPARGE Injection of air below the water table to strip dissolved volatile organic compounds and/or oxygenate groundwater to facilitate aerobic biodegradation of organic compounds.

SPATIAL CONCENTRATION Dry weight of sediment per unit volume of water-sediment mixture in place; ratio of the dry weight of sediment to the total weight of water-sediment mixture in a sample; unit volume of the mixture.

SPAWN To lay eggs (i.e. aquatic animals).

SPAWNING BEDS Places in which eggs of aquatic animals lodge or are placed during or after fertilization.

SPC Storm Prediction Center.

SPCLY Especially.

SPEARHEAD ECHO Radar echo associated with a downburst with a pointed appendage extending toward the direction of the echo motion.

SPECIAL AVALANCHE WARNING
Issued by the National Weather Service when avalanches are imminent or occurring in the mountains.

SPECIAL DRAINAGE WELLS
Class V wells that include potable water tank overflow, construction dewatering, swimming pool drainage, and mine dewatering wells; receive fluids that cannot be classified as agricultural, industrial, or storm water.

SPECIAL EXAMINATION Field review performed on a high- or significant-hazard dam to address an identified visible dam safety deficiency or to investigate significant changes in operating or loading conditions.

SPECIAL FEATURES Area containing ecological, geological, or other features of scientific, educational, scenic, or historical value.

SPECIAL LOCAL-NEEDS REGISTRATION
Registration of a pesticide product by a state agency for a specific use that is not federally registered.

SPECIAL MARINE WARNING (SMW)
Issued for potentially hazardous weather conditions usually of short duration (up to 2 hours) producing sustained marine thunderstorm winds or associated gusts of 34 knots or greater; and/or hail 3/4 inch or more in diameter; and/or waterspouts affecting areas included in a Coastal Waters Forecast, a Nearshore Marine Forecast, or an Great Lakes Open Lakes Forecast that is not adequately covered by existing marine warnings; used for short duration mesoscale events such as a strong cold front, gravity wave, squall line, etc., lasting less than 2 hours and producing winds or gusts of 34 knots or greater.

SPECIAL POPULATIONS Those individuals or groups that may be institutionalized and have needs that require special consideration in emergencies.

SPECIAL REVIEW Regulatory process through which existing pesticides suspected of posing unreasonable risks to human health, non-target organisms, or the environment are referred for review by EPA.

SPECIAL TROPICAL DISTURBANCE STATEMENT Issued by the National Hurricane Center furnishes information on strong and formative non-depression systems.

SPECIAL WASTE Items such as household hazardous waste, bulky wastes (refrigerators, pieces of furniture, etc.) tires, and used oil.

SPECIES Basic category of biological classification intended to designate a single kind of animal or plant; populations of organisms that may interbreed and produce fertile offspring having similar structure, habits, functions and usually designated by a common name.

SPECIES (TAXA) RICHNESS Number of species (taxa) present in a defined area or sampling unit.

SPECIES DIVERSITY Ecological concept that incorporates both the number of species in a particular sampling area and the evenness with which individuals are distributed among the various species.

SPECIFIC AREA MESSAGE ENCODING (SAME) Tone alert system which allows NOAA Weather Radio receivers equipped with the SAME feature to sound an alert for only certain weather conditions or within a limited geographic area such as a county.

SPECIFIC CAPACITY Yield of a well per unit of drawdown; rate of discharge from a well divided by the drawdown of the water level within the well at a specific time since pumping started.

SPECIFIC CONDUCTANCE Rapid method of estimating the dissolved solid content of a water supply by testing its ability to conduct an electrical current.

SPECIFIC DISCHARGE Rate of discharge of groundwater per unit area of a porous medium measured at right angle to the direction of groundwater flow.

SPECIFIC GRAVITY Ratio of the mass of a body to an equal volume of water; comparison of the weight of the chemical to the weight of an equal volume of water; ratio of the density of any substance to the density of water.

SPECIFIC HUMIDITY Ratio of the mass of water vapor to the total mass of the system.

SPECIFIC STORAGE Volume of water released from or taken into storage per unit volume of the porous medium per unit change in head.

SPECIFIC WEIGHT Weight per unit volume.

SPECIFIC YIELD Ratio of the volume of water that will drain freely from the material to the total volume of that material (e.g., saturated rock, aquifer formation); always less than porosity value.

SPECIFIED CONCENTRATION BOUNDARY (CONSTANT CONCENTRATION) Boundary at which the solute concentration is specified.

SPECIFIED FLUX BOUNDARY Model boundary condition in which the groundwater flux or mass flux is specified.

SPECTRAL DENSITY Radar term for the distribution of power by frequency.

SPECTRAL WAVE DENSITY Energy for each frequency bin.

SPECTRAL WAVE DIRECTION Mean wave direction, in degrees from true North, for each frequency bin.

SPECTRUM INTENSITY Integral of the pseudo-velocity response spectrum taken over the range of significant structural vibration periods of the structure being analyzed.

SPECTRUM WIDTH WSR-88D radar product depicts a full 360 degree sweep of spectrum width data indicating a measure of velocity dispersion within the radar sample volume.

SPECTRUM WIDTH CROSS SECTION WSR-88D radar product displays a vertical cross section of spectrum width on a grid with heights up to 70,000 feet on the vertical axis and distance up to 124 nm on the horizontal axis.

SPEED SHEAR Component of wind shear which is due to a change in wind speed with height.

SPENES NESDIS Satellite Precipitation Estimates.

SPENT BRINE RETURN FLOW WELLS Class V wells that dispose of the spent brine that results from the extraction of minerals, halogens, and other compounds from fluids.

SPHERE CALIBRATION Reflectivity calibration of a radar by pointing the dish at a metal sphere of (theoretically) known reflectivity.

SPIGOT Plain end of a cast-iron pipe.

SPILE (FOREPOLE) Plank driven ahead of a tunnel face for roof support.

SPILL PREVENTION CONTROL AND COUNTERMEASURE PLAN (SPCC) Plan prepared by a facility to minimize the likelihood of a spill and to expedite control and cleanup activities should a spill occur.

SPILL PREVENTION, CONTAINMENT, AND COUNTERMEASURES PLAN (SPCP) Plan covering the release of hazardous substances as defined in the Clean Water Act.

SPILLS Water releases that cannot be put to use for project purposes.

SPILLWAY Structure that passes normal and/or flood flows in a manner that protects the structural integrity of the dam; structure over or through which excess or flood flows are discharged where a controlled spillway's flow is controlled by gates and an uncontrolled spillway on control is the elevation of the spillway crest.

SPILLWAY CAPACITY Maximum spillway outflow that a dam can safely pass with the reservoir at its maximum level.

SPILLWAY CHANNEL Open channel or closed conduit conveying water from the spillway inlet downstream.

SPILLWAY CHUTE Steeply sloping spillway channel that conveys discharges at super-critical velocities.

SPILLWAY CREST Lowest level at which water can flow over or through a spillway; elevation of the highest point of a spillway.

SPILLWAY, AUXILIARY Any secondary spillway that is designed to be operated infrequently, possibly in anticipation of some degree of structural damage or erosion to the spillway that would occur during operation.

SPILLWAY, EMERGENCY See SPILLWAY, AUXILIARY.

SPILLWAY, FUSE PLUG Form of auxiliary spillway consisting of a low embankment designed to be overtopped and washed away during an exceptionally large flood.

SPILLWAY, SERVICE Spillway that is designed to provide continuous or frequent regulated or unregulated releases from a reservoir, without significant damage to either the dam or its appurtenant structures.

SPILLWAY, SHAFT Vertical or inclined shaft into which water spills and then is conveyed through, under, or around a dam by means of a conduit or tunnel.

SPINNING RESERVES Available capacity of generating facilities synchronized to the interconnected electric system where it can be called upon for immediate use in response to system problems or sudden load changes.

SPIN-UP Small-scale vortex initiation (i.e., when a gustnado, landspout, or suction vortex forms).

SPIT Small point or low tongue or narrow embankment of land having one end attached to the mainland and the other terminating in open water.

SPLIT FLOW Flow that leaves the main river flow and takes a completely different path from the main river; flow pattern high in the atmosphere characterized by diverging winds.

SPLIT SAMPLE Sample prepared by dividing it into two or more equal volumes, where each volume is considered a separate sample but representative of the entire sample.

SPLITTER WALL Wall or pier parallel to the direction of flow in a channel that separates flows released from different sources as a means of energy dissipation.

SPLITTING STORM Thunderstorm which splits into two storms which follow diverging paths (i.e., a left mover and a right mover).

SPOIL Dirt or rock removed from its original location; excavated material; overburden or other waste material removed in mining, quarrying, dredging, or excavating.

SPORADIC E Phenomenon occurring in the E region of the ionosphere, which significantly affects HF radiowave propagation.

SPOTNIL Spotless disk.

SPOTTING Outbreak of secondary fires as firebrands or other burning materials are carried ahead of the main fire line by winds.

SPRAWL Unplanned development of open land.

SPRAY Ensemble of water droplets torn by the wind from an extensive body of water, generally from the crests of waves, and carried up into the air in such quantities that it reduces the horizontal visibility.

SPRAY IRRIGATION Irrigation method where water is shot from high-pressure sprayers onto crops and some water is lost to evaporation.

SPRAY TOWER SCRUBBER Device that sprays alkaline water into a chamber where acid gases are present to aid in neutralizing the gas.

SPRING Ground water seeping or flowing out of the Earth where the water table intersects the ground surface; issue of water from the Earth; a natural fountain; a source of a reservoir of water; season of the year comprising the transition period from winter to summer occurring when the Sun is approaching the summer solstice.

SPRING MELT/THAW Process whereby warm temperatures melt winter snow and ice.

SPRING TIDE Tide higher than normal which occurs around the time of the new and full Moon.

SPRINGLINE Imaginary horizontal reference line located at mid-height, or halfway point, of a circular conduit, pipe, tunnel, or the point at which the side walls are vertical on a horseshoe-shaped conduit.

SPRINKLER IRRIGATION Method of irrigation in which the water is sprayed, or sprinkled, through the air to the ground surface.

SPS Severe Weather Potential Statement.

SQLN Squall Line.

SQUALL Strong wind characterized by a sudden onset in which the wind speed increases at least 16 knots and is sustained at 22 knots or more for at least one minute; severe local storm considered as a whole, that is, winds and cloud mass and (if any) precipitation, thunder and lightning.

SQUALL LINE (SQLN) Line of active thunderstorms, either continuous or with breaks, including contiguous precipitation areas resulting from the existence of the thunderstorms.

SRH Storm-Relative Helicity.

SRN Southern.

SS Sandstorm.

SSHS Saffir/Simpson Hurricane Scale.

SST Sea Surface Temperature.

ST Stratus.

ST. ELMO'S FIRE Glow on a masthead produced by an extreme buildup of electrical charge where lightning may strike the mast within five minutes after it begins to glow.

STABILITY Persistence of a component of a system; degree of resistance of a layer of air to vertical motion; condition of a structure or a mass of material when it is able to support the applied stress for a long time without suffering any significant deformation or movement that is not reversed by the release of the stress.

STABILITY INDEX Overall stability or instability of a sounding is sometimes conveniently expressed in the form of a single numerical value.

STABILIZATION Conversion of the active organic matter in sludge into inert, harmless material.

STABILIZATION PONDS See LAGOON.

STABLE Atmospheric state with warm air above cold air which inhibits the vertical movement of air.

STABLE AIR Motionless mass of air that holds, instead of dispersing, pollutants.

STABLE BOUNDARY LAYER Stably-stratified layer that forms at the surface and grows upward, usually at night or in winter, as heat is extracted from the atmosphere's base in response to longwave radiative heat loss from the ground; stably-stratified layer that forms when warm air is advected over a cold surface or over melting ice.

STABLE CHANNEL Stream channel that does not change in planform or bed profile during a particular period of time (i.e., years to tens of years).

STABLE CORE Post-sunrise, elevated remnant of the temperature inversion that has built up overnight within a valley.

STABLE ISOTOPE RATIO (PER MIL) Unit expressing the ratio of the abundance of two radioactive isotopes.

STACCATO LIGHTNING Cloud to Ground (CG) lightning discharge which appears as a single very bright, short-duration stroke, often with considerable branching.

STACK Chimney, smokestack, or vertical pipe that discharges used air.

STACK EFFECT Air that moves upward because it is warmer than the ambient atmosphere (e.g. chimney) creating a positive pressure area at the top and negative pressure area at the bottom.

STACK GAS See FLUE GAS.

STAFF GAUGE Vertical staff graduated where a portion of the gauge is in the water at all times that indicates the height of the water.

STAGE Height of a water surface above an established datum at a given location; elevation; depth of water.

STAGE II CONTROLS Systems placed on service station gasoline pumps to control and capture gasoline vapors during refueling.

STAGE-CAPACITY CURVE Graph showing the relation between the surface elevation of the water in a reservoir (ordinate) plotted against the volume below that elevation (abscissa).

STAGE-DISCHARGE CURVE (RATING CURVE) Graph showing the relation between the water height (ordinate) and the volume of water flowing in a channel (abscissa) per unit of time.

STAGNATION Lack of motion in a mass of air or water that holds pollutants in place.

STAIR STEPPING Process of continually updating river forecasts for the purpose of incorporating the effects rain that has fallen since the previous forecast was prepared.

STAKEHOLDER Any organization, governmental entity, or individual that has a stake in or may be impacted by a given approach to environmental regulation, pollution prevention, energy conservation, etc.

STANDARD Document that has been developed and established within the consensus principles of the American Society for Testing and Materials (ASTM) and that meets the approval requirements of ASTM procedures and regulations.

STANDARD ATMOSPHERE Hypothetical vertical distribution of atmospheric temperature, pressure, and density that, by international agreement, is taken to be representative of the atmosphere.

STANDARD DEVIATION Statistical measure of the dispersion or scatter of a series of values; square root of the variance.

STANDARD HYDROLOGIC EXCHANGE FORMAT (SHEF) Documented set of rules for coding data for operational day-to-day use in a form for both visual and computer recognition.

STANDARD INDUSTRIAL CLASSIFICATION (SIC) CODE Four digit code established by the Office of Management and Budget to identify various types of industries.

STANDARD SAMPLE Part of finished drinking water that is examined for the presence of coliform bacteria.

STANDARD STEP METHOD Method where the total distance is divided into reaches by cross sections at fixed locations along the channel and, starting from one control, profile calculations proceed in steps from cross section to cross section to the next control.

STANDARD SYNOPTIC TIMES Times of 0000, 0600, 1200, and 1800 UTC.

STANDARDS Norms that impose limits on the amount of pollutants or emissions produced.

STANDBY RESERVES Unused capacity in an electric system in machines that are not in operation but that are available for immediate use if required.

STANDING OPERATING PROCEDURES (SOP) Comprehensive single-source document covering all aspects of dam and reservoir operation and maintenance and emergency procedures.

STANDPIPE Pipe or tank connected to a closed conduit and extending to or above the hydraulic grade line of the conduit to afford relief from surges of pressure in pipelines; tank used for storage of water in distribution systems.

START OF A RESPONSE ACTION Point in time when there is a guarantee or set-aside of funding by EPA, other federal agencies, states or Principal Responsible Parties in order to begin response actions at a Superfund site.

STATE CLIMATE DIVISION Geographic area in a state based primarily on crop-reporting districts where states can have 2 to 10 climate divisions.

STATE EMERGENCY RESPONSE COMMISSION (SERC) Appointed by the governor of each state for the designation of emergency planning districts, appoint LEPC's, supervise and coordinate their activities, and review local emergency response plans as provided by SARA Title III.

STATE ENVIRONMENTAL GOALS AND INDICATION PROJECT Program to assist state environmental agencies by providing technical and financial assistance in the development of environmental goals and indicators.

STATE FORECAST PRODUCT National Weather Service product intended to give a good general picture of what weather may be expected in the state during the next 5 days.

STATE IMPLEMENTATION PLANS (SIP) EPA approved state plans for the establishment, regulation, and enforcement of air pollution standards.

STATE MANAGEMENT PLAN State management plan required by EPA to allow states, tribes, and U.S. territories the flexibility to design and implement ways to protect groundwater from the use of certain pesticides.

STATE ORGANIZATION State government agency or office having the principal or lead role in emergency planning and preparedness.

STATE WEATHER ROUNDUP National Weather Service tabular product which provides routine hourly observations within the state through the National Weather Wire Service (NWWS) that gives the current weather condition in one word (cloudy, rain, snow, fog, etc.), the temperature and dew point in Fahrenheit, the relative humidity, wind speed and direction, and finally additional information (wind chill, heat index, a secondary weather condition).

STATEMENT OF BASIS Document prepared for every draft NPDES permit for which a fact sheet is not required that briefly describes how permit conditions were derived and the reasons the conditions are necessary for the permit.

STATES Include: the 50 States, the District of Columbia, Guam, the Commonwealth of Puerto Rico, Virgin Islands, American Samoa, the Trust Territory of the Pacific Islands, and the Commonwealth of the Northern Mariana Islands.

STATIC HEAD Difference in elevation between the pumping source and the point of delivery; vertical distance between two points in a fluid.

STATIC WATER DEPTH Vertical distance from the centerline of the pump discharge down to the surface level of the free pool while no water is being drawn from the pool or water table.

STATIC WATER LEVEL Elevation or level of the water table in a well when the pump is not operating; level or elevation to which water would rise in a tube connected to an artesian aquifer or basin in a conduit under pressure.

STATION Any one of a series of stakes or points indicating distance from a point of beginning or reference.

STATION MODEL Specified pattern for plotting, on a weather map, the meteorological symbols that represent the state of the weather at a particular observing station.

STATION PRESSURE Absolute air pressure at a given reporting station.

STATION USE Energy used in a generating plant as necessary in production of electricity.

STATIONARY FRONT Front between warm and cold air masses that is moving very slowly or not at all.

STATIONARY SOURCE Fixed-site producer of pollution, mainly power plants and other facilities using industrial combustion processes.

STATIONARY WHITE NOISE Random energy with statistical characteristics that do not vary with time.

STATISTICS Branch of mathematics dealing with the collection, analysis, interpretation, and presentation of masses of numerical data.

STATOR That portion of a machine which contains the stationary (non-moving) parts that surround the moving parts (rotor).

STATOR WINDINGS Armature or stationary winding of a synchronous generator.

STEADY FLOW In an open channel, flow is steady if the depth of flow does not change over a given time interval; no change occurs with respect to time.

STEADY STATE CONDITION When model input values are nearly constant for a defined period of time; condition in which system inputs and outputs are in equilibrium so that there is no net change in the system with time.

STEADY STATE FLOW Characteristic of a groundwater or vadose zone flow system where the magnitude and direction of specific discharge at any point in space are constant in time.

STEADY STATE MODEL Model in which the variables being investigated do not change with time.

STEAM Water vapor that rises from boiling water.

STEAM FOG Fog formed when water vapor is added to air which is much colder than the source of the vapor.

STEERING CURRENTS See STEERING WINDS.

STEERING WINDS Prevailing synoptic scale flow which governs the movement of smaller features embedded within it.

STEMFLOW Precipitation that makes it to the ground sliding down the stems of plants.

STEMMING Crushed stone, soil, sand, or drill cuttings used to plug the unloaded portion of a drill hole.

STEPPED LEADER Faint, negatively charged channel that emerges from the base of a thunderstorm and propagates toward the ground in a series of steps of about 1 microsecond duration and 50-100 meters in length, initiating a lightning stroke.

STERILIZATION Removal or destruction of all microorganisms, including pathogenic and other bacteria, vegetative forms, and spores.

STERILIZER One of three groups of anti-microbials registered by EPA for public health uses.

STFR Stratus Fractus.

STG Strong.

STICKY LIMIT Lowest moisture content at which a soil will stick to a metal blade drawn across the surface of the soil mass.

STILLING BASIN Basin constructed to dissipate the energy of fast-flowing water (e.g., from a spillway or bottom outlet), and to protect the streambed from erosion.

STILLING POOL Pool, usually lined with reinforced concrete, located below a spillway, gate, or valve into which the discharge dissipates energy to avoid downstream channel degradation.

STILLWATER LEVEL Elevation that a water surface would assume if all wave actions were absent.

STOCKPILE Storage pile of materials.

STONE Concretion of earthy or mineral matter; rock.

STONEY GATE Gate for large openings that bears on a train of rollers in each gate guide.

STOPING Upward erosion/piping action into an embankment or foundation (possibly leading to a breach).

STOPLOGS Large logs, timbers, or steel beams placed on top of each other with their ends held in guides on each side of a channel or conduit so as to provide a cheaper or more easily handled means of temporary closure than a bulkhead gate.

STORAGE Water artificially impounded in surface or underground reservoirs for future use; water naturally detained in a drainage basin, such as ground water, channel storage, and depression storage; temporary holding of waste pending treatment or disposal, as in containers, tanks, waste piles, and surface impoundments.

STORAGE COEFFICIENT Volume of water an aquifer releases from or takes into storage per unit surface are of the aquifer per unit change in head where for a confined aquifer, the storage coefficient is equal to the product of the specific storage and aquifer thickness and for an unconfined aquifer, the storage coefficient is approximately equal to specific yield.

STORAGE EQUATION Equation for the conservation of mass when calculating storage.

STORAGE, BANK See BANK STORAGE.

STORAGE, CONSERVATION See CONSERVATION STORAGE.

STORAGE, DEAD See DEAD STORAGE.

STORAGE, DEPRESSION See DEPRESSION STORAGE.

STORAGE, TOTAL See TOTAL STORAGE.

STORAGE, USABLE See USABLE STORAGE.

STORAGE-REQUIRED FREQUENCY CURVE Graph showing the frequency with which storage equal to or greater than selected amounts will be required to maintain selected rates of regulated flow.

STORATIVITY See STORAGE COEFFICIENT.

STORET EPA's computerized STOrage and RETrieval water quality data base that includes physical, chemical, and biological data measured in water bodies throughout the United States.

STORM Disturbance of the ordinary average conditions of the atmosphere which, unless specifically qualified, may include any or all meteorological disturbances, such as wind, rain, snow, hail, or thunder.

STORM DATA National Climatic Data Center (NCDC) monthly publication documents a chronological listing, by states, of occurrences of storms and unusual weather phenomena

STORM HYDROGRAPH Hydrograph representing the total flow or discharge past a point.

STORM MOTION Speed and direction at which a thunderstorm travels.

STORM RELATIVE Measured relative to a moving thunderstorm, usually referring to winds, wind shear, or helicity.

STORM RELATIVE MEAN RADIAL VELOCITY MAP (SRM) WSR-88D radar product depicts a full 360° sweep of radial velocity data with the average motion of all identified storms subtracted out.

STORM RELATIVE MEAN RADIAL VELOCITY REGI (SRR) WSR-88D radar product depicts a 27 nm by 27 nm region of storm relative mean radial velocity centered on a point which the operator can specify anywhere within a 124 nm radius of the radar.

STORM SCALE Referring to weather systems with sizes on the order of individual thunderstorms.

STORM SEEPAGE That part of precipitation which infiltrates the surface soil, and moves toward the streams as ephemeral, shallow, perched groundwater above the main ground-water level.

STORM SEWER System of pipes (separate from sanitary sewers) that carries surface runoff from buildings and land surfaces, street wash, and snow melt from the land.

STORM SURGE Abnormal and sudden rise of the sea along a shore as a result of the winds of a storm.

STORM TIDE Actual level of sea water resulting from the astronomic tide combined with the storm surge.

STORM TOTAL PRECIPITATION Estimate of accumulated rainfall from radar since the last time there was a one-hour, or more, break in precipitation.

STORM TRACKING INFORMATION WSR-88D radar product displays the previous, current, and projected locations of storm centroids (forecast and past positions are limited to one hour or less).

STORM WARNING Warning of sustained surface winds, or frequent gusts, in the range of 48 knots (55 mph) to 63 knots (73 mph) inclusive, either predicted or occurring, and not directly associated with a tropical cyclone.

STORM WATCH Watch for an increased risk of a storm force wind event for sustained surface winds, or frequent gusts, of 48 knots (55 mph) to 63 knots (73 mph), but its occurrence, location, and/or timing is still uncertain.

STORM WATER DRAINAGE WELLS Shallow Class V wells that are designed for the disposal of rain water and melted snow.

STORMFLOW See DIRECT RUNOFF.

STORMWATER Stormwater runoff, snow melt runoff, and surface runoff and drainage.

STORMWATER DISCHARGE Precipitation that does not infiltrate into the ground or evaporate due to impervious land surfaces but instead flows onto adjacent land or water areas and is routed into drain/sewer systems.

STORMWATER DISCHARGE-RELATED ACTIVITIES Activities that cause, contribute to, or result in stormwater point source pollutant discharges, including excavation, site development, grading, and other surface disturbance activities; measures to control stormwater, including the siting, construction, and operation of BMPs to control, reduce, or prevent stormwater pollution.

STORMWATER INFILTRATION Process through which stormwater runoff penetrates into soil from the ground surface.

STRAIGHT-LINE HODOGRAPH Hodograph with straight line shape caused by a steady increase of winds with height (vertical wind shear).

STRAIGHT-LINE WINDS Any wind that is not associated with rotation; used to differentiate from tornadic winds.

STRAIN Change in length per unit of length in a given direction.

STRAIN METERS (CARLSON TYPE, VIBRATING-WIRE) Instrument that uses electrical principles to measure the strain at the location of the strain meter.

STRATA (STRATUM) Distinct layers of stratified rock; layer of sedimentary rock, visually separable from other layers above and below

STRATEGIC PLAN Written plan outlining a government agency's framework for management.

STRATIFICATION Layering of water in lakes and streams; formation of separate layers (of temperature, plant, or animal life) in a lake or reservoir; subareas that exhibit reasonably homogeneous environmental conditions.

STRATIFIED RESERVOIR Reservoir with several thermal layers of water.

STRATIFORM Having extensive horizontal development, as opposed to the more vertical development characteristic of convection.

STRATIFORM RINGS AND BANDS Occur between the active convective bands of a hurricane outside of the eye wall.

STRATIGRAPHY Geology that deals with the origin, composition, distribution, and succession of strata; study of the formation, composition, and sequence of sediments, whether consolidated or not; description of layered or stratified rocks.

STRATOCUMULUS Low-level clouds, existing in a relatively flat layer but having individual elements.

STRATOPAUSE Boundary between the stratosphere and mesosphere.

STRATOSPHERE Portion of the atmosphere 10-to-25 miles above the Earth's surface; region of the atmosphere extending from the top of the troposphere to the base of the mesosphere.

STRATOSPHERIC OZONE In the stratosphere, ozone forms an ozone shield that prevents dangerous radiation from reaching the Earth's surface.

STRATUS Low, generally gray cloud layer with a fairly uniform base.

STREAM Body of flowing water; natural water course containing water at least part of the year; type of runoff where water flows in a channel; water flowing in a natural channel as distinct from a canal.

STREAM CAPACITY Total volume of water that a stream can carry within the normal high water channel.

STREAM GAGING Process of measuring flow characteristics (depths, areas, velocities, and rates)..

STREAM GAUGE Specific location on a stream where the stage (water level) is read either by eye or measured by a device that measures and records flow characteristics and water surface elevation.

STREAM LINE Imaginary line within the flow which is everywhere tangent to the velocity vector; arrows on a weather chart showing wind speed and direction where the head of the arrow points toward where the wind is blowing and the length of the arrow is proportional to the wind speed. Sometimes shows wind direction and trajectory only.

STREAM MILE Distance of 1 mile along a line connecting the midpoints of the channel of a stream.

STREAM ORDER Method of numbering streams as part of a drainage basin network; ranking of the relative sizes of streams within a watershed based on the nature of their tributaries where the smallest unbranched tributary is called first order, etc.

STREAM POWER Product of bed shear stress and mean cross-sectional velocity at a cross section for a given flow.

STREAM PROFILE Plot of the elevation of a stream bed versus distance along the stream.

STREAM REACH Continuous part of a stream between two specified points.

STREAM SEGMENT Specified portion of a river with an upstream inflow point and with a downstream termination at a control point; surface waters of an approved planning area exhibiting common hydrological, natural, physical, biological, or chemical processes.

STREAM, CONTINUOUS One that does not have interruptions in space.

STREAM, EPHEMERAL One that flows only in direct response to precipitation, and whose channel is at all times above the water table.

STREAM, GAINING Stream or reach of a stream that receives water from the zone of saturation.

STREAM, INTERMITTENT OR SEASONAL One which flows only at certain times of the year when it receives water from springs or from some surface source such as melting snow in mountainous areas.

STREAM, INTERRUPTED One which contains alternating reaches, that are either perennial, intermittent, or ephemeral.

STREAM, LOSING Stream or reach of a stream that contributes water to the zone of saturation.

STREAM, PERENNIAL One which flows continuously.

STREAM-AQUIFER INTERACTIONS Relations of water flow and chemistry between streams and aquifers that are hydraulically connected.

STREAMBED AT THE DAM AXIS Lowest-point elevation in the streambed at the axis or centerline crest of the dam prior to construction.

STREAMFLOW Discharge that occurs in a natural channel; water flowing in the stream channel; actual flow in streams, whether or not it is affected by diversion or regulation.

STREAMFLOW DEPLETION Sum of the water that flows into the stream channel minus the water that flows out.

STREAM-GAGING STATION Gaging station where a record of discharge of a stream is obtained.

STREAMLINE Line on a map that is parallel to the direction of fluid flow and shows flow patterns.

STRESS Force per unit area.

STRESSOR INDICATOR Characteristic measured to quantify a natural process, an environmental hazard, or a management action that results in changes in exposure and habitat (EMAP).

STRESSORS Physical, chemical, or biological entities that can induce adverse effects on ecosystems or human health.

STRFM Stratiform.

STRIATIONS Scratches or grooves in bedrock caused by rocks within a glacier grinding the Earth's surface as the glacier moves; grooves or channels in cloud formations, arranged parallel to the flow of air and therefore depicting the airflow relative to the parent cloud.

STRIKE Direction taken by a bedding or fault plane as it intersects the horizontal; to be aligned or to trend in a direction at right angles to the direction of the dip.

STRIP CROPPING Growing crops in a systematic arrangement of strips or bands that serve as barriers to wind and water erosion.

STRIP-MINING Process that uses machines to scrape soil or rock away from mineral deposits just under the Earth's surface.

STRIPPING Removal of a surface layer or deposit for the purpose of excavating other material beneath it.

STRUCTURAL DEFORMATION Distortion in walls of a tank after liquid has been added or removed.

STRUCTURAL HEIGHT Distance between the lowest point in the excavated foundation (excluding narrow fault zones) and the top of dam.

STRUT Inside brace.

STUD Bolt having one end firmly anchored.

STUDY UNIT Major hydrologic system of the United States in which NAWQA studies are focused; geographically defined by a combination of ground- and surface-water features and generally encompass more than 4,000 square miles of land area.

STUDY-UNIT SURVEY Broad assessment of the water-quality conditions of the major aquifer systems of each NAWQA Study Unit.

SUBBASE Layer used in a pavement system between the subgrade and base course, or between the subgrade and portland cement concrete pavement.

SUBCHRONIC Of intermediate duration; describes studies or periods of exposure lasting between 5 and 90 days.

SUBCHRONIC EXPOSURE Multiple or continuous exposures lasting for approximately ten percent of an experimental species lifetime.

SUBCRITICAL FLOW State of flow where water depth is above critical depth and velocities are less than critical; gravity forces dominate inertial forces; flow is described as slow or tranquil.

SUBDRILLING Overdrilling; drilling below final grade.

SUBGRADE Soil prepared and compacted to support a structure or pavement system.

SUBGRADE SURFACE Surface of the earth or rock prepared to support a structure or a pavement system.

SUBIRRIGATION Applying irrigation water below the ground surface either by raising the water table within or near the root zone, or by use of a buried perforated or porous pipe system which discharge directly into the root zone.

SUBLETHAL Stimulus below the level that causes death.

SUBLIMATION Process of transformation directly between a solid and a gas, without passing through an intermediate liquid phase.

SUBMEANDER Small meander contained with banks of main channel, associated with relatively low discharges.

SUBMERGED AQUATIC VEGETATION (SAV) Vegetation that lives at or below the water surface; vegetation (i.e. sea grasses) that cannot withstand excessive drying and therefore live with their leaves at or below the water surface; provides an important habitat for young fish and other aquatic organisms.

SUBMERSED PLANT Plant that lies entirely beneath the water surface, except for flowering parts in some species.

SUB-ORGANIZATION Any organization such as agencies, departments, offices, or local jurisdictions having a supportive role in emergency planning and preparedness.

SUBREFRACTION Bending of the radar beam in the vertical which is less than under standard refractive conditions.

SUBSIDENCE gradual downward settling or sinking of the Earth's surface due to underground excavation (e.g., removal of groundwater, gas, oil) with little or no horizontal motion; compression of soft aquifer materials in a confined aquifer due to pumping of water from the aquifer; descending motion of air in the atmosphere occurring over a rather broad area.

SUBSIDENCE CONTROL WELLS Class V wells that are used to control land subsidence caused by groundwater withdrawal or over pumping of oil and gas.

SUBSIDENCE INVERSION Temperature inversion that develops aloft as a result of air gradually sinking over a wide area and being warmed by adiabatic compression, usually associated with subtropical high pressure areas.

SUBSTATION Facility equipment that switches, changes, or regulates electric voltage; location where observations are taken or other services are furnished by people not located at National Weather Service offices who do not need to be certified to take observations.

SUBSTRATE Physical surface upon which an organism lives.

SUBSTRATE EMBEDDEDNESS CLASS Visual estimate of riffle streambed substrate larger than gravel that is surrounded or covered by fine sediment (¡2 mm, sand or finer).

SUBSTRATE SIZE Diameter of streambed particles such as clay, silt, sand, gravel, cobble and boulders.

SUBSURFACE DRAIN Shallow drain installed in an irrigated field to intercept the rising ground-water level and maintain the water table at an acceptable depth below the land surface.

SUBSURFACE FLUID DISTRIBUTION SYSTEM Assemblage of perforated pipes, drain tiles, or other similar mechanisms intended to distribute fluids below the surface of the ground.

SUBSURFACE IRRIGATION SYSTEM Irrigation by means of underground porous tile or its equivalent.

SUB-SURFACE LAYER Composed of well mixed sediments brought up from the inactive layer plus sediment which has deposited from the water column.

SUBSURFACE RUNOFF See STORM SEEPAGE.

SUBSURFACE STORM FLOW (INTERFLOW) Lateral motion of water through the upper layers until it enters a stream channel; usually takes longer to reach stream channels than runoff.

SUBTIDAL Continuously submerged; area affected by ocean tides.

SUBTROPICAL ANTICYCLONE Semi-permanent anticyclone located, on the average, over oceans near 30° North and 30° South latitude.

SUBTROPICAL CYCLONE Non-frontal low pressure system that has characteristics of both tropical and extratropical cyclones.

SUBTROPICAL DEPRESSION Subtropical cyclone in which the maximum 1-minute sustained surface wind is 33 knots (38 mph) or less.

SUBTROPICAL JET (STJ) Jet stream is usually found between 20° and 30° latitude at altitudes between 12 and 14 km.

SUBTROPICAL STORM Subtropical cyclone in which the maximum 1-minute sustained surface wind is 34 knots (39 mph) or more.

SUBWATERSHED Topographic perimeter of the catchment area of a stream tributary.

SUCCESSION Directional, orderly process of community change in which the community modifies the physical environment to eventually establish an ecosystem which is as stable as possible at the site in question.

SUCTION VORTEX Small but very intense vortex within a tornado circulation.

SUDDEN COMMENCEMENT (SC)
Abrupt increase or decrease in the northward component of the geomagnetic field, which marks the beginning of a geomagnetic storm.

SUDDEN IMPULSE (SI+ OR SI-)
Sudden perturbation of several gammas in the northward component of the low-latitude geomagnetic field, not associated with a following geomagnetic storm.

SUDDEN IONOSPHERIC DISTURBANCE (SID)
HF propagation anomalies due to ionospheric changes resulting from solar flares, proton events and geomagnetic storms.

SULFATE ATTACK Damage to concrete caused by the effects of a chemical reaction between sulfates in soils or groundwater and hydrated lime and hydrated calcium aluminate in cement paste that results in considerable expansion and disruption of paste.

SULFUR DIOXIDE Gas that causes acid rain; pungent, colorless, gas formed primarily by the combustion of fossil fuels; pollutant when present in large amounts.

SULFUR HEXAFLUORIDE Potent greenhouse gas primarily used as an electrical insulator in high voltage equipment.

SUMMATION PRINCIPLE States that the sky cover at any level is equal to the summation of the sky cover of the lowest layer plus the additional sky cover provided at all successively higher layers up to and including the layer in question.

SUMMER Typically the warmest season of the year during which the Sun is most nearly overhead.

SUMMER SOLSTICE Time at which the Sun is farthest north in the Northern Hemisphere, on or around June 21.

SUMP Pit, pool or tank for draining, collecting, or storing water for drainage or disposal; chamber located at the entrance to the pump which provides water to the pump.

SUMP PUMP Pump used for removing collected water from a sump.

SUN DOG See PARHELION.

SUN PILLAR Bright column above or below the Sun produced by the reflection of sunlight from ice crystals.

SUN POINTING Alignment of the radar antenna by locating the position of the Sun in the sky, which has an exactly known position given the radar's location and the present time.

SUNNY When there are no opaque (not transparent) clouds; same as Clear.

SUNRISE Phenomenon of the Sun's daily appearance on the eastern horizon as a result of the Earth's rotation.

SUNSET Phenomenon of the Sun's daily disappearance below the western horizon as a result of the Earth's rotation.

SUNSPOT Area seen as a dark spot on the photosphere of the Sun; concentrations of magnetic flux, typically occurring in bipolar clusters or groups.

SUNSPOT GROUP CLASSIFICATION A: small single unipolar sunspot or very small group of spots without penumbra; B: bipolar sunspot group with no penumbra; C: elongated bipolar sunspot group where one sunspot must have penumbra; D: elongated bipolar sunspot group with penumbra on both ends of the group; E: elongated bipolar sunspot group with penumbra on both ends where longitudinal extent of penumbra exceeds 10 degrees but not 15 degrees; F: elongated bipolar sunspot group with penumbra on both ends where longitudinal extent of penumbra exceeds 15 degrees; H: unipolar sunspot group with penumbra.

SUNSPOT NUMBER Daily index of sunspot activity.

SUPER TYPHOON Typhoon having maximum sustained winds of 130 knots (150 mph) or greater.

SUPERCELL See SUPERCELL THUNDERSTORM.

SUPERCELL THUNDERSTORM Thunderstorm consisting of one quasi-steady to rotating updraft which may exist for several hours; potentially the most dangerous of the convective storm types; possess structure to generate the vast majority of long-lived strong and violent (F2-F5) tornadoes, as well as downburst damage and large hail.

SUPERCHLORINATION Chlorination with doses that are deliberately selected to produce water free of combined residuals so large as to require dechlorination.

SUPERCOOL To cool a liquid below its freezing point without solidification or crystallization.

SUPERCOOLED LIQUID WATER Liquid water can survive at temperatures colder than 0 °C.

SUPERCRITICAL FLOW State of flow where the water depth is below the critical depth and velocities are greater than critical; inertial forces dominate the gravitational forces; flow is described as rapid or shooting.

SUPERCRITICAL FLUID Fluid above its critical temperature and critical pressure.

SUPERCRITICAL WATER Type of thermal treatment using moderate temperatures and high pressures to enhance the ability of water to break down large organic molecules into smaller, less toxic ones.

SUPERFUND Program operated that funds and carries out EPA solid waste emergency and long-term removal and remedial activities.

SUPERFUND INNOVATIVE TECHNOLOGY EVALUATION (SITE) PROGRAM EPA program to promote development and use of innovative treatment and site characterization technologies in Superfund site cleanups.

SUPERPOSITION PRINCIPLE Addition or subtraction of two or more different solutions of a governing linear partial differential equation (PDE) to obtain a composite solution of the PDE.

SUPERREFRACTION Bending of the radar beam in the vertical which is greater than sub-standard refractive conditions.

SUPERVISORY CONTROL System used to monitor conditions and operate mechanical features associated with a facility from a location other than at the site.

SUPPLEMENTAL REGISTRATION Arrangement whereby a registrant licenses another company to market its product under the second company's registration.

SUPPLIER OF WATER Any person who owns or operates a public water supply.

SUPPRESSED WEIR Rectangular weir that has only the crest far removed from the channel bottom, the sides are coincident with the sides of the approach channel, so no lateral contraction of water passing through the weir is possible.

SURCHARGE To fill or load to excess; any storage above the full pool; volume or space in a reservoir between the controlled retention water level and the maximum water level.

SURCHARGE CAPACITY (SURCHARGE STORAGE) Reservoir capacity provided for use in passing the inflow design flood through the reservoir; reservoir capacity between the maximum water surface elevation and the highest of the following elevations: top of exclusive flood control capacity, top of joint use capacity, or top of active conservation capacity; temporary storage.

SURF ZONE Area of water between the high tide level on the beach and the seaward side of the breaking waves.

SURF ZONE FORECAST (SRF) National Weather Service routine or event driven forecast product geared toward non-boating marine users issued for an area extending from the area of water between the high tide level on the beach and the seaward side of the breaking waves.

SURFACE AREA OF A LAKE Area encompassed by the boundary of the lake as shown on maps or photographs.

SURFACE ENERGY BUDGET Energy or heat budget at the Earth's surface, considered in terms of the fluxes through a plane at the Earth-atmosphere interface including radiative, sensible, latent and ground heat fluxes.

SURFACE IMPOUNDMENT Indented area in the land's surface, such as a pit, pond, or lagoon; treatment, storage, or disposal of liquid hazardous wastes in ponds.

SURFACE PUMP Mechanism for removing water or wastewater from a sump or wet well.

SURFACE RUNOFF Runoff that travels overland to the stream channel; precipitation, snow melt, or irrigation in excess of what can infiltrate the soil surface and be stored; part of the runoff of a drainage basin that has not passed beneath the surface since precipitation.

SURFACE SOIL Upper part of the soil ordinarily moved in tillage, or its equivalent in uncultivated soils, about 10 to 20 cm in thickness.

SURFACE TENSION Attraction of molecules to each other on a liquid's surface.

SURFACE URANIUM MINES Strip mining operations for removal of uranium-bearing ore.

SURFACE VIBRATION Method of compacting soil using a vibrating plate or vibrating smooth drum roller used on the surface of soil placed.

SURFACE WATER Water on the surface of the Earth; open body of water, such as in a stream, river, lake, or reservoir; all water naturally open to the atmosphere (rivers, lakes, reservoirs, streams, impoundments, seas, estuaries, etc.).

SURFACE WAVES Waves that travel along or near the surface; include Rayleigh and Love Waves of an earthquake.

SURFACE WEATHER CHART Analyzed synoptic chart of surface weather observations that shows the distribution of sea-level pressure (i.e., position of highs, lows, ridges and troughs) and the location and nature of fronts and air masses.

SURFACE-BASED CONVECTION Convection occurring within a surface-based layer (i.e., a layer in which the lowest portion is based at or very near the Earth's surface.).

SURFACE-WATER TREATMENT RULE Rule that specifies maximum contaminant level goals for Giardia lamblia, viruses, and Legionella and promulgates filtration and disinfection requirements for public water systems using surface-water or ground-water sources under the direct influence of surface water; specify water quality, treatment, and watershed protection criteria under which filtration may be avoided.

SURFACING ACM Asbestos-containing material that is sprayed or troweled on or otherwise applied to surfaces.

SURFACING MATERIAL Material sprayed or troweled onto structural members for fire protection; material sprayed or troweled onto ceilings or walls for fireproofing, acoustical or decorative purposes.

SURFACTANT Detergent compound that promotes lathering.

SURFICIAL BED MATERIAL Upper surface (0.1 to 0.2 foot) of the bed material that is sampled using U.S. Series Bed-Material Samplers.

SURGE Rapid increase in the depth of flow; jet of material from active regions that reaches coronal heights and then either fades or returns into the chromosphere along the trajectory of ascent.

SURGE CHAMBER Chamber or tank connected to a pipe and located at or near a valve that may quickly open or close or a pump that may suddenly start or stop to allow water to flow into or out of the pipe and minimize any sudden positive or negative pressure waves or surges in the pipe.

SURGE IRRIGATION Irrigation technique wherein flow is applied to furrows (or less commonly, borders) intermittently during a single irrigation set.

SURROGATE Analyte that behaves similarly to a target analyte, but that is highly unlikely to occur in a sample.

SURROGATE DATA Data from studies of test organisms or a test substance that are used to estimate the characteristics or effects on another organism or substance.

SURVEILLANCE SYSTEM Series of monitoring devices designed to check on environmental conditions.

SURVEY Measure external vertical and horizontal movement on the surface; sampling of a representative number of sites during a given hydrologic condition.

SUSCEPTIBILITY ANALYSIS Analysis to determine whether a Public Water Supply is subject to significant pollution from known potential sources.

SUSPECT MATERIAL Building material suspected of containing asbestos (e.g., surfacing material, floor tile, ceiling tile, thermal system insulation).

SUSPENDED State of floating in water rather than being dissolved in it; amount (concentration) of undissolved material in a water-sediment mixture.

SUSPENDED BED MATERIAL LOAD Portion of the suspended load that is composed of particle sizes found in the bed material.

SUSPENDED LOAD (SUSPENDED SEDIMENT) Sediment that is supported by the upward components of turbulence, carried with the flow of water and that stays in suspension for an appreciable length of time.

SUSPENDED SEDIMENT Sediment (e.g., particles of rock, sand, soil, and organic detritus) that is transported in suspension in water and remains in suspension for a considerable period of time without contact with the bottom due to the upward components of turbulence and currents.

SUSPENDED SOLIDS Small particles of solid pollutants that float on the surface of, or are suspended in, sewage or other liquids; solids that are not in true solution and that can be removed by filtration; defined in waste management as small particles of solid pollutants that resist separation by conventional methods.

SUSPENDED, RECOVERABLE Amount of a given constituent that is in solution after the part of a representative water-suspended sediment sample that is retained on a 0.45-micrometer filter has been extracted or digested.

SUSPENDED, TOTAL Total amount of a given constituent in the part of a water-sediment sample that is retained on a 0.45-micrometer membrane filter.

SUSPENDED-SEDIMENT CONCENTRATION Ratio of the mass of dry sediment in a water-sediment mixture to the mass of the water-sediment mixture; velocity-weighted concentration of suspended sediment in the sampled zone (from the water surface to a point approximately 0.3 foot above the bed).

SUSPENDED-SEDIMENT DISCHARGE Quantity of suspended sediment passing a cross section in a unit of time; rate of sediment transport, as measured by dry mass or volume, that passes a cross section in a given time.

SUSPENDED-SEDIMENT LOAD Given characteristic of the material in suspension that passes a point during a specified period of time.

SUSPENSION Method of sediment transport in which air or water turbulence supports the weight of the sediment particles preventing them from settling out or being deposited; suspending the use of a pesticide when EPA deems it necessary to prevent an imminent hazard resulting from its continued use.

SUSPENSION CULTURE Cells growing in a liquid nutrient medium.

SUSTAINED OVERDRAFT Long-term withdrawal from the aquifer of more water than is being recharged.

SUSTAINED WIND Wind speed determined by averaging observed values over a two-minute period.

SVR Severe.

SVRL Several.

SW Southwest; snow showers.

SWALE Low place in a tract of land; wide, shallow ditch, usually grassed or paved; wide open drain with a low center line; slight depression, sometimes filled with water, in the midst of generally level land.

SWAMP Type of wetland dominated by woody vegetation but without appreciable peat deposits; area intermittently or permanently covered with water, and having trees and shrubs.

SWE Snow Water Equivalent.

SWELL Increase of bulk in soil or rock when excavated; wind-generated waves that have travelled out of their generating area.

SWELL DIRECTION Direction from which the swells are propagating.

SWELL DIRECTION (SWD) Compass direction from which the swell wave are coming from.

SWELL HEIGHT (SWH) Vertical distance between any swell crest and the succeeding swell wave trough.

SWELL PERIOD SWP Time that it takes successive swell wave crests or troughs pass a fixed point.

SWELL PRESSURE Pressure required to maintain zero expansion.

SWITCHING STATION Facility equipment used to tie together two or more electric circuits through switches.

SWLY Southwesterly.

SWRN Southwestern.

SWS Severe Weather Statement.

SWWD Southwestward.

SX Stability Index.

SXN Section.

SYMMETRIC DOUBLE EYE Concentrated ring of convection that develops outside the eye wall in symmetric, mature hurricanes that propagates inward and leads to a double-eye.

SYNCHRONOUS CONDENSERS Synchronous machine running without mechanical load and supplying or absorbing reactive power.

SYNCHRONOUS DETECTION Radar processing that retains the received signal amplitude and phase but that removes the intermediate frequency carrier.

SYNCLINE Fold in rocks in which the strata dip inward from both sides toward the axis; troughlike downward sag or fold in rock layers; opposite of anticline.

SYNERGISM Interaction of two or more chemicals that results in an effect greater than the sum of their separate effects.

SYNERGISTIC EFFECT Biological response to exposure to multiple chemicals which is greater than the sum of the effects of the individual agents.

SYNOPSIS Broad discussion of the weather pattern expected across any given area, generally confined to the 0-48 hour time frame.

SYNOPTIC (SYNOP) Relating to the general weather pattern over a wide region, such as areas of high and low pressure or frontal boundaries, as opposed to mesoscale or smaller features such as a thunderstorm.

SYNOPTIC CODE Rules and procedures established by the World Meteorological Organization (WMO) for encoding weather observations.

SYNOPTIC SCALE Spatial scale of the migratory high and low pressure systems of the lower troposphere, with wavelengths of 1000 to 2500 km.

SYNOPTIC SITES Sites sampled during a short-term investigation of specific water-quality conditions during selected seasonal or hydrologic conditions to provide improved spatial resolution for critical water-quality conditions.

SYNOPTIC STUDIES Short-term investigations of specific water-quality conditions during selected seasonal or hydrologic periods to provide improved spatial resolution for critical water-quality conditions.

SYNOPTIC TRACK Weather reconnaissance mission flown to provide vital meteorological information in data sparse ocean areas as a supplement to existing surface, radar, and satellite data.

SYNTHETIC EARTHQUAKE Earthquake time history records developed from mathematical models that use white noise, filtered white noise, and stationary and nonstationary filtered white noise, or theoretical seismic source models of failure in the fault zone.

SYNTHETIC ORGANIC CHEMICALS (SOCS) Man-made (anthropogenic) organic chemicals.

SYSTEM Physically connected generation, transmission, and distribution facilities operated as an integrated unit under one central management, or operating supervision.

SYSTEM WITH A SINGLE SERVICE CONNECTION System that supplies drinking water to consumers via a single service line.

SYSTEMIC PESTICIDE Chemical absorbed by an organism that interacts with the organism and makes the organism toxic to pests.

SYSTEMIC TOXICANTS Chemical compounds that affect entire organ systems, often operating far from the original site of entry.

SYZYGY Instance when the Earth, Sun, and Moon (new moon or full moon) are all in a straight line.

T

T Thunderstorm.

TABLETOP EXERCISE Informal activity involving discussions of actions to be taken on described emergency situations.

TAF Terminal Aerodrome Forecast.

TAFB Tropical Analysis and Forecast Branch.

TAGOUT Placement of a tagout device on an energy isolating device, in accordance with an established procedure, to indicate that the energy isolating device and the equipment being controlled may not be operated until the tagout device is removed.

TAGOUT DEVICE Prominent warning device, such as a tag and a means of attachment, which can be securely fastened to an energy isolating device in accordance with an established procedure, to indicate that the energy isolating device and the equipment being controlled may not be operated until the tagout device is removed.

TAIL CLOUD Horizontal, tail-shaped cloud (not a funnel cloud) at low levels extending from the precipitation cascade region of a supercell toward the wall cloud.

TAIL WATER See TAILWATER.

TAIL-END CHARLIE Slang for the thunderstorm at the southernmost end of a squall line or other line or band of thunderstorms.

TAILINGS Rock that remains after processing ore to remove the valuable minerals; second grade or waste material separated from pay material during screening or processing.

TAILINGS DAM See MINE TAILINGS DAM.

TAILPIPE STANDARDS Emissions limitations applicable to mobile source engine exhausts.

TAILRACE See AFTERBAY.

TAILWATER Water surface elevation immediately downstream from a structure (i.e., dam, weir or drop structure); runoff of irrigation water from the lower end of an irrigated field; water in the natural stream immediately downstream from a dam; elevation of water varies with discharge from the reservoir.

TAILWATER HEIGHT Height of water immediately downstream of the dam.

TAINTER GATE Term used by the U.S. Army Corps of Engineers to describe radial gates.

TALUS Sloping accumulation of rock debris; rock fragments at the base of a cliff as the result of slides or falls; rock fragments mixed with soil at the foot of a natural slope from which they have been separated; accumulation of broken rocks or boulders at the base of a cliff.

TAMP To pound or press soil to compact; to firmly compact earth during backfilling.

TAMPER Tool for compacting soil in spots not accessible to rollers.

TAMPERING Adjusting, negating, or removing pollution control equipment on a motor vehicle.

TAMPING Method of compacting soil using the impact of a power or hand tamper on the surface of the soil placed.

TAMPING ROLLER One or more steel drums, fitted with projection feet, used to densify soil (e.g. sheepsfoot roller).

TANDEM Pair in which one part follows the other.

TANGENT Line that touches a circle and is perpendicular to its radius at the point of contact.

TANK Artificial reservoir for stock water.

TAPROOT Big root that grows downward from the base of a tree.

TARGET Precipitation or other phenomena which produces echoes on a radar display.

TARN Relatively small and deep, steep-sided lake or pool occupying an ice-gouged basin amid glaciated mountains.

TAXA (SPECIES) RICHNESS Number of species (taxa) present in a defined area or sampling unit.

TAXON (PLURAL TAXA) Any identifiable group of taxonomically related organisms.

TAXONOMY Division of biology concerned with the classification and naming of organisms.

TCU Towering Cumulus Clouds.

TD Tropical Depression.

TDA Today.

TDS Total Dissolved Solids.

TDWR Terminal Doppler Weather Radar.

Technical Support Document (TSD) Contains procedures for water quality-based limitation development.

TECHNICAL-GRADE ACTIVE INGREDIENT (TGA) Pesticide chemical in pure form as it is manufactured prior to being formulated into an end-use product (e.g., wettable powders, granules, emulsifiable concentrates).

TECHNOLOGY-BASED EFFLUENT LIMIT Permit limit for a pollutant that is based on the capability of a treatment method to reduce the pollutant to a certain concentration.

TECHNOLOGY-BASED LIMITATIONS Industry-specific effluent limitations based on best available preventive technology applied to a discharge when it will not cause a violation of water quality standards at low stream flows.

TECHNOLOGY-BASED STANDARDS Industry-specific effluent limitations applicable to direct and indirect sources which are developed on a category-by-category basis using statutory factors, not including water-quality effects.

TECTONIC ACTIVITY Movement of the Earth's crust resulting in the formation of ocean basins, continents, plateaus, and mountain ranges.

TECTONICS Science that deals with the structure of the Earth's crust.

TEE Pipe fitting that has two threaded openings in line, and a third at right angles to them.

TELECONNECTION Linkage between changes in atmospheric circulation occurring in widely separated parts of the globe.

TELESCOPE To slide one piece inside another.

TEMP Temperature.

TEMPERATURE (TEMP) Measure of the internal energy that a substance contains.

ΔT (TEMPERATURE CHANGE) Mean lapse rate within a layer of the atmosphere, obtained by calculating the difference between observed temperatures at the bottom and top of the layer; difference in temperature between the surface of a lake and 850mb, typically used to determine lake effect snow potential.

TEMPERATURE INVERSION (SURFACE-BASED OR ELEVATED) Layer of the atmosphere in which air temperature increases with height.

TEMPERATURE RECOVERY Change in temperature over a given period of time.

TEMPORARY IRRIGATION SERVICE LAND Irrigable land for which a water supply is delivered under temporary arrangements.

TEMPORARY STRUCTURE Any structure that can be readily and completely dismantled and removed from the site between periods of actual use.

TENSIOMETER Instrument, consisting of a porous cup filled with water and connected to a manometer or vacuum gauge, used for measuring the soil water matrix potential.

TERATOGEN Material that produces a physical defect in a developing embryo causing birth defects.

TERATOGENESIS Introduction of non-hereditary birth defects in a developing fetus by exogenous factors such as physical or chemical agents acting in the womb to interfere with normal embryonic development.

TERMINAL AERODROME FORECAST National Weather Service aviation product is a concise statement of the expected meteorological conditions at an airport during a specified period (usually 24 hours).

TERMINAL MORAINE End moraine extending across a glacial plain or valley as an arcuate or crescent ridge that marks the farthest advance or maximum extent of a glacier.

TERMINAL STRUCTURE Portion of spillway downstream from chute, tunnel or conduit, which generally dissipates or stills releases; concrete portion of an outlet works downstream from a conduit, tunnel, or control structure; structure dissipates or stills releases.

TERRACE Ridge, a ridge and hollow, or a flat bench built along a ground contour; surface form of a high sediment deposit having a relatively flat surface and steep slope facing the river; berm or discontinuous segments of a berm, in a valley at some height above the flood plain, representing a former abandoned flood plain of the stream; broad channel, bench, or embankment constructed across the slope to intercept runoff and detain or channel it to protected outlets, thereby reducing erosion from agricultural areas.

TERRACING Dikes built along the contour of sloping farm land that hold runoff and sediment to reduce erosion.

TERRAIN Ground surface.

TERRAIN FORCED FLOW Airflow that is modified or channeled as it passes over or around mountains or through gaps in a mountain barrier.

TERRANE Area or surface over which a particular rock type or group of rock types is prevalent.

TERRESTRIAL Pertaining to, consisting of, or representing the Earth; living or growing on land; land-based.

TERRITORIAL COMMUNITIES Populations that function within a particular geographic area.

TERTIARY TREATMENT Advanced cleaning of wastewater that goes beyond the secondary or biological stage, removing nutrients such as phosphorus, nitrogen, and most BOD and suspended solids.

TERTIARY WASTEWATER TREATMENT Selected biological, physical, and chemical separation processes to remove organic and inorganic substances that resist conventional treatment practices; additional treatment of effluent beyond that of primary and secondary treatment methods to obtain a very high quality of effluent.

TERTIARY-TREATED SEWAGE Third phase of treating sewage that removes nitrogen and phosphorus before it is discharged.

TEST PIT Pit dug for geologic investigation or inspection and testing of earthwork placement.

TEXAS HOOKER Low pressure systems that originate in the panhandle region of Texas and Oklahoma which initially move east and then "hook" or recurve more northeast toward the upper Midwest or Great Lakes region.

THALWEG Deepest part of a river channel in a cross section of a river profile; path of deepest flow; line connecting the deepest points along a riverbed; lowest thread along the axial part of a valley; middle or chief navigable channel of a waterway; part that has the maximum velocity and causes cut banks and channel migration.

THE "ACT" Clean Water Act.

THEODOLITE Instrument used in surveying to measure horizontal and vertical angles with a small telescope that can move in the horizontal and vertical planes.

THEORETICAL MAXIMUM RESIDUE CONTRIBUTION Theoretical maximum amount of a pesticide in the daily diet of an average person.

THEORY Possible or plausible explanation of phenomena based on available evidence.

THERAPEUTIC INDEX Ratio of the dose required to produce toxic or lethal effects to the dose required to produce non-adverse or therapeutic response.

THERMAL Relatively small-scale, rising air current produced when the Earth's surface is heated.

THERMAL BELT Zone of high nighttime temperatures (and relatively low humidities) that is often experienced within a narrow altitude range on valley sidewalls, especially evident during clear weather with light winds.

THERMAL HIGH Area of high pressure that is shallow in vertical extent and produced primarily by cold surface temperatures.

THERMAL LOADING Amount of waste heat discharged to a water body.

THERMAL LOW Area of low pressure that is shallow in vertical extent and produced primarily by warm surface temperatures.

THERMAL POLLUTION Discharge of heated water from industrial processes that can kill or injure aquatic organisms; reduction in water quality caused by increasing its temperature, often due to disposal of waste heat from industrial or power generation processes.

THERMAL SAND Sand used to dissipate heat away from buried electrical cables.

THERMAL STRATIFICATION Formation of layers of different temperatures in a lake or reservoir.

THERMAL SYSTEM INSULATION (TSI) Asbestos-containing material applied to pipes, fittings, boilers, breeching, tanks, ducts, or other interior structural components to prevent heat loss or gain or water condensation.

THERMAL TREATMENT Use of elevated temperatures to treat hazardous wastes.

THERMAL WIND Theoretical wind that blows parallel to the thickness lines, for the layer considered, analogous to how the geostrophic wind blows parallel to the height contours where the closer the thickness isopleths, the stronger the thermal wind.

THERMAL-ELECTRIC POWERPLANT Generating plant which uses heat to create steam driven electricity.

THERMALLY DRIVEN CIRCULATION Diurnally reversing closed cellular wind current resulting from horizontal temperature contrasts caused by different rates of heating or cooling over adjacent surfaces.

THERMISTOR Resistor whose resistance changes with temperature.

THERMOCLINE Middle layer of a lake, separating the upper, warmer portion (epilimnion) from the lower, colder portion (hypolimnion); middle layer in a thermally stratified lake or reservoir.

THERMOCOUPLE Heat-sensing device made of two conductors of different metals joined at their ends.

THERMODYNAMIC CHART Chart containing contours of pressure, temperature, moisture, and potential temperature, all drawn relative to each other such that basic thermodynamic laws are satisfied.

THERMODYNAMIC DIAGRAM See THERMODYNAMIC CHART.

THERMODYNAMICS Relationships between heat and other properties (such as temperature, pressure, density, etc.).

THERMOELECTRIC POWER Electrical power generated by use of fossil-fuel (coal, oil, or natural gas), geothermal, or nuclear energy.

THERMOELECTRIC POWER WATER USE Water used in the process of the generation of thermoelectric power.

THERMOGRAPH Instrument that measures and continuously records variations of temperature.

THERMOKARST Irregular land surface formed in a permafrost region by melting ground ice and a subsequent settling of the ground.

THERMOMETER Instrument for measuring temperature.

THERMOSPHERE Atmospheric shell extending from the top of the mesosphere to outer space.

THETA-E (EQUIVALENT POTENTIAL TEMPERATURE) Temperature a parcel of air would have if a) it was lifted until it became saturated, b) all water vapor was condensed out, and c) it was returned adiabatically (i.e., without transfer of heat or mass) to a pressure of 1000 millibars.

THETA-E RIDGE Axis of relatively high values of theta-e.

THICK ARCH DAM Arch dam with a base thickness to structural height ratio of 0.3 or greater (previously defined as 0.5 or greater).

THICKNESS OR WIDTH OF DAM Thickness or width of a dam as measured horizontally between the upstream and downstream faces and normal to the axis or centerline crest of the dam.

THIN ARCH DAM Arch dam with a base thickness to structural height ratio of 0.2 or less (previously defined as 0.3 or less).

THIN LINE ECHO Narrow, elongated, non-precipitating echo.

THIXOTROPHY Property of a material that enables it to stiffen in a relatively short time on standing but, upon agitation or manipulation, to change to a very soft consistency or to a fluid of high viscosity, the process being completely reversible.

THK Thick/Thickness.

THN Thin.

THREATENED Legal classification for a species which is likely to become endangered within the foreseeable future.

THREATENED SPECIES Any species which has potential of becoming endangered in the near future.

THREATENED WATERS Waters that fully support their designated uses, but may not support uses in the future unless pollution-control action is taken because of anticipated sources or adverse pollution trends.

THREE-HOUR RAINFALL RATE WSR-88D Radar product displays

precipitation total (in inches) of the current and past two clock hours as a graphical image.

THRESHOLD Lowest dose of a chemical at which a specified measurable effect is observed and below which it is not observed; dose or exposure level below which a significant adverse effect is not expected.

THRESHOLD EFFECTS Result from chemicals that have a safe level (i.e., acute, subacute, or chronic human health effects).

THRESHOLD LEVEL Time-weighted average pollutant concentration values, exposure beyond which is likely to adversely affect human health.

THRESHOLD LIMIT VALUE (TLV) Concentration of an airborne substance to which an average person can be repeatedly exposed without adverse effects.

THRESHOLD ODOR See ODOR THRESHOLD.

THRESHOLD PLANNING QUANTITY Quantity designated for each chemical on the list of extremely hazardous substances that triggers notification by facilities that such facilities are subject to emergency planning requirements.

THRESHOLD RUNOFF Runoff in inches from a rain of specified duration that causes a small stream to slightly exceed bankfull (flood stage).

THRFTR Thereafter.

THROPIC LEVELS Functional classification of species that is based on feeding relationships (e.g. generally aquatic and terrestrial green plants comprise the first thropic level, and herbivores comprise the second.).

THROUGH CUT Excavation between parallel banks that begins and ends at original grade.

THROUGHFALL Excess precipitation that falls from foliage.

THRU Through.

THRUST BLOCK (ANCHOR BLOCK) Massive block of concrete built to withstand a thrust or pull; mass of concrete or similar material appropriately placed around a pipe to prevent movement when the pipe is carrying water.

THRUT Throughout.

THSD Thousand.

THUNDER Sound caused by rapidly expanding gases in a lightning discharge.

THUNDERSTORM Local storm produced by a cumulonimbus cloud and accompanied by lightning and thunder.

TIDAL CYCLE Periodic changes in the intensity of tides caused primarily by the varying relations between the Earth, Moon, and Sun.

TIDAL FLAT Extensive, nearly horizontal, tract of land that is alternately covered and uncovered by the tide and consists of unconsolidated sediment.

TIDAL MARSH Low, flat marshlands traversed by channels and tidal hollows, subject to tidal inundation.

TIDAL PILING Occurs when unusually high water levels occur as the result of an accumulation of successive incoming tides that do not completely drain due to opposing strong winds and/or waves.

TIDAL WAVE See TSUNAMI.

TIDE Periodic (occurring at regular intervals) variations in the surface water level of the Great Lakes, oceans, bays, gulfs, and inlets from the gravitational attraction between the Sun and the Moon on the Earth; water level in feet above or below Mean Lower Low Water (MLLW).

TIDE ANOMALY Actual water level minus the prediction.

TIDE PREDICTION Computation of tidal highs and lows at a given location resulting from the gravitational interactions between the Sun and the Moon on the Earth.

TIE LINES Transmission line connecting two or more power systems.

TIER 1 Bureau of Reclamation public protection guideline dealing with loss of life (LOL) considerations.

TIER 1 SEDIMENT GUIDELINE Threshold concentration above which there is a high probability of adverse effects on aquatic life from sediment contamination, determined using modified USEPA (1996) procedures.

TIER 2 Bureau of Reclamation public protection guideline dealing with public trust responsibilities based on the annual failure probability of the structure.

TIERED PERMIT LIMITS Permit limits that only apply to the discharge when a certain threshold (e.g. production level), specific circumstance (e.g. batch discharge), or timeframe (e.g. after 6 months) triggers their use.

TIERED TESTING Any of a series of tests that are conducted as a result of a previous test's findings.

TIERING Coverage of general matters in a broad document with subsequent narrowly focused documents.

TIGHT Soil or rock formations lacking veins of weakness.

TIL Until.

TILE Pipe made of baked clay.

TILE DRAIN Buried perforated pipe designed to remove excess water from soils.

TILL Deposit of sediment formed under a glacier, consisting of an unlayered mixture of clay, silt, sand, and gravel ranging widely in size and shape.

TILLAGE Plowing, seedbed preparation, and cultivation practices.

TILT Storm in which a line connecting the centroid of a mid-level storm component to the centroid of the lowest storm component is to the right or the rear of the direction of motion.

TILT SEQUENCE Radar term indicating that the radar antenna is scanning through a series of antenna elevations in order to obtain a volume scan.

TILTED STORM Thunderstorm or cloud tower which is not purely vertical but instead exhibits a slanted or tilted character; sign of vertical wind shear.

TILTED UPDRAFT See TILTED STORM.

TILTH Soil condition in relation to lump or particle size.

TILTMETERS Instrument that monitors the horizontal or vertical tilt of structures and rock masses.

TIMBER Wood beams larger than 4 inches by 6 inches.

TIMBERING Wood bracing in a tunnel or excavation.

TIMBERING SET Tunnel support consisting of a roof beam or arch, and two posts.

TIME OF CONCENTRATION Time required for storm runoff to flow from the most remote point of a catchment or drainage area to the outlet or point under consideration; travel time from the hydraulically furthermost point in a watershed to the outlet; time from the end of rainfall excess to the inflection point on the recession curve.

TIME OF RISE Time from the start of rainfall excess to the peak of the hydrograph.

TIME TO PEAK Time from the center of mass of the rainfall excess to the peak of the hydrograph.

TIME-WEIGHTED AVERAGE (TWA) Computed by multiplying the number of days in the sampling period by the concentrations of individual constituents for the corresponding period and dividing the sum of the products by the total number of days.

TINAJA Pocket of water developed below a waterfall; temporary pool.

TIPPING-BUCKET RAIN GAUGE Precipitation gauge where collected water is funneled into a two compartment bucket where rain fills one compartment, tips the bucket actuating an electric circuit (emptying into a reservoir) and moves the second compartment beneath the funnel.

TIRE PROCESSOR Intermediate operating facility where recovered tires are processed in preparation for recycling.

TISSUE STUDY Assessment of concentrations and distributions of trace elements and certain organic contaminants in tissues of aquatic organisms.

TKE Turbulent Kinetic Energy.

TMW Tomorrow.

TNDCY Tendency.

TNGT Tonight.

T-NUMBER System used to subjectively estimate tropical cyclone intensity based solely on visible and infrared satellite images.

TOE DRAIN System of pipe and/or pervious material along the downstream toe of a dam used to collect seepage from the foundation and embankment and convey it to a free outlet.

TOE OF DAM Point of intersection between the bottom of a slope or the upstream or downstream face of a dam and the natural ground; junction of the face of a dam with the ground surface.

TOE WEIGHT Additional material placed at the toe of an embankment dam to increase its stability.

TOLERANCE PETITION Formal request to establish a new tolerance or modify an existing one.

TOLERANCES Permissible residue levels for pesticides in raw agricultural produce and processed foods.

TOLERANT SPECIES Those species that are adaptable to (tolerant of) human alterations to the environment and often increase in number when human alterations occur.

TONNAGE Amount of waste that a landfill accepts, usually expressed in tons per month; rate at which a landfill accepts waste is limited by the landfill's permit.

TONS PER ACRE-FOOT (T/ACRE-FT) Dry mass (tons) of a constituent per unit volume (acre-foot) of water; equivalent to 2,000 pounds per day or 0.9072 metric ton per day.

TOP Cloud Top.

TOP OF ACTIVE CONSERVATION CAPACITY Reservoir water surface elevation at the top of the capacity allocated to the storage of water for conservation purposes only.

TOP OF DAM See CREST.

TOP OF DEAD CAPACITY Lowest elevation in the reservoir from which water can be drawn by gravity.

TOP OF EXCLUSIVE FLOOD CONTROL CAPACITY Reservoir water surface elevation at the top of the reservoir capacity allocated to exclusive use for the regulation of flood inflows to reduce damage downstream.

TOP OF INACTIVE CAPACITY Reservoir water surface elevation below which the reservoir will not be evacuated under normal conditions.

TOP OF JOINT USE CAPACITY Reservoir water surface elevation at the top of the reservoir capacity allocated to joint use.

TOP OF SURCHARGE CAPACITY Maximum water surface of a reservoir.

TOP THICKNESS (TOP WIDTH) Thickness or width of a dam at the level of the top of dam (excluding corbels or parapets).

TOPOGRAPHIC MAP Map indicating surface elevation and slope; detailed graphic delineation (representation) of natural and man-made features of a region with particular emphasis on relative position and elevation.

TOPOGRAPHY Physical features of the ground surface area including relative elevations and the position of natural and man-made (anthropogenic) features; shape of the land.

TOPSOIL Topmost layer of soil, usually containing organic matter; soil containing humus which is capable of supporting plant growth.

TOR Tornado; Tornado Warning.

TORNADO Violently rotating column of air, usually pendant to a cumulonimbus, with circulation reaching the ground.

TORNADO EMERGENCY Issued when there is a severe threat to human life and catastrophic damage from an imminent or ongoing tornado.

TORNADO FAMILY Series of tornadoes produced by a single supercell, resulting in damage path segments along the same general line.

TORNADO VORTEX SIGNATURE (TVS) Image of a tornado on the Doppler radar screen that shows up as a small region of rapidly changing wind speeds inside a mesocyclone.

TORNADO WARNING Issued by the National Weather Service when a tornado is indicated by the WSR-88D radar or sighted by spotters.

TORNADO WATCH Issued by the National Weather Service when conditions are favorable for the development of tornadoes in and close to the watch area.

TOTAL Amount of a given constituent in a representative whole-water (unfiltered) sample, regardless of the constituents physical or chemical form.

TOTAL CAPACITY Reservoir capacity below the highest of the elevations representing either the top of exclusive flood control capacity, the top of joint use capacity, or the top of active conservation capacity.

TOTAL COLIFORM BACTERIA Particular group of bacteria that are used as indicators of possible sewage pollution.

TOTAL CONCENTRATION Concentration of a constituent regardless of its form (dissolved or bound) in a sample.

TOTAL DDT Sum of DDT and its metabolites (breakdown products), including DDD and DDE.

TOTAL DISCHARGE Quantity of a given constituent, measured as dry mass or volume, that passes a stream cross section per unit of time.

TOTAL DISSOLVED PHOSPHOROUS Total phosphorous content of all material that will pass through a filter, which is determined as orthophosphate without prior digestion or hydrolysis.

TOTAL DISSOLVED SOLIDS (TDS) Quantitative measure of the residual mineral dissolved in water that remains after the evaporation of a solution; all material that passes the standard glass river filter.

TOTAL DYNAMIC HEAD (TDH) When a pump is lifting or pumping water, the vertical distance from the elevation of the energy grade line on the suction side of the pump to the elevation of the energy grade line on the discharge side of the pump.

TOTAL GROSS RESERVOIR CAPACITY Total amount of storage capacity available in a reservoir for all purposes from the streambed to the normal water or normal water or normal pool surface level.

TOTAL HEAD Height above a datum plane of a column of water; in a ground-water system, it is composed of elevation head and pressure head.

TOTAL LENGTH (FISH) Straight-line distance from the anterior point of a fish specimens snout, with the mouth closed, to the posterior end of the caudal (tail) fin, with the lobes of the caudal fin squeezed together.

TOTAL LOAD See TOTAL SEDIMENT LOAD.

TOTAL MAXIMUM DAILY LOAD (TMDL) Estimates of the amount of specific pollutants that a body of water can safely take without threatening beneficial uses and safely meet water quality standards; sum of the individual wasteload allocations (WLAs) for point sources and load allocations (LAs) for non-point sources and natural background; total allowable pollutant load to a receiving water such that any additional loading will produce a violation of water-quality standards.

TOTAL ORGANIC CARBON (TOC) Measures the amount of organic carbon in water.

TOTAL ORGANISM COUNT Number of organisms collected and enumerated in any particular sample.

TOTAL PETROLEUM HYDROCARBONS (TPH) Measure of the concentration or mass of petroleum hydrocarbon constituents present in a given amount of soil or water.

TOTAL RECOVERABLE Amount of a given constituent in a whole-water sample after a sample has been digested by a method (usually using a dilute acid solution) that results in dissolution of only readily soluble substances.

TOTAL RECOVERED PETROLEUM HYDROCARBON Method for measuring petroleum hydrocarbons in samples of soil or water.

TOTAL SEDIMENT DISCHARGE Total rate at which sediment passes a given point on the stream (tons/day); mass of suspended sediment plus bed-load transport, measured as dry weight, that passes a cross section in a given time.

TOTAL SEDIMENT LOAD (TOTAL LOAD) All of a constituent in transport; sediment in transport including bed load, suspended bed material load, and wash load.

TOTAL STORAGE Volume (including dead storage) of storage below the maximum designed water surface level.

TOTAL SUSPENDED PARTICLES (TSP) Method of monitoring airborne particulate matter by total weight.

TOTAL SUSPENDED SOLIDS (TSS) Measure of the filterable solids present in a sample; measure of the suspended solids in wastewater, effluent, or water bodies.

TOTALIZER (INTEGRATOR) Device or meter that continuously measures and calculates (adds) total flow.

TOTALLY ENCAPSULATING CHEMICAL PROTECTIVE (TECP) SUIT Special protective suits made of material that prevents toxic or corrosive substances or vapors from coming in contact with the body; gas and vapor tight suit.

TOTAL-TOTALS INDEX Stability index and severe weather forecast tool, equal to the temperature at 850 mb plus the dew point at 850 mb, minus twice the temperature at 500 mb.

TOWERING CUMULUS (TCU) Large cumulus cloud with great vertical development, usually with a cauliflower-like appearance, but lacking the characteristic anvil of a towering cumulus.

TOXAPHENE Chemical that causes adverse health effects in domestic water supplies and is toxic to fresh water and marine aquatic life.

TOXIC Relating to harmful effects to biota caused by a substance or contaminant.

TOXIC CHEMICAL RELEASE FORM
Information form required of facilities that manufacture, process, or use (in quantities above a specific amount) chemicals listed.

TOXIC CHEMICAL USE SUBSTITUTION
Replacing toxic chemicals with less harmful chemicals in industrial processes.

TOXIC CLOUD Airborne plume of gases, vapors, fumes, or aerosols containing toxic materials.

TOXIC CONCENTRATION Concentration at which a substance produces a toxic effect.

TOXIC DOSE Dose level at which a substance produces a toxic effect.

TOXIC POLLUTANT Material that causes death, disease, behavioral abnormalities, cancer, genetic mutations, physiological malfunctions, (including malfunctions in reproduction) or physical deformations in organisms that ingest or absorb them.

TOXIC RELEASE INVENTORY
Database of toxic releases in the United States.

TOXIC SUBSTANCE Chemical or mixture that may present an unreasonable risk of injury to health or the environment.

TOXIC UNIT ACUTE (TUA) Reciprocal of the effluent concentration that causes 50 percent of the organisms to die by the end of the acute exposure period.

TOXIC UNIT CHRONIC (TUC)
Reciprocal of the effluent concentration that causes no observable effect on the test organisms by the end of the chronic exposure period.

TOXIC UNITS (TUS) Measure of toxicity in an effluent as determined by the acute toxicity units or chronic toxicity units measured.

TOXIC WASTE Waste that can produce injury if inhaled, swallowed, or absorbed through the skin.

TOXICANT Harmful substance or agent that may injure an exposed organism.

TOXICITY Degree of danger posed by a substance to animal or plant life

TOXICITY ASSESSMENT Characterization of the toxicological properties and effects of a chemical, with special emphasis on establishment of dose-response characteristics.

TOXICITY REDUCTION EVALUATION (TRE)
Site-specific study conducted in a stepwise process designed to identify the causative agent(s) of effluent toxicity, isolate the sources of toxicity, evaluate the effectiveness of toxicity control options, and then confirm the reduction in effluent toxicity.

TOXICITY TEST Procedure to determine the toxicity of a chemical or an effluent using living organisms.

TOXICOLOGICAL PROFILE Examination, summary, and interpretation of a hazardous substance to determine levels of exposure and associated health effects.

TOXICOLOGY Study of the adverse effects of chemicals on biological systems, and the assessment of the probability of their occurrence.

TOXIN Poisonous substance, generally from a plant or animal.

TPW Total Precipitable Water.

TRACE Hydrograph or similar plot for an extended-range time horizon showing one of many scenarios generated through an ensemble forecast process.

TRACE ELEMENT Chemical element that is present in minute quantities in a substance.

TRACE OF PRECIPITATION Rainfall amount less than 0.01 of an inch.

TRACER Stable, easily detected substance or a radioisotope added to a material to follow the location of the substance in the environment or to detect any physical or chemical changes that it undergoes.

TRACK Path that a storm or weather system follows.

TRACTIVE FORCE Force developed in the direction of water flow in a channel.

TRACTOR GATE Rectangular gate supported and sealed similar to the paradox gate.

TRADABLE RENEWABLE CREDITS See RENEWABLE ENERGY CERTIFICATES.

TRADE WINDS Persistent tropical winds that blow from the subtropical high pressure centers towards the equatorial low; system of easterly winds that dominate most of the tropics.

TRADITIONAL CULTURAL PROPERTY (TCP) Site or resource that is eligible for inclusion in the National Register of Historic Places because of its association with cultural practices or beliefs of a living community.

TRAFFIC CONTROL POINT Location staffed to ensure the continued movement of traffic inside or outside an area of risk during an emergency or disaster.

TRAINING Repeated areas of rain, typically associated with thunderstorms, that move over the same region in a relatively short period of time and are capable of producing excessive rainfall totals.

TRAINING WALL Wall built to confine or guide the flow of water.

TRANQUIL FLOW Exists when Froude number is less than one; control of flow depth is always downstream; surface waves propagate upstream as well as downstream.

TRANSBOUNDARY POLLUTANTS Air pollution that travels from one jurisdiction to another, often crossing state or international boundaries.

TRANSCRIBED WEATHER BROADCASTS (TWEB) National Weather Service aviation product that forecast sky cover (height and amount of cloud bases), cloud tops, visibility (including vertical visibility), weather, and obstructions to vision are described for a corridor 25 miles either side of the route.

TRANSDUCER Device that senses some varying condition and converts it to an electrical signal for transmission to some other device (a receiver) for processing or decision making.

TRANSECT Line across a stream perpendicular to the flow and along which measurements are taken, so that morphological and flow characteristics along the line are described from bank to bank.

TRANSFER POINT See TURNING POINT.

TRANSFER STATION Facility where solid waste is transferred from collection vehicles to larger trucks or rail cars for longer distance transport.

TRANSFORMATION Chemical alteration of a compound by processes such as reaction with other compounds or breakdown into component elements.

TRANSFORMED FLOW NET Flow net whose boundaries have been properly modified (transformed) so that a net consisting of curvilinear squares can be constructed to represent flow conditions in an anisotropic porous medium.

TRANSFORMER Device which through electromagnetic induction transforms alternating electric energy in one circuit into energy of similar type on another circuit, commonly with altered values of voltage and current.

TRANSIENT CONDITIONS Condition in which system inputs and outputs are not in equilibrium so that there is a net change in the system with time.

TRANSIENT WATER SYSTEM Non-community water system that does not serve 25 of the same non-residents per day for more than six months per year.

TRANSIENT, NON-COMMUNITY WATER SYSTEM Water system which provides water in a place such as a gas station or campground where people do not remain for long periods of time.

TRANSIT Surveying instrument that can measure both vertical and horizontal angles.

TRANSITION ZONE (SEMIPERVIOUS ZONE) Substantial part of the cross section of an embankment dam comprising material whose grading is of intermediate size between that of an impervious zone and that of a permeable zone.

TRANSMISSIBILITY Capacity of a rock to transmit water under pressure.

TRANSMISSION Act or process of transporting electric energy in bulk.

TRANSMISSION LINE Facility for transmitting electrical energy at high voltage from one point to another point; pipeline that transports raw water from its source to a water treatment plant, then to the distribution grid system.

TRANSMISSIVE BOUNDARY Boundary (cross section) that will allow sediment that reaches it to pass without changing that cross section.

TRANSMISSIVITY Ability of an aquifer to transmit water; volume of water at the existing kinetic viscosity that will move in a unit time under a unit hydraulic gradient through a unit width of the aquifer; equals hydraulic conductivity multiplied by aquifer thickness.

TRANSMITTER Radar equipment used for generating and amplifying a radio frequency (RF) carrier signal, modulating the carrier signal with intelligence, and feeding the modulated carrier to an antenna for radiation into space as electromagnetic waves.

TRANSPIRATION Process by which water vapor escapes from the living plant, principally the leaves, and enters the atmosphere; evaporation of water through the leaves of plants; water discharged into the atmosphere from plant surfaces.

TRANSPORT Hydrological, atmospheric, or other physical processes that convey pollutants through and across media from source to receptor.

TRANSPORT CAPACITY Ability of the river to transport a given volume or weight of sediment material of a specific size per time for a given flow condition.

TRANSPORT POTENTIAL Rate at which a river could transport sediment of a given grain size for given hydraulic conditions if the bed and banks were composed entirely of material of that size.

TRANSPORT WIND Average wind over a specified period of time within a mixed layer near the surface of the Earth.

TRANSPORTATION (SEDIMENT) Complex processes of moving sediment particles from place to place where the principal transporting agents are flowing water and wind.

TRANSPORTATION CONTROL MEASURES (TCMS) Steps taken by a locality to reduce vehicular emission and improve air quality by reducing or changing the flow of traffic.

TRANSPORTER Hauling firm that picks up properly packaged and labeled hazardous waste from generators and transports it to designated facilities for treatment, storage, or disposal.

TRANSVERSE Pertaining to or extending along the short axis, or width, of a structure; perpendicular to or across the long axis, or length, of a structure.

TRANSVERSE BANDS Bands of clouds oriented perpendicular to the flow in which they are embedded.

TRANSVERSE ROLLS (T ROLLS) Elongated low-level clouds, arranged in parallel bands and aligned parallel to the low-level winds but perpendicular to the mid-level flow.

TRAP EFFICIENCY Proportion of sediment inflow to a stream reach that is retained within that reach.

TRAPEZOIDAL WEIR See CIPOLLETTI WEIR.

TRAPPER Valley or basin in which cold air becomes trapped or pooled.

TRASH Material considered worthless or offensive that is thrown away.

TRASHRACK Screen located at an intake to prevent floating or submerged debris from entering the intake.

TRASHRAKE Device that is used to remove debris which has collected on a trash rack to prevent blocking the associated intake.

TRASH-TO-ENERGY PLAN Burning trash to produce energy.

TRAVEL TIME Time required for a flood wave to travel from one location to a subsequent location downstream.

TREATABILITY MANUAL Five-set library of EPA guidance manuals that contain information related to the treatability of many pollutants.

TREATABILITY STUDIES Tests of potential cleanup technologies conducted in a laboratory.

TREATED REGULATED MEDICAL WASTE Medical waste treated to substantially reduce or eliminate its pathogenicity, but that has not yet been destroyed.

TREATED WASTEWATER Wastewater that has been subjected to one or more physical, chemical, and biological processes to reduce its potential of being health hazard.

TREATMENT Any method, technique, or process designed to remove solids and/or pollutants from solid waste, waste-streams, effluents, and air emissions; methods used to change the biological character or composition of any regulated medical waste so as to substantially reduce or eliminate its potential for causing disease.

TREATMENT PLANT Structure built to treat wastewater before discharging it into the environment.

TREATMENT TECHNIQUE Required process intended to reduce the level of a contaminant in drinking water.

TREATMENT WORKS TREATING DOMESTIC SEWAGE (TWTDS) Includes all POTWs and other facilities that treat domestic wastewater, and facilities that do not treat domestic wastewater, but that treat or dispose of sewage sludge.

TREATMENT, STORAGE, AND DISPOSAL FACILITY Site where a hazardous substance is treated, stored, or disposed of.

TREMIE Device used to place concrete or grout under water.

TRENCH See DITCH.

TREND Statistical term referring to the direction or rate of increase or decrease in magnitude of the individual members of a time series of data when random fluctuations of individual members are disregarded.

TRESTLE Bridge, usually of timber or steel, that has a number of closely spaced supports between the abutments.

TRIAL BURN Incinerator test in which emissions are monitored for the presence of specific organic compounds, particulates, and hydrogen chloride.

TRIAXIAL SHEAR TEST Test in which a specimen of soil or rock encased by an impervious membrane is subjected to a confining pressure and then loaded axially to failure.

TRIAZINE HERBICIDE Class of herbicides containing a symmetrical triazine ring (a nitrogen-heterocyclic ring composed of three nitrogens and three carbons in an alternating sequence).

TRIBE Describes any Indian Tribe, band, group, or community recognized by the Secretary of the Interior and exercising governmental authority over a Federal Indian reservation.

TRIBUTARY River or stream flowing into a larger river, stream or lake; river segment other than the main stem in which sediment transport is calculated.

TRICHLOROETHYLENE (TCE) Stable, low boiling-point colorless liquid, toxic if inhaled, used as a solvent or metal degreasing agent, and in other industrial applications.

TRICKLE IRRIGATION Method in which water drips to the soil from perforated tubes or emitters.

TRICKLING FILTER Coarse treatment system in which wastewater is trickled over a bed of stones or other material covered with bacteria that break down the organic waste and produce clean water.

TRIHALOMETHANE (THM) By-products of chlorination of drinking water that contains organic material.

TRIPLE POINT Intersection point between two boundaries (dry line, outflow boundary, cold front, etc.), often a focus for thunderstorm development.

TRIPOD Three-legged support for a surveying instrument.

TRITIUM Radioactive form of hydrogen with atoms of three times the mass of ordinary hydrogen that can be used to determine the age of water.

TROF Trough.

TROP Tropopause.

TROPHIC LEVEL Place of an animal in the food chain.

TROPICAL ADVISORY Official information issued by tropical cyclone warning centers describing all tropical cyclone watches and warnings in effect along with details concerning tropical cyclone locations, intensity and movement, and precautions that should be taken.

TROPICAL ANALYSIS AND FORECAST BRANCH One of three branches of the Tropical Prediction Center (TPC) that provides year-round products involving marine forecasting, aviation forecasts and warnings (SIGMETs), and surface analyses.

TROPICAL CYCLONE Cyclone that originates over the tropical oceans; warm-core, non-frontal synoptic-scale cyclone, originating over tropical or subtropical waters with organized deep convection and a closed surface wind circulation about a well-defined center.

TROPICAL CYCLONE PLAN OF THE DAY Coordinated mission plan that tasks operational weather reconnaissance requirements during the next 1100 to 1100 UTC day or as required, describes reconnaissance flights committed to satisfy both operational and research requirements, and identifies possible reconnaissance requirements for the succeeding 24-hour period.

TROPICAL CYCLONE POSITION ESTIMATE National Hurricane Center issues a position estimate between scheduled advisories whenever the storm center is within 200 nautical miles of U.S. land-based weather radar and if sufficient and regular radar reports are available to the hurricane center.

TROPICAL CYCLONE UPDATE Brief statement issued by the National Hurricane Center in lieu of or preceding special advisories to inform of significant changes in a tropical cyclone or the posting or cancellation of watches and warnings.

TROPICAL DEPRESSION Tropical cyclone in which the maximum 1-minute sustained surface wind is 33 knots (38 mph) or less.

TROPICAL DISTURBANCE Discrete tropical weather system of apparently organized convection originating in the tropics or subtropics, having a non-frontal migratory character and maintaining its identity for 24 hours or more.

TROPICAL PREDICTION CENTER (TPC) NCEP center which produces marine offshore and high seas forecasts south of 30° N in the Eastern Pacific, Gulf of Mexico and Caribbean.

TROPICAL STORM Tropical cyclone in which the maximum 1-minute sustained surface wind ranges from 34 to 63 knots (39 to 73 mph) inclusive.

TROPICAL STORM WARNING Announcement that tropical storm conditions (sustained winds of 39 to 73 mph) are expected somewhere within the specified coastal area within 36 hours.

TROPICAL STORM WATCH Announcement that tropical storm conditions (sustained winds of 39 to 73 mph) are possible within the specified coastal area within 48 hours.

TROPICAL UPPER-TROPOSPHERIC TROUGH (TUTT LOW) Semi-permanent trough extending east-northeast to west-southwest from about 35° N in the eastern Pacific to about 15° to 20° N in the central west Pacific.

TROPICAL WAVE Trough or cyclonic curvature maximum in the trade wind easterlies.

TROPICAL WEATHER DISCUSSION Messages issued 4 times daily by the Tropical Analysis and Forecast Branch (TAFB) to describe significant synoptic weather features in the tropics.

TROPICAL WEATHER OUTLOOK Covers the tropical and subtropical waters, discussing the weather conditions, emphasizing any disturbed and suspicious areas which may become favorable for tropical cyclone development within the next day to two.

TROPICAL WEATHER SUMMARY National Hurricane Center issues a monthly summary of tropical weather is included at the end of the month or as soon as feasible thereafter, to describe briefly the past activity or lack thereof and the reasons why.

TROPICS Areas of the Earth within 20° North and South of the equator.

TROPOPAUSE Upper boundary of the troposphere, usually characterized by an abrupt change in lapse rate from positive (decreasing temperature with height) to neutral or negative (temperature constant or increasing with height).

TROPOPAUSE JET Type of jet stream found near the tropopause (e.g., subtropical and polar fronts).

TROPOSPHERE Layer of the atmosphere closest to the Earth's surface; lowest 6 to 12 miles of the atmosphere, characterized by a general decrease in temperature with height, vertical wind motion, appreciable water content, and sensible weather (clouds, rain, etc.).

TROUGH Elongated area of relatively low atmospheric pressure; opposite of ridge; elongated depression in a potentiometric surface.

TROUGH OF WARM AIR ALOFT (TROWAL)
"Tongue" of relatively warm/moist air aloft that wraps around to the north and west of a mature cyclone.

TRPCL Tropical.

TRRN Terrain.

TRUE WIND Wind relative to a fixed point on the Earth.

TRUNNION Heavy horizontal hinge.

TRUST FUND (CERCLA) Fund set up under the Comprehensive Environmental Response, Compensation and Liability Act (CERCLA) to help pay for cleanup of hazardous waste sites and for legal action to force those responsible for the sites to clean them up.

TS Tropical Storm.

TSRA Thunderstorms with rain.

TSTM Thunderstorm.

TSUNAMI Series of long-period waves (on the order of tens of minutes) that are usually generated by an impulsive disturbance (e.g., earthquake, underwater volcanic activity, landslide) that displaces massive amounts of water.

TSUNAMI ADVISORY Third highest level of tsunami alert issued to coastal populations within areas not currently in either warning or watch status when a tsunami warning has been issued for another region of the same ocean.

TSUNAMI INFORMATION STATEMENT
Issued to inform emergency management officials and the public that an earthquake has occurred, or that a tsunami warning, watch or advisory has been issued for another section of the ocean.

TSUNAMI WARNING Highest level of tsunami alert issued due to the imminent threat of a tsunami from a large undersea earthquake or following confirmation that a potentially destructive tsunami is underway.

TSUNAMI WATCH Second highest level of tsunami alert issued based on seismic information without confirmation that a destructive tsunami is underway.

TUBE SETTLER Device using bundles of tubes to let solids in water settle to the bottom for removal by conventional sludge collection means; sometimes used in sedimentation basins and clarifiers to improve particle removal.

TUBE VALVE Valve which is opened or closed by mechanically moving a tube upstream or downstream by an actuating screw.

TUBERCULATION Development or formation of small mounds of corrosion products on the inside of iron pipe.

TUBING Small-diameter pipe installed inside the casing of a well.

TUFF Igneous rock formed from hardened volcanic ash.

TULE FOG Radiation fog that forms during night and morning hours in late fall and winter months following the first significant rainfall.

TUNDRA Type of treeless ecosystem dominated by lichens, mosses, grasses, and woody plants found at high latitudes (arctic tundra) and high altitudes (alpine tundra).

TUNNEL Covered portion of spillway between the gate or crest structure and the terminal structure, where open channel flow and/or pressure flow conditions may exist; portion of an outlet works between upstream and downstream portals, excluding the gate chamber; generally located in the dam abutments, and are concrete lined or concrete/steel lined; enclosed channel that is constructed by excavating through natural ground; can convey water or house conduits or pipes; long underground excavation with two or more openings to the surface, usually having a uniform cross section used for access, conveying flows, etc.

TURBID Having a cloudy or muddy appearance.

TURBIDIMETER Device that measures the amount of suspended solids in a liquid; a measure of the quantity of suspended solids.

TURBIDITY Measure of extent to which light passing through water is reduced due to suspended materials (e.g., suspended silt, organic matter, etc.); thickness or opaqueness of water caused by the suspension of matter; haziness in air caused by the presence of particles and pollutants.

TURBINE Machine for generating rotary mechanical power from the energy of a stream of fluid (such as water, steam, or hot gas); convert ;kinetic energy of fluids to mechanical energy through the principles of impulse and reaction, or a mixture of the two.

TURBULENCE Irregular motion of medium (e.g., flowing fluid, atmosphere, etc.).

TURBULENT FLOW Open channel flow characterized by random fluid motion; inertial forces are very much greater than the viscous forces and the Reynolds number is large; type of flow in which any water particle may move in any direction with respect to any other particle, and in which the head loss is approximately proportional to the second power of the velocity.

TURKEY TOWER Narrow, individual cloud tower that develops and falls apart rapidly.

TURN ANGLES To measure the angle between directions with a surveying instrument.

TURNING POINT (TRANSFER POINT) Temporary point whose elevation is determined by additions and subtractions of backsights and foresights respectively.

TURNOUT Structure used to divert water from a supply channel to a smaller channel.

TWENTY FOUR HOUR WARNING POINT (24-HOUR WARNING POINT) Communication facility at a state or local level, operating 24 hours a day, which has the capability to receive alerts and warnings, plus activate the public warning system in its area of responsibility.

TWILIGHT Average time of civil twilight, which is the time between civil dawn and sunrise in the morning, and between sunset and civil dusk in the evening.

TWISTER Colloquial term for a tornado.

TWO-RIBBON FLARE Flare that has developed as a pair of bright strands (ribbons) on both sides of the main inversion ("neutral") line of the magnetic field of the active region.

TYPHOON tropical cyclone in the Western Pacific Ocean in which the maximum 1-minute sustained surface wind is 64 knots (74 mph) or greater.

TYPHOON SEASON Part of the year having a relatively high incidence of tropical cyclones.

U

U BURST Fast radio burst spectrum of a flare.

U.S. ARMY CORPS OF ENGINEERS (USACE)
Federal agency, part of the Department of the Army, has authority for approval of dredge and fill permits in navigable waters and tributaries thereof; enforces wetlands regulations, and constructs and operates a variety of water resources projects, mostly notably levee, dams and locks.

U.S. BUREAU OF RECLAMATION (USBR)
Federal agency whose mandate was to reclaim the arid west of the United States; operating in 17 western states, this agency builds, operates and maintains a variety of irrigation, power, and flood control projects.

U.S. GEOLOGICAL SURVEY (USGS)
Federal agency chartered to classify public lands, and to examine the geologic structure, mineral resources, and products of the national domain; provides information and data on the Nation's rivers and streams that are useful for mitigation of hazards associated with floods and droughts.

UCP See UNIT CONTROL POSITION.

UIC PROGRAM DIRECTOR Chief administrative officer of any state or tribal agency or EPA Region that has been delegated to operate an approved UIC program.

ULJ Upper Level Jet.

ULTIMATE BEARING CAPACITY
Load per unit of area required to produce failure by rupture of a supporting soil or rock mass.

ULTRA CLEAN COAL (UCC) Coal that is washed, ground into fine particles, then chemically treated to remove sulfur, ash, silicone, and other substances; usually briquetted and coated with a sealant made from coal.

ULTRA HIGH FREQUENCY (UHF)
Those radio frequencies exceeding 300 MHz.

ULTRAVIOLET (UV) ABSORBANCE (ABSORPTION) AT 254 OR 280 NANOMETERS
Measure of the aggregate concentration of the mixture of UV absorbing organic materials dissolved in the analyzed water, such as lignin, tannin, humic substances, and various aromatic compounds.

ULTRAVIOLET INDEX Index provides important information to help you plan your outdoor activities in ways that prevent overexposure to the Sun's rays; computed using forecasted ozone levels, a computer model that relates ozone levels to UV incidence on the ground, forecasted cloud amounts, and the elevation of the forecast cities.

ULTRAVIOLET RADIATION Electromagnetic radiation of shorter wavelength than visible radiation but longer than x-rays.

ULTRAVIOLET RAYS Radiation from the Sun that can be useful or potentially harmful.

UMBRA Dark core or cores (umbrae) in a sunspot with penumbra; sunspot lacking penumbra.

UNBALANCED HEAD See DIFFERENTIAL HEAD.

UNBUNDLING (FUNCTIONAL UNBUNDLING) Electric service is traditionally provided on a bundled basis, meaning that generation, transmission, and distribution services are provided as a single package.

UNCERTAINTY Describes situations where potential outcomes cannot be estimated based on historical events.

UNCERTAINTY FACTOR One of several factors used in calculating the reference dose from experimental data.

UNCLASSIFIED EXCAVATION Excavation paid for at a fixed price per yard, regardless of whether it is common or rock excavation.

UNCOMPACTED BACKFILL Material used in refilling an excavation without the material being compacted.

UNCONFINED AQUIFER Aquifer containing water that is not under pressure; water level in a well is the same as the water table outside the well; aquifer whose upper surface is a water table free to fluctuate under atmospheric pressure; aquifer that discharges and recharges with an upper surface that is the water table.

UNCONFORMITY Break or gap in the geologic record where rock layers were eroded or never deposited.

UNCONSOLIDATED DEPOSIT Deposit of loosely bound sediment that typically fills topographically low areas.

UNDERCURRENT Current below the upper currents or surface of a fluid body.

UNDERFLOW Lateral motion of water through the upper layers until it enters a stream channel; downstream flow of water through the permeable deposits that underlie a stream and that are more or less limited by rocks of low permeability.

UNDERGROUND INJECTION CONTROL (UIC) Program under the Safe Drinking Water Act that regulates the use of wells to pump fluids into the ground.

UNDERGROUND INJECTION WELLS Steel- and concrete-encased shafts into which hazardous waste is deposited by force and under pressure.

UNDERGROUND SOURCE OF DRINKING WATER (USDW) Aquifer or portion of an aquifer that supplies any public water system or that contains a sufficient quantity of groundwater to supply a public water system, and currently supplies drinking water for human consumption, or that contains fewer than 10,000 mg/l total dissolved solids and is not an exempted aquifer.

UNDERGROUND STORAGE TANK (UST) Tank located at least partially underground and designed to hold gasoline or other petroleum products or chemicals.

UNDERGROUND WATER Subsurface water in the unsaturated and saturated zones.

UNDERSTORY Vegetation underneath the trees; foliage layer lying beneath and shaded by the main canopy of a forest.

UNDERSUN Optical effect seen by an observer above a cloud deck when looking toward the Sun, as sunlight is reflected upwards off the faces of ice crystals in the cloud deck.

UNDERTAKING Any project, activity, or program that could change the character or use of historic properties.

UNDERTOW Relatively small-scale surf-zone current moving away from the beach.

UNDISTURBED SAMPLE Soil sample that has been obtained by methods in which every precaution has been taken to minimize disturbance to the sample.

UNFILTERED Pertains to the constituents in an unfiltered, representative water-suspended sediment sample.

UNFILTERED, RECOVERABLE Amount of a given constituent in a representative water-suspended sediment sample that has been extracted or digested.

UNIFIED SOIL CLASSIFICATION SYSTEM Method of grouping and describing soils according to their engineering properties.

UNIFORM FLOW Open channel flow where the depth and discharge remain constant with respect to space; velocity at a given depth is the same everywhere.

UNIMODAL Distribution having only one localized maximum (i.e. only one peak).

UNIONIZED Neutral form of an ionizable compound (such as an acid or a base).

UNIONIZED AMMONIA Neutral form of ammonia-nitrogen in water.

UNIT CONTROL POSITION WSR-88D radar operator uses this to control the entire radar system.

UNIT HYDROGRAPH (UNITGRAPH) Direct runoff hydrograph resulting from a unit depth of excess rainfall produced by a storm of uniform intensity and specified duration; discharge hydrograph from one inch of surface runoff distributed uniformly over the entire basin for a given time period.

UNIT HYDROGRAPH DURATION Time over which one inch of surface runoff is distributed for unit hydrograph theory.

UNIT HYDROGRAPH THEORY States that surface runoff hydrographs for storm events of the same duration will have the same shape, and the ordinates of the hydrograph will be proportional to the ordinates of the unit hydrograph (e.g. hydrograph for $1/2''$ of storm runoff will be half that of that from the unit hydrograph).

UNIT TIME May be either the hour or day, depending on the incubation period.

UNIT WEIGHT OF WATER Weight per unit volume of water.

UNIVERSAL COORDINATED TIME Local time at the prime meridian that passes through Greenwich, England; known as Coordinated Universal Time, Universal Time Coordinated (UTC), "Z time" or "Zulu Time".

UNIVERSAL GEOGRAPHIC CODE (UGC) National Weather Service product that specifies the affected geographic area of the event, typically by state, county (or parish), or unique National Weather Service zone (land and marine).

UNIVERSAL TIME (UT) See UNIVERSAL COORDINATED TIME.

UNIVERSAL TYPE WEIGHTING AND RECORDING GAUGE Gage which collects precipitation and then converts the weight onto an inked pen movement which traces on graph paper fixed to a clock driven drum.

UNLISTED ITEMS Line item used in an appraisal estimate for design changes and to estimate pay items that have little influence on the total cost.

UNLOAD-RELOAD CYCLE Loading sequence resulting in rebound and recompression.

UNPAID FEDERAL INVESTMENT Year-end amount owed to the Treasury which remains to be repaid.

UNSATURATED ZONE Subsurface zone between land surface and water table where the pores contain both water and air that may include the capillary fringe.

UNSBL Unseasonable.

UNSETTLED Colloquial term used to describe a condition in the atmosphere conducive to precipitation; with regard to geomagnetic levels, a descriptive word specifically meaning that 8 is less than or equal to the Ap Index which is less than or equal to 15.

UNSTABLE AIR Air that is able to rise easily, and has the potential to produce clouds, rain, and thunderstorms.

UNSTBL Unstable.

UNSTEADY FLOW Velocity at a point varies with time.

UNSUITABLE MATERIAL Those soils that cannot be compacted in embankment or backfill or where excavated to finished grade result in unstable material.

UNWATERING Interception and removal of groundwater outside of excavations and the removal of ponded or flowing surface water from within excavations.

UPDRAFT Small-scale current of rising air.

UPGRADIENT Of or pertaining to the place(s) from which groundwater originated or traveled through before reaching a given point in an aquifer.

UPLAND General term for non-wetland; elevated land above low areas along streams or between hills; any elevated region from which rivers gather drainage.

UPLIFT Upward pressure in the pores of a material (interstitial pressure) on the base of a structure; upward force on a structure caused by frost heave or wind force; upward water pressure on a structure.

UPLIFT PRESSURE See PORE-WATER PRESSURE.

UPPER DETECTION LIMIT Largest concentration that an instrument can reliably detect.

UPPER LEVEL Portion of the atmosphere that is above the lower troposphere, generally 850 hPa and above.

UPPER LEVEL DISTURBANCE Disturbance in the upper atmospheric flow pattern which is usually associated with clouds and precipitation.

UPPER LEVEL SYSTEM General term for any large-scale or mesoscale disturbance capable of producing upward motion (lift) in the middle or upper parts of the atmosphere.

UPPER-AIR WEATHER CHART Weather maps that are produced for the portion of the atmosphere above the lower troposphere, generally at and above 850 mb.

UPR Upper.

UPRATE Increase of greater than 15 percent in power plant unit output at an existing facility through modifications or replacement of equipment.

UPSET Exceptional incident in which there is unintentional and temporary non-compliance with the permit limit because of factors beyond the reasonable control of the permittee.

UPSLOPE FLOW Air that flows toward higher terrain, and hence is forced to rise.

UPSLOPE FOG Fog that forms when moist, stable air is carried up a mountain slope.

UPSLP Upslope.

UPSTREAM Towards the source of flow; located in the area from which the flow is coming.

UPSTREAM BLANKET Impervious blanket placed on the reservoir floor and abutments upstream of a dam.

UPSTREAM FACE Inclined surface of the dam that is in contact with the reservoir.

UPSTREAM SLOPE Part of the dam which is in contact with the reservoir water.

UPSTRM Upstream.

UP-VALLEY WIND Diurnal thermally driven flow directed up a valley's axis, usually occurring during daytime; part of the along-valley wind system.

UPWELLING Upward motion of subsurface water toward the surface of the ocean.

URANIUM (U) Heavy silvery-white metallic element, highly radioactive and easily oxidized.

URANIUM MILL TAILINGS PILES Former uranium ore processing sites that contain leftover radioactive materials (wastes), including radium and unrecovered uranium.

URANIUM MILL-TAILINGS WASTE PILES Licensed active mills with tailings piles and evaporation ponds created by acid or alkaline leaching processes.

URBAN AND SMALL STREAM FLOOD ADVISORY Advisory alerts the public to flooding which is generally only an inconvenience (not life-threatening) to those living in the affected area.

URBAN AND SMALL STREAM FLOODING Flooding of small streams, streets, and low-lying areas, such as railroad underpasses and urban storm drains.

URBAN AREA Predominantly cities, towns or developed areas where the population is significant.

URBAN FLASH FLOOD GUIDANCE Guidance which estimates the average amount of rain needed over an urban area during a specified period of time to initiate flooding on small, ungauged streams in the urban area.

URBAN FLOODING Flooding of streets, underpasses, low lying areas, or storm drains.

URBAN HEAT ISLAND Increased air temperatures in urban areas in contrast to cooler surrounding rural areas.

URBAN RUNOFF Storm water from city streets and adjacent domestic or commercial properties that carries pollutants of various kinds into the sewer systems and receiving waters.

URBAN SITE Site that has greater than 50 percent urbanized and less than 25 percent agricultural area.

URBANIZATION To become urban in nature or character; residential, commercial, and industrial development.

UREA-FORMALDEHYDE FOAM INSULATION Material once used to conserve energy by sealing crawl spaces, attics, etc.

USABLE STORAGE Quantity of groundwater of acceptable quality that can be economically withdrawn from storage.

USDW Underground Source of Drinking Water.

USE ATTAINABILITY ANALYSIS (UAA) Structured scientific assessment of the factors affecting the attainment of the use which may include physical, chemical, biological, and economic factors.

USE CLUSTER Set of competing chemicals, processes, and/or technologies that can substitute for one another in performing a particular function.

USE VALUE Economic benefit associated with the physical use of a resource, usually measured by the consumer surplus or net economic value associated with such use.

USED OIL Spent motor oil from passenger cars and trucks collected at specified locations for recycling.

USER DAY Participation in a recreation activity at a given resource during a 24-hour period by one person.

USER FEE Fee collected from only those persons who use a particular service, as compared to one collected from the public in general.

USFS U.S. Forest Service.

UTC Local time at the prime meridian that passes through Greenwich, England; known as Coordinated Universal Time, "Z time" or "Zulu Time".

UTILITY Regulated entity which exhibits the characteristics of a natural monopoly.

UTILITY LOAD Total electricity demand for a utility district.

UVM Upward Vertical Motion.

UVV Upward Vertical Velocity.

UWNDS Upper Winds.

V

V NOTCH Radar reflectivity signature seen as a V-shaped notch in the downwind part of a thunderstorm echo.

VAAC Volcanic Ash Advisory Centers.

VAD Velocity Azimuth Display.

VAD WIND PROFILE (VWP) Radar plot of horizontal winds, derived from VAD data, as a function of height above a Doppler Radar.

VADOSE ZONE Portion of the soil profile above the saturation zone; zone between land surface and the water table within which the moisture content is less than saturation (except in the capillary fringe), pressure is less than atmospheric and soil pores may either contain air or water.

VADOSE ZONE FLOW SYSTEM Aggregate of rock, in which both water and air enters and moves and which is bounded by rock that does not allow any water movement, and by zones of interaction with the Earth's surface, atmosphere and surface water systems.

VALDRIFT Air pollution transport and diffusion model developed to determine pesticide drift from aerial spraying operations in valleys.

VALID TIME Period of time during which a forecast or warning, until it is updated or superseded by a new forecast issuance, is in effect.

VALID TIME EVENT CODE (VTEC) Always is used in conjunction with, and provides supplementary information to, the Universal Geographic Code (UGC), to further aid in the automated delivery of National Weather Service text products to users.

VALIDATION MONITORING Determines if predictive model coefficients are adequately protecting the targeted resources.

VALLEY EXIT JET Strong elevated down-valley air current issuing from a valley above its intersection with the adjacent plain.

VALLEY VOLUME EFFECT Reduction in volume of a valley (or basin) as compared to an equal depth volume with a horizontal floor.

VALUED ENVIRONMENTAL ATTRIBUTES/COMPONENTS Those aspects (i.e., components, processes, functions) of ecosystems, human health, and environmental welfare considered to be important and potentially at risk from human activity or natural hazards.

VALVE Device fitted to a pipeline or orifice in which the closure member is either rotated or moved in some way as to control or stop flow in a conduit, pipe, or tunnel.

VANE SHEAR TEST In-place shear test in which a rod with thin radial vanes at the end is forced into the soil and the resistance to rotation of the rod is determined.

VAPOR Gas given off by substances that are solids or liquids at ordinary atmospheric pressure and temperatures.

VAPOR CAPTURE SYSTEM Any combination of hoods and ventilation system that captures or contains organic vapors so they may be directed to an abatement or recovery device.

VAPOR DENSITY Measure of the heaviness of a chemicals vapor as compared to the weight of a similar amount of air.

VAPOR DISPERSION Movement of vapor clouds in air due to wind, thermal action, gravity spreading, and mixing.

VAPOR PLUMES Flue gases visible because they contain water droplets.

VAPOR PRESSURE Measure of a substance's propensity to evaporate; partial pressure of water vapor in an air-water system; force per unit area exerted by vapor in an equilibrium state with surroundings at a given pressure.

VAPOR RECOVERY SYSTEM System by which the volatile gases from gasoline are captured instead of being released into the atmosphere.

VARIABLE AIR VOLUME (VAV) HVAC system strategy through which the volume of air delivered to conditioned spaces is varied as a function of ventilating needs, energy needs, or both.

VARIABLE COSTS Input costs that change as the nature of the production activity of its circumstances change.

VARIABLE FREQUENCY DRIVE (VFD) Specific type of adjustable-speed drive that controls the rotational speed of an alternating current (AC) electric motor by controlling the frequency of the electrical power supplied to the motor.

VARIABLE WIND DIRECTION Condition when the wind direction fluctuates by 60° or more during the 2-minute evaluation period and the wind speed is greater than 6 knots; condition when the wind direction is variable and the wind speed is less than 6 knots.

VARIANCE Government permission for a delay or exception in the application of a given law, ordinance, or regulation; measure of variability.

VARVED Sedimentary bed or lamination that is deposited within one year's time.

VASCULAR PLANT Plant composed of or provided with vessels or ducts that convey water or sap.

VCNTY Vicinity.

VEBE TIME Time for concrete to fully consolidate in a Vebe cylinder.

VECTOR Quantity having both direction and magnitude; organism, often an insect or rodent, that carries disease; plasmids, viruses, or bacteria used to transport genes into a host cell.

VEERING Clockwise shift in wind direction.

VEGETATIVE CONTROLS Nonpoint source pollution control practices that involve plants (vegetative cover) to reduce erosion and minimize the loss of pollutants.

VEHICLE MILES TRAVELLED (VMT)
Measure of the extent of motor vehicle operation; total number of vehicle miles travelled within a specific geographic area over a given period of time.

VEIN Layer, seam, or narrow irregular body of material different from surrounding formations.

VELOCITY Vector quantity that has magnitude and direction; rate of flow (e.g., feet per second, miles per hour); time rate of displacement of a fluid particle from one point to another.

VELOCITY AZIMUTH DISPLAY
WSR 88-D product which shows the radar derived wind speeds at various heights.

VELOCITY CROSS SECTION
WSR-88D radar product displays a vertical cross section of velocity on a grid with heights up to 70,000 feet on the vertical axis and distance up to 124 nm on the horizontal axis.

VELOCITY ZONES Areas within the floodplain subject to potential high damage.

VELOCITY, AVERAGE INTERSTITIAL
Average rate of groundwater flow to interstices expressed as the product of hydraulic conductivity and hydraulic gradient divided by the effective porosity.

VELOCITY, DARCIAN See SPECIFIC DISCHARGE.

VENTILATION INDEX Product of the mixing depth and transport wind speed; measure of the potential of the atmosphere to disperse airborne pollutants from a stationary source.

VENTILATION RATE Rate at which indoor air enters and leaves a building expressed as the number of changes of outdoor air per unit of time (air changes per hour (ACH) or the rate at which a volume of outdoor air enters in cubic feet per minute (CFM).

VENTILATION/SUCTION Act of admitting fresh air into a space in order to replace stale or contaminated air.

VENTURI Pressure jet that draws in and mixes air.

VENTURI EFFECT Speedup of air through a constriction due to the pressure rise on the upwind side of the constriction and the pressure drop on the downwind side as the air diverges to leave the constriction.

VENTURI SCRUBBERS Air pollution control devices that use water to remove particulate matter from emissions.

VER HIGH FREQUENCY (VHF)
That portion of the radio frequency spectrum from 30 to 300 MHz.

VERNAL POOL Small lake or pond that is filled with water for only a short time during the spring.

VERNIER Device permitting finer measurement or control than standard markings or adjustments.

VERTICAL CURVE Meeting of different gradients in a road or pipe.

VERTICAL DATUM See DATUM.

VERTICAL LIFT GATE All rectangular gates set in vertical guides within which the gate moves vertically in its own plane.

VERTICAL VELOCITY Component of velocity (motion) in the vertical.

VERTICAL WIND SHEAR Change in the wind's direction and speed with height.

VERTICALLY STACKED SYSTEM Low-pressure system, usually a closed low or cutoff low, which is not tilted with height (i.e. located similarly at all levels of the atmosphere).

Vertically-Integrated Liquid (VIL) water Property computed by RADAP II and WSR-88D units that takes into account the three-dimensional reflectivity of an echo.

VERY LOW FREQUENCY (VLF) That portion of the radio frequency spectrum from 3 to 30 kHz.

VERY WINDY 30 to 40 mph winds.

VESICULAR (VESICLES) Containing many small cavities formed by the expansion of a gas bubble or steam when the rock solidifies; tiny holes in volcanic rock caused by gas bubbles trapped in lava when it cooled.

VFR Visual Flight Rules.

VIABLE POPULATION Population that has the estimated numbers and distribution of reproductive individuals to ensure the continued existence of the species throughout its existing range in the planning area.

VIBRATING SCREEN Screen which is vibrated to separate and move pieces resting on it.

VINYL CHLORIDE Chemical compound, used in producing some plastics, that is believed to be oncogenic.

VIOLATION Failure to meet any state or federal regulation.

VIRGA Streaks or wisps of precipitation falling from a cloud but evaporating before reaching the ground.

VIRGIN COMPRESSION Compression corresponding to stresses greater than the preconsolidation stress.

VIRGIN COMPRESSION CURVE Portion of the compression curve corresponding to virgin compression.

VIRGIN COMPRESSION LINE Straight line approximating the virgin compression curve.

VIRGIN MATERIALS Resources extracted from nature in their raw form, such as timber or metal ore.

VIRTUAL POTENTIAL TEMPERATURE Temperature a parcel at a specific pressure level and virtual temperature would have if it were lowered or raised to 1000 mb.

VIRTUAL TEMPERATURE Temperature a parcel which contains no moisture would have to equal the density of a parcel at a specific temperature and humidity.

VIS Visible Satellite Imagery; Visible; Visibility.

VISCOSITY Resistance of a fluid to flow; molecular friction within a fluid that produces flow resistance.

VISIBILITY Distance at which a given standard object can be seen and identified with the unaided eye.

VISIBILITY PROTECTION PROGRAM Program specified by the Clean Air Act to achieve a national goal of remedying existing impairments to visibility and preventing future visibility impairment throughout the United States.

VISIBLE SATELLITE IMAGERY Uses reflected sunlight (i.e. reflected solar radiation) to see things in the atmosphere and on the Earth's surface.

VISITOR DAY Twelve visitor hours which may be aggregated by one or more persons in single or multiple visits.

VISITOR USE Visitor use of recreation and wilderness resource for inspiration, stimulation, solitude, relaxation, education, pleasure, or satisfaction.

VISUAL SPECTRUM Portion of the electromagnetic spectrum to which the eye is sensitive (i.e. light with wavelengths between 0.4 and 0.7 micrometers).

VLCTY Velocity.

VLY Valley.

V-NOTCH WEIR Weir that is V-shaped, with its apex downward, used to accurately measure small rates of flow.

VOID Space in a soil or rock mass not occupied by solid mineral matter but rather occupied by air, water, or other gaseous or liquid material.

VOID RATIO Ratio of the volume of void space to the volume of solid particles in a given soil mass.

VOLATILE Any substance that evaporates readily.

VOLATILE LIQUIDS Liquids which easily vaporize or evaporate at room temperature.

VOLATILE ORGANIC COMPOUND (VOC) Any organic compound that participates in atmospheric photochemical reactions, with high vapor pressure relative to their water solubility, that are emitted as gases from certain solids or liquids.

VOLATILE SOLIDS Those solids in water or other liquids that are lost on ignition of the dry solids at 550 °C.

VOLATILE SYNTHETIC ORGANIC CHEMICALS Chemicals that tend to volatilize or evaporate.

VOLATILIZATION Entry of contaminants into the atmosphere by evaporation from soil or water.

VOLCANIC ASH Fine particles of mineral matter from a volcanic eruption which can be dispersed long distances by winds aloft.

VOLCANIC ROCK Rock that forms from the solidification of molten rock or magma at the Earth's surface (extrusive igneous rock).

VOLT (V) Unit of measurement of electromotive force; equivalent to the force required to produce a current of 1 ampere through a resistance of 1 ohm.

VOLTAGE (E) Electrical pressure (i.e. the force which causes current to flow through an electrical conductor).

VOLTAGE REDUCTION Any intentional reduction of system voltage by 3 percent or greater for reasons of maintaining the continuity of service of the bulk electric power supply system.

VOLT-AMPERE (VA) Unit of apparent power in an AC circuit containing reactance; equal to the potential in volts multiplied by the current in amperes, without taking phase into consideration.

VOLT-AMPERES REACTIVE (VARS) Unit of measure for reactive power.

VOLUME COVERAGE PATTERN (VCP) Volumetric sampling procedure designed for the surveillance of one or more particular meteorological phenomena.

VOLUME MEDIAN DIAMETER (VMD) Statistical measure of the average droplet size in a spray cloud, such that fifty percent of the volume of sprayed material is composed of droplets smaller in diameter than the VMD.

VOLUME OF CONCRETE Total space occupied by concrete forming the dam structure computed between abutments and from the top to the bottom of the dam.

VOLUME OF DAM See VOLUME OF CONCRETE.

VOLUME REDUCTION Processing waste materials to decrease the amount of space they occupy, usually by compacting, shredding, incineration, or composting.

VOLUME SCAN Radar scanning strategy in which sweeps are made at successive antenna elevations (i.e. a tilt sequence), and then combined to obtain the three-dimensional structure of the echoes.

VOLUME VELOCITY PROCESSING (VVP) Way to guess the large-scale 2-dimensional winds, divergence and fall speeds from one-dimensional radial velocity data.

VOLUMETRIC SHRINKAGE Decrease in volume, expressed as a percentage of the soil mass when dried, of a soil mass when the moisture content is reduced from a given percentage to the shrinkage limit.

VOLUMETRIC TANK TEST Volume of fluid in the tank is measured directly or calculated from product-level changes.

VOLUNTARY OBSERVING SHIP PROGRAM (VOS) International voluntary marine observation program under the auspices of the World Meteorological Organization (WMO).

VORTEX Revolving mass of water (whirlpool) in which the streamlines are concentric circles and in which the total head is the same; water rotating about an axis; whirling mass of air in the form of a column or spiral.

VORTICITY (VOT) Measure of the rotation of air in a horizontal plane.

VORTICITY MAXIMUM (VORT MAX) Slang for center, or maximum, in the vorticity field of a fluid.

VR Veer.

VRT MOTN Vertical Motion.

VSB Visible Satellite Imagery.

VSBY Visibility.

VULNERABILITY ANALYSIS Assessment of elements in the community that are susceptible to damage if hazardous materials are released.

VULNERABILITY ASSESSMENT Evaluation of source quality and its vulnerability to contamination.

VULNERABLE ZONE Area over which the concentration of a chemical accidentally released could reach the level of concern.

W

W West.

WAKE Region of turbulence immediately to the rear of a solid body caused by the flow of air over or around the body.

WALL CLOUD Localized, persistent, often abrupt lowering from a rain-free base.

WALL FRICTION Frictional resistance mobilized between a wall and the soil or rock in contact with the wall.

WARM ADVECTION See WARM AIR ADVECTION.

WARM AIR ADVECTION (WAA) Movement of warm air into a region by horizontal winds.

WARM CORE LOW Low pressure area which is warmer at its center than at its periphery.

WARM FRONT Transition zone between a mass of warm air and the colder air it is replacing.

WARM OCCLUSION Frontal zone formed when a cold front overtakes a warm front and, finding colder air ahead of the warm front, leaves the ground and rises up and over this denser air.

WARM SECTOR Region of warm surface air between a cold front and a warm front.

WARM-WATER FISHERY Water or water system that has an environment suitable for species of fish other than salmonids.

WARNING Fourth of five Early Warning System components consisting of the processes (including the media) and equipment necessary to make the public aware of potential, probable, or imminent danger or risk; issued when a hazardous weather or hydrologic event is occurring, is imminent, or has a very high probability of occurring; used for conditions posing a threat to life or property.

WARNING STAGE Depth of water in a river at which the National Weather Service (NWS) reviews basin conditions for potential flooding.

WARNING TIME (WT) Amount of time between detection of failure or incipient failure and arrival of dam failure flood.

WASH LOAD That part of the suspended load that is finer than the bed material.

WASTE Digging, hauling and dumping of valueless material to get it out of the way; valueless material; unwanted materials left over from a manufacturing process; refuse from places of human or animal habitation.

WASTE CHARACTERIZATION Identification of chemical and microbiological constituents of a waste material.

WASTE EXCHANGE Arrangement in which companies exchange their wastes for the benefit of both parties.

WASTE FEED Continuous or intermittent flow of wastes into an incinerator.

WASTE GENERATION Weight or volume of materials and products that enter the waste stream before recycling, composting, landfilling, or combustion takes place; amount of waste generated by a given source or category of sources.

WASTE LOAD ALLOCATION (WLA) Maximum load of pollutants each discharger of waste is allowed to release into a particular waterway; portion of a stream's total assimilative capacity assigned to an individual discharge.

WASTE MINIMIZATION Measures or techniques that reduce the amount of wastes generated during industrial production processes; recycling and other efforts to reduce the amount of waste going into the waste stream.

WASTE PILES Non-containerized, lined or unlined accumulations of solid, non-flowing waste.

WASTE REDUCTION Using source reduction, recycling, or composting to prevent or reduce waste generation.

WASTE STREAM Total flow of solid waste from homes, businesses, institutions, and manufacturing plants that is recycled, burned, or disposed of in landfills, or segments thereof such as the "residential waste stream" or the "recyclable waste stream."

WASTE TREATMENT LAGOON Impoundment made by excavation or earth fill for biological treatment of wastewater.

WASTE TREATMENT PLANT Facility containing a series of tanks, screens, filters and other processes by which pollutants are removed from water.

WASTE TREATMENT STREAM Continuous movement of waste from generator to treater and disposer.

WASTEBANK Bank made of excess or unstable material excavated from a construction site.

WASTE-HEAT RECOVERY Recovering heat discharged as a byproduct of one process to provide heat needed by a second process.

WASTELOAD ALLOCATION (WLA) Proportion of a receiving water's total maximum daily load that is allocated to one of its existing or future point sources of pollution.

WASTE-TO-ENERGY FACILITY/MUNICIPAL-WASTE COMBUSTOR Facility where recovered municipal solid waste is converted into a usable form of energy, usually via combustion.

WASTEWATER Spent or used water from a homes, businesses, and industries, communities, farms, or industries that contains wastes (i.e., dissolved or suspended matter); water that has been used and is not for reuse unless it is treated.

WASTEWATER INFRASTRUCTURE Plan or network for the collection, treatment, and disposal of sewage in a community.

WASTEWATER OPERATIONS AND MAINTENANCE Actions taken after construction to ensure that facilities constructed to treat wastewater will be operated, maintained, and managed to reach prescribed effluent levels in an optimum manner.

WASTEWATER TREATMENT Processing of wastewater for the removal or reduction of contained solids or other undesirable constituents.

WASTEWATER TREATMENT PLAN Facility containing a series of tanks, screens, filters, and other processes by which pollutants are removed from water.

WASTEWATER-TREATMENT RETURN FLOW Water returned to the environment by wastewater-treatment facilities.

WASTEWAY Waterway used to drain excess irrigation water dumped from the irrigation delivery system.

WATCH Used when the risk of a hazardous weather or hydrologic event has increased significantly, but its occurrence, location, and/or timing is still uncertain; intended to provide enough lead time so that those who need to set their plans in motion can do so.

WATCH BOX Slang for severe weather watch (Tornado or Severe Thunderstorm).

WATCH CANCELLATION Issued to let the public know when a severe weather watch (Tornado or Severe Thunderstorm) has been canceled early.

WATCH REDEFINING STATEMENT Tells the public which counties/parishes are included in the severe weather watch (Tornado or Severe Thunderstorm).

WATCH STATUS REPORTS Lets the NWFO know of the status of the current severe weather watch (Tornado or Severe Thunderstorm).

WATER BALANCE See HYDROLOGIC BUDGET.

WATER BUDGET (WATER BALANCE MODEL) Accounting of the inflow, outflow, and storage changes of water in a hydrologic unit over a fixed period.

WATER COLUMN Imaginary vertical column of water extending through a water body from its floor to its surface.

WATER COLUMN STUDIES Investigations of physical and chemical characteristics of surface water, which include suspended sediment, dissolved solids, major ions, and metals, nutrients, organic carbon, and dissolved pesticides, in relation to hydrologic conditions, sources, and transport.

WATER CONTENT See MOISTURE CONTENT.

WATER CONTENT OF SNOW See WATER EQUIVALENT OF SNOW.

WATER CONVEYANCE EFFICIENCY Ratio of the volume of irrigation water delivered by a distribution system to the water introduced into the system.

WATER CONVEYANCE STRUCTURE Any structure that conveys water from one location to another.

WATER CROP See WATER YIELD.

WATER CYCLE Circuit of water movement from the oceans to the atmosphere and to the Earth and return to the atmosphere through various stages or processes such as precipitation, interception, runoff, infiltration, percolation, storage, evaporation, and transportation; movement of water from the air to and below the Earth's surface and back into the air.

WATER DELIVERY SYSTEM Reservoirs, canals, ditches, pumps, and other facilities to move water.

WATER DEMAND Water requirements for a particular purpose, such as irrigation, power, municipal supply, plant transpiration, or storage.

WATER EQUIVALENT Amount of water, in inches, obtained by melting a snow sample; liquid content of solid precipitation that has accumulated on the ground (snow depth).

WATER EQUIVALENT OF SNOW Amount of liquid water in the snow at the time of observation.

WATER EXPORTS Artificial transfer (by pipes or canals) of freshwater from one region or subregion to another.

WATER GAP Deep, narrow pass in a mountain ridge through which a stream flows.

WATER HAMMER (HYDRAULIC TRANSIENT) Pressure fluctuations caused by a sudden increase or decrease in flow velocity, usually associated with a rapid closure or opening of a valve in a pipeline.

WATER HOLDING CAPACITY Amount of soil water available to plants; smallest value to which the moisture content of a soil can be reduced by gravity drainage.

WATER IMPORTS Artificial transfer (by pipes or canals) of freshwater to one region or subregion from another.

WATER LOSS See EVAPOTRANSPIRATION.

WATER MANAGEMENT PLAN Plan developed during construction to help assure water quality compliance for both point and non-point pollution sources.

WATER MASS BALANCE Inventory of the difference source and sinks of water in a hydrogeologic system.

WATER POLLUTION Alteration of the constituents of a body of water by man to such a degree that the water loses its value as a natural resource.

WATER PURVEYOR Agency or person that supplies water (e.g. drinking) water to customers.

WATER QUALITY Condition of water as it relates to impurities; describes the chemical, physical, and biological characteristics of water, usually in respect to its suitability for a particular purpose.

WATER QUALITY ASSESSMENT Evaluation of the condition of a water body using biological surveys, chemical-specific analyses of pollutants in water bodies, and toxicity tests.

WATER QUALITY CRITERIA Levels of water quality expected to render a body of water suitable for its designated use based on specific levels of pollutants that would make the water harmful if used for drinking, swimming, farming, fish production, or industrial processes.

WATER QUALITY LIMITED SEGMENT Any segment where it is known that water quality does not meet applicable water quality standards and/or is not expected to meet applicable water quality standards even after application of technology-based effluent limitations.

WATER QUALITY STANDARD (WQS) Law or regulation that consists of the beneficial use or uses of a waterbody, the numeric and narrative water quality criteria that are necessary to protect the use or uses of that particular waterbody, and an anti-degradation statement; state-adopted and EPA-approved ambient standards for water bodies.

WATER QUALITY-BASED EFFLUENT LIMIT (WQBEL) Value determined by selecting the most stringent of the effluent limits calculated using all applicable water quality criteria (e.g., aquatic life, human health, and wildlife) for a specific point source to a specific receiving water for a given pollutant.

WATER QUALITY-BASED LIMITATIONS Effluent limitations applied to dischargers when mere technology-based limitations would cause violations of water quality standards.

WATER QUALITY-BASED PERMIT Permit with an effluent limit more stringent than one based on technology performance.

WATER RIGHTS Legal rights to the use of water.

WATER SATURATION That point at which a material will no longer absorb water.

WATER SIDE ECONOMIZER Reduces energy consumption in cooling mode by allowing the chiller to be turned off when the cooling tower alone can produce water at the desired chilled water set point.

WATER SOLUBILITY Maximum possible concentration of a chemical compound dissolved in water.

WATER STORAGE POND Impound for liquid wastes designed to accomplish some degree of biochemical treatment.

WATER SUPPLIER Person who owns or operates a public water system.

WATER SUPPLY OUTLOOK Seasonal volume forecast, generally for a period centered around the time of spring snowmelt (e.g. April-July).

WATER SUPPLY OUTLOOK (ESS) PRODUCT Public product issued by a Forecast Office which contains narrative and numeric information on current and extended water supply conditions.

WATER SUPPLY SYSTEM Collection, treatment, storage, and distribution of potable water from source to consumer.

WATER SURFACE ELEVATION (STAGE) Elevation of a water surface above or below an established reference level, such as sea level.

WATER TABLE Surface of underground, gravity-controlled water; level of ground water; level below the Earth's surface at which the ground becomes saturated with water (zone of saturation) and where the ground is filled with water and air (zone of aeration); upper surface of the zone of saturation of groundwater above an impermeable layer of soil or rock (through which water cannot move) as in an unconfined aquifer; surface of a groundwater body at which the water pressure equals atmospheric pressure.

WATER TRANSFER Selling or exchanging water or water rights among individuals or agencies; artificial conveyance of water from one area to another.

WATER TREATMENT LAGOON Impound for liquid wastes designed to accomplish some degree of biochemical treatment.

WATER USE Water that is actually used for a specific purpose, such as for domestic use, irrigation, or industrial processing; pertains to human's interaction with and influence on the hydrologic cycle, and includes elements such as water withdrawal, delivery, consumptive use, wastewater release, reclaimed wastewater, return flow, and instream use.

WATER USER Any individual, district, association, government agency, or other entity that uses water supplied.

WATER VAPOR PLUME Appears in the water vapor satellite imagery as a plume-like object that extends from the Intertropical Convergence Zone (ITCZ) northward or southward into the higher latitudes.

WATER WELL Excavation where the intended use is for location, acquisition, development, or artificial recharge of ground water.

WATER YEAR (WY) Time period form October 1 through September 30.

WATER YIELD (WATER CROP OR RUNOUT) Runoff from the drainage basin, including ground-water outflow that appears in the stream plus groundwater outflow that bypasses the gaging station and leaves the basin underground; precipitation minus evapotranspiration.

WATERBORNE DISEASE OUTBREAK Significant occurrence of acute illness associated with drinking water from a public water system that is deficient in treatment, as determined by appropriate local or state agencies.

WATERCOURSE Any surface flow such as a river, stream, tributary.

WATER-EFFECT RATIO (WER) Appropriate measure of the toxicity of a material obtained in a site water divided by the same measure of the toxicity of the same material obtained simultaneously in a laboratory dilution water.

WATER-QUALITY CRITERIA Specific levels of water quality which, if reached, are expected to render a body of water unsuitable for its designated use.

WATER-QUALITY DATA Chemical, biological, and physical measurements or observations of the characteristics of surface and ground waters, atmospheric deposition, potable water, treated effluents, and waste water and of the immediate environment in which the water exists.

WATER-QUALITY GUIDELINES Specific levels of water quality which, if reached, may adversely affect human health or aquatic life.

WATER-QUALITY LIMITED SEGMENT Stretch or area of surface water where technology-based controls are not sufficient to prevent violations of water-quality standards.

WATER-QUALITY MONITORING Integrated activity for evaluating the physical, chemical, and biological character of water in relation to human health, ecological conditions, and designated water uses.

WATER-QUALITY STANDARD Law or regulation that consists of the beneficial designated use or uses of a water body, the numerical and narrative water-quality criteria that are necessary to protect the use or uses of that particular water body, and an anti-degradation statement.

WATER-RESOURCE QUALITY Condition of water or some water-related resource as measured by biological surveys, habitat-quality assessments, chemical-specific analyses of pollutants in water bodies, and toxicity tests; condition of water or some water-related resource as measured by the following: habitat quality, energy dynamics, chemical quality, hydrological regime, and biotic factors.

WATER-RESOURCES REGION Natural drainage basin or hydrologic area that contains either the drainage area of a major river or the combined areas of a series of rivers; in the United States, there are 21 regions of which 18 are in the conterminous United States, and one each in Alaska, Hawaii, and the Caribbean.

WATER-RESOURCES SUBREGION Designated water-resources regions (21) of the United States are subdivided into 222 subregions where each subregion includes that area drained by a river system, a reach of a river and its tributaries in that reach, a closed basin(s), or a group of streams forming a coastal drainage system.

WATERS OF THE UNITED STATES All waters that are currently used, were used in the past, or may be susceptible to use in interstate or foreign commerce, including all waters subject to the ebb and flow of the tide.

WATERSHED (DRAINAGE AREA) Land area from which water drains toward a common watercourse in a natural basin; land area from which water drains into a stream, river, or reservoir; area of land that contributes runoff to one specific delivery point; part of the surface of the Earth that is occupied by a drainage system, which consists of a surface stream or a body of impounded surface water together with all tributary surface streams and bodies of impounded surface water; topographically defined area drained by a river/stream or system of connecting rivers/streams such that all outflow is discharged through a single outlet.

WATERSHED APPROACH Coordinated framework for environmental management that focuses public and private efforts on the highest priority problems within hydrologically-defined geographic areas taking into consideration both ground and surface water flow.

WATERSHED DIVIDE Boundary between catchment (drainage) areas.

WATER-SOLUBLE PACKAGING Packaging that dissolves in water.

WATER-SOURCE HEAT PUMP Heat pump that uses wells or heat exchangers to transfer heat from water (i.e. groundwater) to the inside of a building.

WATERSPOUT Tornado occurring over water.

WATERSTAGE RECORDER Motor-driven (spring wound or electric) instrument for monitoring water surface elevation.

WATERSTOP (WATER BAR) Continuous strip of waterproof material placed at concrete joints designed to control cracking and limit moisture penetration.

WATERWAYS Spillways and outlet works.

WATT Power of a current of one ampere flowing across a potential difference of one volt; equals one Joule per second.

WATTHOUR (WH) Electrical energy unit of measure equal to one watt of power supplied to, or taken from, an electrical circuit steadily for one hour.

WAVE CREST Highest part of a wave.

WAVE HEIGHT Distance from wave trough to wave crest.

WAVE PERIOD Time, in seconds, between the passage of consecutive wave crests or wave troughs past a fixed point.

WAVE PROTECTION Riprap, concrete, or other armoring on the upstream face of an embankment dam to protect against scouring or erosion due to wave action.

WAVE RUNUP Vertical height above the stillwater level to which water from a specific wave will run up the face of a structure or embankment.

WAVE SPECTRUM Distribution of wave energy with respect to wave frequency or period.

WAVE STEEPNESS Ratio of wave height to wave length and is an indicator of wave stability.

WAVE TROUGH Lowest part of the wave.

WAVE WALL See PARAPET WALL.

WAVELENGTH Distance between crests or troughs of a wave.

WAVEWATCH III One of the operational forecast models run at NCEP that is run four times daily, with forecast output out to 126 hours.

WBND Westbound.

WC/ATWC West Coast and Alaska Tsunami Warning Center.

WCM Warning Coordination Meteorologist.

WDIR Wind direction (the direction the wind is coming from in degrees clockwise from true N).

WDLY Widely.

WDSPRD Widespread.

WEAK ECHO REGION (WER) WSR-88D radar product that displays reflectivity for up to 8 elevation angles for a radar operator selected location as a set presentation of a storm.

WEATHER State of the atmosphere at any particular time and place with respect to wind, temperature, cloudiness, moisture, pressure, etc.

WEATHER FORECAST OFFICE (WFO) Responsible for issuing advisories, warnings, statements, and short term forecasts for its county warning area.

WEATHERFAX See RADIOFACSIMILE.

WEATHERING Process whereby earthy or rocky materials are changed in color, texture, composition, or form (with little or no transportation) by exposure to atmospheric agents.

WEDGE Piece that tapers from a thick end to a chisel point.

WEDGE TORNADO Slang for a large tornado with a condensation funnel that is at least as wide (horizontally) at the ground as it is tall (vertically) from the ground to cloud base.

WEEP HOLE Drain embedded in a concrete or masonry structure intended to relieve pressure caused by seepage behind the structure.

WEFAX System for transmitting weather charts and imagery via satellite.

WEIGHING-TYPE PRECIPITATION GAUGE Rain gauge that weighs the rain or snow which falls into a bucket set on a platform of a spring or lever balance.

WEIGHT OF SCIENTIFIC EVIDENCE Considerations in assessing the interpretation of published information about toxicity - quality of testing methods, size and power of study design, consistency of results across studies, and biological plausibility of exposure-response relationships and statistical associations.

WEIGHTED MEAN Value obtained by multiplying each of a series of values by its assigned weight and dividing the sum of these products by the sum of the weights.

WEIGHTING OF A SLOPE Additional material placed on the slope of an embankment.

WEIR Notch of regular form through which water flows; low dam built across a stream to raise the upstream water level (fixed-crest weir when uncontrolled); structure built across a stream or channel for the purpose of measuring flow (measuring or gaging weir); structure built into a levee or river bank that allows water to flow from the main river channel into a bypass channel during time of high flows; wall or obstruction used to control flow from settling tanks and clarifiers to ensure a uniform flow rate and avoid short-circuiting.

WEIR, MEASURING Device for measuring the rate of flow of water where the height of water above the weir crest is used to determine the rate of flow.

WELD To build up or fasten together metals by bonding on molten metal.

WELL Hole or shaft drilled into the Earth to get water or other underground substances; bored, drilled, or driven shaft, or a dug hole, whose depth is greater than the largest surface dimension and whose purpose is to reach underground water supplies or oil, or to store or bury fluids below ground.

WELL FIELD Area containing one or more wells that produce usable amounts of water or oil.

WELL GRADED Good representation in the material of all particle sizes present from the largest to smallest.

WELL INJECTION Subsurface emplacement of fluids into a well.

WELL MONITORING Measurement by on-site instruments or laboratory methods of well water quality.

WELL PLUG Watertight, gastight seal installed in a bore hole or well to prevent movement of fluids.

WELL POINT Hollow vertical tube, rod, or pipe with a driving point and a fine mesh screen used to remove underground water; complete set of equipment for drying up ground including wellpoints, connecting pipe and pump.

WELLHEAD PROTECTION AREA
Protected surface and subsurface zone surrounding a well or well field supplying a public water system to keep contaminants from reaching the well water.

WEST AFRICAN DISTURBANCE LINE
Line of convection about 300 miles long, similar to a squall line, that forms over west Africa north of the equator and south of 15 degrees North latitude.

WEST WALL Coast side boundary of the Gulf Stream, typically south of Cape Hatteras.

WESTERLIES Prevailing winds that blow from the west in the mid-latitudes.

WESTERN AREA POWER ADMINISTRATION (WAPA)
Markets and delivers reliable, cost-based hydroelectric power and related services within a 15-state region of the central and western U.S. WAPA's transmission system carries electricity from 55 hydropower plants, including those operated by the Bureau of Reclamation.

WESTERN SYSTEMS COORDINATING COUNCIL (WSCC)
Voluntary industry association created to enhance reliability among western utilities.

WET BULB ZERO (WBZ) Height where the wet-bulb temperature goes below 0 °C.

WET FLOODPROOFING Approach to flood proofing which usually is a last resort where flood waters are intentionally allowed into the building to minimize water pressure on the structure.

WET LABORATORY Laboratory where chemicals, drugs, or other material or biological matter is tested and analyzed, and which requires water, direct ventilation, and specialized piped utilities.

WET MICROBURST Microburst accompanied by heavy precipitation at the surface.

WET UNIT WEIGHT Unit weight of solids plus water per unit volume, irrespective of the degree of saturation.

WET WEATHER GREEN INFRASTRUCTURE Infrastructure associated with stormwater management and low impact development that encompasses approaches and technologies to infiltrate, evapotranspire, capture, and reuse stormwater to maintain or restore natural hydrologies.

WET-BULB TEMPERATURE Lowest temperature that can be obtained by evaporating water into the air.

WETLAND Area that is regularly wet or flooded and has a water table that stands at or above the land surface for at least part of the year.

WETLAND FUNCTION Process or series of processes that take place within a wetland that are beneficial to the wetland itself, the surrounding ecosystems, and people.

WETLANDS Ecosystems whose soil is saturated for long periods seasonally or continuously, including marshes, swamps, ephemeral ponds, bogs, and similar areas such as wet meadows, river overflows, mudflats, and natural ponds; area characterized by periodic inundation or saturation, hydric soils, and vegetation adapted for life in saturated soil conditions; any number of tidal and nontidal areas characterized by saturated or nearly saturated soils most of the year that form an interface between terrestrial and aquatic environments including freshwater marshes around ponds and channels, and brackish and salt marshes.

WETTABILITY Relative degree to which a fluid will spread into or coat a solid surface in the presence of other immiscible fluids.

WETTABLE POWDER Dry formulation that must be mixed with water or other liquid before it is applied.

WETTED PERIMETER Distance along the bottom and sides of a stream, creek, or channel in contact with the water; length of the wetted contact between a conveyed liquid and the open channel or closed conduit conveying it, measured in a plane at right angles to the direction of flow.

WFO National Weather Service Weather Forecast Office.

WFP Warm Front Passage.

WHALER Horizontal beam in a bracing structure.

WHEELING Transmission of electricity by an entity that does not own or directly use the power it is transmitting.

WHIRLWIND Small, rotating column of air.

WHITE LIGHT (WL) Sunlight integrated over the visible portion of the spectrum (4000 - 7000 angstroms) so that all colors are blended to appear white to the eye.

WHITE LIGHT FLARE Major flare in which small parts become visible in white light.

WHITE NOISE Random energy containing all frequency components in equal proportions.

WHITECAP Breaking crest of a wave, usually white and frothy.

WHOLE-EFFLUENT TOXICITY (WET) Total toxic effect of an effluent measured directly with a toxicity test.

WHOLE-EFFLUENT-TOXICITY TESTS Tests to determine the toxicity levels of the total effluent from a single source as opposed to a series of tests for individual contaminants.

WHOLESALE CUSTOMERS Any entity that purchases electricity at the wholesale level, including municipal utilities, private utilities, rural electric cooperatives, or government-owned utility districts.

WHOLESALE POWER MARKET Purchase and sale of electricity from generators to resellers (who sell to retail customers) along with the ancillary services needed to maintain reliability and power quality at the transmission level.

WHOLESALE TRANSMISSION SERVICES Transmission of electric energy sold, or to be sold, at wholesale in interstate commerce.

WHOLESALE WHEELING Process of sending electricity from one utility to another wholesale purchaser over the transmission lines of an intermediate utility.

WICKET GATE Gate which pivots open around the periphery of a turbine or pump to allow water to enter.

WIDESPREAD Areal coverage of non-measurable, non-convective weather and/or restrictions to visibility affecting more than 50 percent of a forecast zone(s).

WILD AND SCENIC RIVERS ACT (PUBLIC LAW 90-542) Selects certain rivers possessing remarkable scenic, recreational, geologic, fish and wildlife, historic, or other similar values, for preservation in free-flowing conditions.

WILDERNESS Tract or region of land uncultivated and uninhabited by human beings, or unoccupied by human settlements.

WILDERNESS RESOURCE Resources identified in officially designated wilderness areas on Forest Service or Bureau of Land Management administered land.

WILDFIRE Any free burning uncontainable wildland fire not prescribed for the area which consumes the natural fuels and spreads in response to its environment.

WILDLANDS Any non-urbanized land not under extensive agricultural cultivation (e.g., forests, grasslands, rangelands).

WILDLIFE REFUGE Area designated for the protection of wild animals, within which hunting and fishing are either prohibited or strictly controlled.

WILLINGNESS TO PAY Method of estimating the value of activities, services, or other goods, where value is defined as the maximum amount a consumer would be willing to pay for the opportunity rather than do without.

WILLOW CARR Pool or wetland dominated by willow trees or shrubs.

WILLY-WILLY Slang for a dust devil; formerly used to denote a tropical cyclone.

WILTING POINT Soil water content below which plants growing in that soil will remain wilted even when transpiration is nearly eliminated.

WINCH Drum that can be rotated so as to exert a strong pull while winding in a line.

WIND Horizontal motion of the air past a given point due to differences in air pressures where air moves from high to low pressure areas.

WIND ADVISORY Sustained winds 25 to 39 mph and/or gusts to 57 mph.

WIND CHILL ADVISORY National Weather Service issued product when the wind chill could be life threatening if action is not taken.

WIND CHILL FACTOR Increased wind speeds accelerate heat loss from exposed skin.

WIND CHILL WARNING National Weather Service issued product when the wind chill is life threatening.

WIND COUPLET Area on the radar display where two maximum wind speeds are blowing in opposite directions.

WIND DIRECTION True direction from which the wind is blowing at a given location (i.e. wind blowing from the east to the west is an east wind).

WIND FIELD Three-dimensional spatial pattern of winds.

WIND GUST Rapid fluctuations in the wind speed with a variation of 10 knots or more between peaks and lulls.

WIND RADII Largest radii of that wind speed found in that quadrant.

WIND ROSE Diagram, for a given locality or area, showing the frequency and strength of the wind from various directions.

WIND SETUP Vertical rise in the still water level at the face of a structure or embankment caused by wind stresses on the surface of the water.

WIND SHEAR Rate at which wind velocity changes from point to point in a given direction.

WIND SHEAR PROFILE Change in wind speed and/or direction usually in the vertical.

WIND SHIFT Change in wind direction of 45 degrees or more in less than 15 minutes with sustained wind speeds of 10 knots or more throughout the wind shift.

WIND SHIFT LINE Long, but narrow axis across which the winds change direction (usually veer).

WIND SOCK Tapered fabric shaped like a cone that indicates wind direction by pointing away from the wind.

WIND SPEED Rate at which air is moving horizontally past a given point.

WIND WAVES Local, short period waves generated from the action of wind on the water surface (as opposed to swell); waves generated by the local wind blowing at the time of observation.

WINDROW Ridge of loose material.

WINDWARD Side toward the wind.

WINDY 20 to 30 mph winds.

WING WALL Wall that guides a water into a conveyance structure.

WINT Winter.

WINTER Typically the coldest season of the year during which the Sun is farthest from overhead.

WINTER POOL Pool, or height of the water surface, of a reservoir during the winter.

WINTER SOLSTICE Time at which the Sun is farthest south in the Southern Hemisphere, on or around December 21.

WINTER STORM WARNING Product issued by the National Weather Service when a winter storm is producing or is forecast to produce heavy snow or significant ice accumulations.

WINTER STORM WATCH Product issued by the National Weather Service when there is a potential for heavy snow or significant ice accumulations, usually at least 24 to 36 hours in advance.

WINTER WEATHER ADVISORY Product issued by the National Weather Service when a low pressure system produces a combination of winter weather (snow, freezing rain, sleet, etc.) that present a hazard, but does not meet warning criteria.

WINTERS DOCTRINE Provides that the establishment of a Indian Reservation impliedly reserves the amount of water necessary for the purposes of the reservation.

WIRE WEIGHT GAUGE River gauge comprised of a weight which is lowered to the water level.

WIRE-TO-WATER EFFICIENCY (OVERALL EFFICIENCY) Efficiency of a pump and motor together.

WITHDRAWAL Water removed from the ground or diverted from a surface-water source for use; process of taking water from a source and conveying it to a place for a particular type of use; act or process of removing.

WLY Westerly.

WMC World Meteorological Center(s).

WMO World Meteorological Organization.

WOLF NUMBER Historic term for Sunspot Number.

WOOD PACKAGING Wood products such as pallets, crates, and barrels.

WOOD TREATMENT FACILITY Industrial facility that treats lumber and other wood products for outdoor use.

WOOD-BURNING-STOVE POLLUTION Air pollution caused by emissions of particulate matter, carbon monoxide, total suspended particulates, and polycyclic organic matter from wood-burning stoves.

WORK PLAN Plans that are prepared which detail the scope, direction, and purpose of a proposed plan.

WORKERS RIGHT-TO-KNOW Legislation mandating communicating of information to employees.

WORKING LEVEL (WL) Unit of measure for documenting exposure to radon decay products, the so-called "daughters."

WORKING LEVEL MONTH (WLM) Unit of measure used to determine cumulative exposure to radon.

WRAPPING GUST FRONT Gust front which wraps around a mesocyclone, cutting off the inflow of warm moist air to the mesocyclone circulation and resulting in an occluded mesocyclone.

WRCC Western Regional Climate Center.

WRN Western.

WRNG Warning.

WSFO Weather Service Forecast Office.

WSHFT Wind Shift.

WSPD Wind speed (m/s) averaged over an eight-minute period for buoys and a two-minute period for land stations.

WSR-57 National Weather Service Weather Surveillance Radar designed in 1957 that was replaced by WSR-88D units.

WSR-74 National Weather Service Weather Surveillance Radar designed in 1974 that was replaced by WSR-88D units.

WSR-88D Weather Surveillance Radar - 1988 Doppler; NEXRAD unit.

WSR-88D SYSTEM Summation of all hardware, software, facilities, communications, logistics, staffing, training, operations, and procedures specifically associated with the collection, processing, analysis, dissemination and application of data from the WSR-88D unit.

WSW Winter Storm Message.

WTMP Sea Surface Temperature (Celsius).

WTR Water Equivalent.

WVHT Significant Wave Height.

WWH Wind Wave Height.

WWP Wind Wave Period.

WX Weather.

WYE BRANCH (Y BRANCH) Section of pipe that joins the main run of pipe at an angle; fitting that makes the joint is in the shape of the letter Y.

X

X-BAND Frequency band of microwave radiation in which radars operate.

XBT Expendable Bathythermograph.

XCITED Excited.

XCPT Expecting.

XENOBIOTA Any biotum displaced from its normal habitat; chemical foreign to a biological system.

XERISCAPE Landscaping that does not require a lot of water.

XERISCAPING Landscaping based on native, water-efficient plants to minimize the need for irrigation; method of landscaping that uses plants that are well adapted to the local area and are drought-resistant.

XEROPHYTE Plant adapted for growth under dry conditions.

XPC Expect.

X-RAY BACKGROUND Daily average background X-ray flux in the 1 to 8 angstrom range.

X-RAY BURST Temporary enhancement of the X-ray emission of the Sun.

X-RAY FLARE CLASS Rank of a flare based on its X-ray energy output.

X-RAYS Very energetic electromagnetic radiation with wavelengths intermediate between 0.01 and 10 nanometers (0.1-100 Angstroms) or between gamma rays and ultraviolet radiation.

XSEC Cross Section.

Y

YARD WASTE Part of solid waste composed of grass clippings, leaves, twigs, branches, and other garden refuse.

YEAR Period during which the Earth completes one revolution around the Sun.

YELLOW SNOW Snow given a golden or yellow appearance by the presence in it of pine, cypress pollen, or anthropogenic material or animal-produced material.

YELLOW-BOY Iron oxide flocculant (clumps of solids in waste or water) usually observed as orange-yellow deposits in surface streams with excess iron content.

YIELD Total water runout or crop that includes runoff plus underflow; quantity of water (expressed as a rate of flow or total quantity per year) that can be collected for a given use from surface or groundwater sources; mass per unit time per unit area (e.g., mass of material or constituent transported by a river in a specified period of time divided by the drainage area of the river basin.).

YOUNG-OF-YEAR Refers to young (usually fish) produced in one reproductive year.

Z

Z Zulu time; See UNIVERSAL COORDINATED TIME (UTC).

Z/R RELATIONSHIP Empirical conversion relationship between radar reflectivity and precipitation rate.

ZERO AIR Atmospheric air purified to contain less than 0.1 ppm total hydrocarbons.

ZERO AIR VOIDS CURVE Curve showing the relationship between dry unit weights and corresponding moisture contents, assuming that all of the voids are completely filled with water.

ZERO DATUM Reference "zero" elevation for a stream or river gauge.

ZFP Zone Forecast Product.

ZL Freezing Drizzle.

ZNS Zones.

ZONAL FLOW Large-scale atmospheric flow in which the east-west component (i.e. latitudinal) is dominant.

ZONE OF AERATION Zone above the water table; comparatively dry soil or rock located between the ground surface and the top of the water table where soil pores may either contain air or water; vadose zone.

ZONE OF SATURATION Zone in which all the interstices between particles of geologic material or all of the joints, fractures, or solution channels in a consolidated rock unit are filled with water under pressure greater than that of the atmosphere; locus of soil or rock points located below the top of the groundwater table that is saturated with water; phreatic zone.

ZONED EARTHFILL See ZONED EMBANKMENT DAM.

ZONED EMBANKMENT DAM Embankment dam which is comprised of zones of selected materials having different degrees of porosity, permeability and density.

ZONING Identification of areas of specified uses or restrictions.

ZOOPLANKTON Small (often microscopic), free-floating aquatic animals (e.g. protozoans), found in lakes and reservoirs; animal part of the plankton.

ZR Freezing Rain.

ZULU (Z) TIME See UNIVERSAL COORDINATED TIME.

ZURICH SUNSPOT CLASSIFICATION Sunspot classification system that has been modified for SESC use.

CPSIA information can be obtained
at www.ICGtesting.com
Printed in the USA
BVHW092106200319
543214BV00008B/279/P

9 780983 908500